Lecture Notes of the Institute for Computer Sciences, Social-Informatics and Telecommunications Engineering   35

Editorial Board

Ozgur Akan
  Middle East Technical University, Ankara, Turkey
Paolo Bellavista
  University of Bologna, Italy
Jiannong Cao
  Hong Kong Polytechnic University, Hong Kong
Falko Dressler
  University of Erlangen, Germany
Domenico Ferrari
  Università Cattolica Piacenza, Italy
Mario Gerla
  UCLA, USA
Hisashi Kobayashi
  Princeton University, USA
Sergio Palazzo
  University of Catania, Italy
Sartaj Sahni
  University of Florida, USA
Xuemin (Sherman) Shen
  University of Waterloo, Canada
Mircea Stan
  University of Virginia, USA
Jia Xiaohua
  City University of Hong Kong, Hong Kong
Albert Zomaya
  University of Sydney, Australia
Geoffrey Coulson
  Lancaster University, UK

Thomas Phan   Rebecca Montanari
Petros Zerfos (Eds.)

Mobile Computing, Applications, and Services

First International ICST Conference, MobiCASE 2009
San Diego, CA, USA, October 26-29, 2009
Revised Selected Papers

Volume Editors

Thomas Phan
Microsoft Corporation
One Microsoft Way
Redmond, WA 98052, USA
E-mail: thomas.phan@microsoft.com

Rebecca Montanari
DEIS - University of Bologna
Via Risorgimento, 2
40137 Bologna, Italy
E-mail: rebecca.montanari@unibo.it

Petros Zerfos
IBM T.J. Watson Research Center
19 Skyline Drive
Hawthorne, NY 10532, USA
E-mail: pzerfos@us.ibm.com

Library of Congress Control Number: 2010925232

CR Subject Classification (1998): C.2, H.4, I.2, D.2, H.3, H.5

ISSN	1867-8211
ISBN-10	3-642-12606-5 Springer Berlin Heidelberg New York
ISBN-13	978-3-642-12606-2 Springer Berlin Heidelberg New York

This work is subject to copyright. All rights are reserved, whether the whole or part of the material is concerned, specifically the rights of translation, reprinting, re-use of illustrations, recitation, broadcasting, reproduction on microfilms or in any other way, and storage in data banks. Duplication of this publication or parts thereof is permitted only under the provisions of the German Copyright Law of September 9, 1965, in its current version, and permission for use must always be obtained from Springer. Violations are liable to prosecution under the German Copyright Law.

springer.com

© ICST Institute for Computer Science, Social Informatics and Telecommunications Engineering 2010
Printed in Germany

Typesetting: Camera-ready by author, data conversion by Scientific Publishing Services, Chennai, India
Printed on acid-free paper   SPIN: 06/3180   5 4 3 2 1 0

Preface

This proceedings volume includes the full research papers presented at the First International Conference on Mobile Computing, Applications, and Services (MobiCASE) held in San Diego, California, during October 26-29, 2009. It was sponsored by ICST and held in conjunction with the First Workshop on Innovative Mobile User Interactivity (WIMUI).

MobiCASE highlights state-of-the-art academic and industry research work in domain topics above the OSI transport layer with an emphasis on complete end-to-end systems and their components. Its vision is largely influenced by what we see in the consumer space today: high-end mobile phones, high-bandwidth wireless networks, novel consumer and enterprise mobile applications, scalable software infrastructures, and of course an increasingly larger user base that is moving towards an almost all-mobile lifestyle.

This year's program spanned a wide range of research that explored new features, algorithms, and infrastructure related to mobile platforms. We received submissions from many countries around the world with a high number from Europe and Asia in addition to the many from North America. Each paper received at least three independent reviews from our Technical Program Committee members during the Spring of 2009, with final results coming out in July. As a result of the review process, we selected 15 high-quality papers and complemented them with six invited submissions from leading researchers, reaching the final count of 21 papers in the program. The papers were presented at the conference in single-track sessions, and they covered a number of fascinating topics, including healthcare, transportation, mobile assistants, rich mobile media, and system software. The WIMUI workshop featured six papers on context and interactive games.

The participation throughout the MobiCASE 2009 conference and workshop was strong from both industry and academia, and the attendees were enthusiastic about the main theme and topics of the conference. It was often noted that the technical program had direct relevance and importance to the research and product development efforts of several companies, whose researchers and engineers attended the sessions.

MobiCASE further featured a lively demonstration and poster session with eight software and hardware demonstrations and five posters. Additionally, the conference held an industry panel discussion that included Victoria Coleman, VP of the Samsung Computer Science Lab, and Monica Lam and Nick Bambos of Stanford University. The panel discussion covered challenges and advantages in adopting open source as a mobile device platform and novel server and middleware services to support cloud computing paradigms in a mobile ecosystem.

Many people contributed to the organization of the MobiCASE. The conference could not be held without the tremendous work done by the TPC members, each of whom reviewed an average of five papers and provided in-depth critique and analysis of the submissions. Additionally, thanks go to Guang Yang for handling the invited papers, Benjamin Greenstein for chairing the workshop, Angela Nicoara and Jatinder Pal Singh for organizing the invited panel, Angela Dalton for managing the demo/poster session, and Yafei Yang for taking care of the local arrangements. Finally, particular thanks and appreciation must be put forth for the MobiCASE Steering Committee Chair, Imrich Chlamtac, whose vision has steered and shaped this effort.

Thomas Phan
Rebecca Montanari
Petros Zerfos

Organization

General Co-chair

Rebecca Montanari — University of Bologna, Italy
Petros Zerfos — IBM T.J. Watson Research Center, USA

Steering Committee Chair

Imrich Chlamtac — Create-Net

Technical Program Chair

Thomas Phan — Microsoft, USA

Industry Track Co-chair

Jatinder Pal Singh — Deutsche Telekom, Inc. R&D and Stanford University, USA
Angela Nicoara — Deutsche Telekom Inc., R&D Lab, USA

Demo/Poster Chair

Angela Dalton — Johns Hopkins University, USA

Workshop Chair

Benjamin Greenstein — Intel Research, USA

Local Arrangements Chair

Yafei Yang — Qualcomm, USA

Web/Publicity Chair

Alessandra Toninelli — University of Bologna, Italy

Technical Program Committee

Murali Annavaram	USC
Trevor Armstrong	Microsoft
Paolo Bellavista	University of Bologna
Nina Bhatti	HP Labs
Lawrence Brakmo	Google
Tim Brecht	University of Waterloo
Jerry Cheng	Yahoo!
Saumitra Das	Qualcomm
Rajit Gadh	UCLA
Hani Jamjoom	IBM T.J.Watson Research Center
Yiming Ji	UCSB
Minkyong Kim	IBM T.J.Watson Research Center
Haiyun Luo	Azalea Networks
April Mitchell	HP Labs
Trevor Pering	Intel Research
Nischal Piratla	Deutsche Telekom R&D Labs
Calicrates Policroniades	Telenor R&D
Karim Seada	Nokia Research
Timothy Sohn	Nokia Research
Andreas Terzis	Johns Hopkins University
Alexander Varshavsky	AT&T Labs
Pablo Vidales	Deutsche Telekom R&D Labs
Guang Yang	Nokia Research

IMUI Workshop Chair

Ben Greenstein	Intel Labs Seattle

IMUI Workshop Program Committee

Murali Annavaram	USC
Nina Bhatti	HP Labs
Ben Greenstein	Intel Labs Seattle
Yiming Ji	UCSB
Thomas Phan	Microsoft
Andreas Terzis	Johns Hopkins University
Alexander Varshavsky	AT&T Labs
Guang Yang	Nokia Research
Pablo Vidales	Deutsche Telekom R&D Labs
Petros Zerfos	IBM T.J.Watson Research Center

Table of Contents

MobiCASE 2009 - Session 1: To Your Health

Nutrition Monitor: A Food Purchase and Consumption Monitoring Mobile System 1
    Kyle Dorman, Marjan Yahyanejad, Ani Nahapetian, Myung-kyung Suh, Majid Sarrafzadeh, William McCarthy, and William Kaiser

A Personalised Body Motion Sensitive Training System Based on Auditive Feedback 12
    Gerold Hoelzl

MobiCASE 2009 - Session 2: System Software I

Context-Aware Authentication Framework 26
    Diwakar Goel, Eisha Kher, Shriya Joag, Veda Mujumdar, Martin Griss, and Anind K. Dey

The Tradeoff between Energy Efficiency and User State Estimation Accuracy in Mobile Sensing 42
    Yi Wang, Bhaskar Krishnamachari, Qing Zhao, and Murali Annavaram

Dynamic Migration of Computation through Virtualization of the Mobile Platform 59
    Shivani Sud, Roy Want, Trevor Pering, Kent Lyons, Barbara Rosario, and Michelle X. Gong

MobiCASE 2009 - Session 3: On the Go!

Intelligent Telemetry for Freight Trains 72
    Johnathan M. Reason, Han Chen, Riccardo Crepaldi, and Sastry Duri

OneBusAway: A Transit Traveler Information System 92
    Brian Ferris, Kari Watkins, and Alan Borning

MobiCASE 2009 - Session 4: Industry Track

WebCall – A Rich Context Mobile Research Framework 107
    Zhigang Liu, Hawk Yin Pang, Jun Yang, Guang Yang, and Péter Boda

Lively Mashups for Mobile Devices 123
  Feetu Nyrhinen, Arto Salminen, Tommi Mikkonen, and
  Antero Taivalsaari

Ads Go Mobile: Assessing the Opportunities and Challenges of
Personalised Ads in a Mobile Search Service 142
  Sigmund Akselsen, Bente Evjemo, and Calicrates Policroniades

Mobile Visual Analytics for Datacenter Power and Cooling
Management 160
  Ratnesh Sharma, Ming Hao, Ravigopal Vennelakanti, Manish Gupta,
  Umeshwar Dayal, Cullen Bash, Chandrakant Patel, Deepa Naik,
  A. Jayakumar, Sairabanu Z. Ganihar, Ramesh Munusamy, and
  Vani Mohan

MobiCASE 2009 - Session 5: System Software II

RFID-based Distributed Memory for Mobile Applications 172
  Michel Simatic

A Mobile Application to Detect Abnormal Patterns of Activity 190
  Omar Abdul Baki, Joy Zhang, Martin Griss, and Tony Lin

Energy-Efficient Localization via Personal Mobility Profiling............ 203
  Ionut Constandache, Shravan Gaonkar, Matt Sayler,
  Romit Roy Choudhury, and Landon Cox

MobiCASE 2009 - Session 6: Mobile Assistants

Multi-agent Meeting Scheduling Using Mobile Context................. 223
  Kathleen Yang, Neha Pattan, Alejandro Rivera, and Martin Griss

Bayesian Networks-Based Interval Training Guidance System for
Cancer Rehabilitation 236
  Myung-kyung Suh, Kyujoong Lee, Alfred Heu, Ani Nahapetian, and
  Majid Sarrafzadeh

Mobile Context-Aware Personal Messaging Assistant 254
  Senaka Buthpitiya, Deepthi Madamanchi,
  Sumalatha Kommaraju, and Martin Griss

MobiCASE 2009 - Session 7: Rich Mobile Media

OCRdroid: A Framework to Digitize Text Using Mobile Phones........ 273
  Mi Zhang, Anand Joshi, Ritesh Kadmawala, Karthik Dantu,
  Sameera Poduri, and Gaurav S. Sukhatme

Gradient Domain Image Blending and Implementation on Mobile
Devices 293
  Yingen Xiong and Kari Pulli

A Comparative Evaluation of HTML5 as a Pervasive Media Platform... 307
  Tom Melamed and Ben Clayton

MobiCASE 2009 - Demonstrations and Posters Session

Study of Usability of Security and Privacy in Context Aware Mobile
Applications... 326
  Neha Pattan and Deepthi Madamanchi

Friendlee: A Mobile Application for Your Social Life................ 331
  *Anupriya Ankolekar, Gabor Szabo, Yarun Luon, and
  Bernardo A. Huberman*

A Prototype for Resource Optimized Context Determination in
Pervasive Care Environments 335
  Nirmalya Roy, Christine Julien, Archan Misra, and Sajal K. Das

RFID-based Distributed Shared Memory for Pervasive Games 339
  Michel Simatic and Annie Gentès

IMUI Workshop - Session 1: Context and Constraints

Coarse In-Building Localization with Smartphones 343
  *Avinash Parnandi, Ken Le, Pradeep Vaghela, Aalaya Kolli,
  Karthik Dantu, Sameera Poduri, and Gaurav S. Sukhatme*

Delay Analysis of Large-Scale Wireless Sensor Networks 355
  Jun Yin, Yun Wang, and Xiaodong Wang

ProVer: A Secure System for the Provision of Verified Location
Information 366
  Michelle Graham and David Gray

IMUI Workshop - Session 2: Games and Guides

Designing Mobility: Pervasiveness as the Enchanting Tool of Mobility... 374
  Annie Gentes, Camille Jutant, Aude Guyot, and Michel Simatic

Context-Aware Recommendations in Decentralized, Item-Based
Collaborative Filtering on Mobile Devices 383
  *Wolfgang Woerndl, Henrik Muehe, Stefan Rothlehner, and
  Korbinian Moegele*

deSCribe: A Personalized Tour Guide and Navigational Assistant 393
  Dheeraj Kota, Neha Laumas, Urmila Shinde, Saurabh Sonalkar,
  Karthik Dantu, Sameera Poduri, and Gaurav S. Sukhatme

Author Index 405

Nutrition Monitor:
A Food Purchase and Consumption Monitoring Mobile System

Kyle Dorman[1], Marjan Yahyanejad[1], Ani Nahapetian[1,2], Myung-kyung Suh[1],
Majid Sarrafzadeh[1,2], William McCarthy[3,4], and William Kaiser[2,5]

[1] UCLA Computer Science Department
[2] UCLA Wireless Health Institute
[3] UCLA Department of Health Services, School of Public Health
[4] UCLA Department of Psychology
[5] UCLA Electrical Engineering Department
{kdorman,marjan,ani,dmksuh,majid}@cs.ucla.edu,
wmccarth@ucla.edu, kaiser@ee.ucla.edu

Abstract. The challenge of monitoring food intake can be facilitated by the truly transformational power of mobile phones. Mobile phones provide a pervasive and fairly ubiquitous infrastructure, which we leverage to provide cost-effective, high quality aids to behavior monitoring and modification. Additionally, the technology allows public health messages to reach certain target groups, such as youth and members of low-income communities, which may not otherwise be practical. Our system leverages the existing mobile phone infrastructure. We use the highly capable computational and data-gathering platform of mobile phones to facilitate the collection, transmission and processing of data for purposes of monitoring in the field, behavior and activity classification, and timely behavioral cuing. The nature of mobile phones coupled with a web-interface also allow for customization and personalization, retrieval of nutrition information on demand, as well as the ability to truly monitor the user's consumption trends.

1 Introduction

The development and the incorporation of wireless technologies to promote healthy lifestyle behavior, specifically healthy eating and weight control, has the potential to address our ultimate goal of enabling healthier lifestyle choices and behavior modifications needed to prevent obesity and obesity-related diseases.

The obesity epidemic can be ameliorated by the truly transformational power of mobile phones [22] and other wireless embedded technologies that have become well-incorporated into our lives. Mobile phones provide a pervasive and fairly ubiquitous infrastructure, which we can leverage to provide cost-effective, high quality aids to behavior monitoring, evaluation, and modification. Additionally, the technology can be used to provide personalized services that may otherwise be unavailable to certain target groups, youth, for example, and members of low-income communities.

This wireless and mobile technology, specifically, provides a highly capable computational and networked platform, which we are using in combination with wearable sensors for the collection, processing, and retrieval of data for purposes such as monitoring behavior in the field, behavior and activity classification, and timely behavioral cuing. The nature of mobile phones allows for customization and personalization, as well as the ability to dynamically and automatically adapt to the user's environment.

Today's American population lives in an obesigenic environment that puts them all at a high risk of obesity. Age appropriate, culturally aware and gender specific solutions will have a higher likelihood of success in this environment, particularly for achieving prevention. Consider, for example, that while older adults are motivated by reducing their risk of disease, the youth are more health-enhancement oriented and also have a higher potential for social contagion. Bearing in mind this aspect of youth culture, we are incorporating both mobile phone technology and social networking websites into one specific approach for healthy behavior change.

This project involves an interdisciplinary effort to address a national crisis of obesity and preventable obesity-related chronic diseases. It leverages the most personal and beloved of all technologies to have come about in this generation, the mobile phone.

We have developed a system that leverages mobile phones for monitoring food purchases, food consumption and providing timely and appropriate informational cues for making healthier eating choices. This can be coupled with a web-based scheme, to provide self-monitoring and progress comparison with peers.

This food purchases and food intake monitoring system allows the entry of names of foods through the use of manual keyed entry or, more conveniently, the use of the mobile phone camera for scanning the UPC code on packaged foods. The system connects with various databases to provide ingredient lists, nutritional information as is commonly seen on packaged foods, as well as nutrients and nutrition information that typically are not described on food packaging, such as food water content and satiety value. The information is stored by the system for later examination on the web.

The system can cue the user if food choices inconsistent with adherence to the Dietary Guidelines for Americans are being made. The information is also be used for subtle reminders such as "you are 50% over your daily energy density intake goal" or "you are consuming 'trans fats'." Additionally, the system allows users to obtain information about how well each scanned product could help the purchaser to adhere to the Dietary Guidelines, to help in choices made during grocery shopping. Upon availability, suggestions for other related foods choices, which would be in stricter adherence to these guidelines, can also be provided. Finally, the long term tracking of a person's food intake can be useful for motivating longer-term adherence to the Dietary Guidelines.

In addition to interfacing with mobile phones, patients can go online to personalize the system, obtain more information about their food choices, and view detailed summaries of their food purchases and consumption habits. Using the online interface is common with weight management programs, such as Weight Watchers [21]. It allows for a more thorough examination of past activities, as well as different data presentation formats. Additionally, it enables a social networking framework where patients monitoring their food intake can compare notes, compete, and develop a support system.

The system provides increased accuracy of monitoring and ease of use and hence has the potential for pervasive use. Due to the nature of the technology, the system reduces bias by relying less on memory by collecting data in real time, and avoids the difficult or inexact calculations of food portions.

2 Obesity Crisis

Overweight adults are defined as those who have a Body Mass Index (kg/m2) (BMI) between 25 and 29.9. Obese adults are those who have a BMI of 30 or higher. Figure 1 shows obesity trends from 1985 to 1997; this does not even include overweight adults and are based on self-reported weight, and yet the numbers are still unbelievably high, reaching 22.6% in California alone.

Obesity increases the risk of many negative health consequences, such as coronary heart disease, type 2 diabetes, and hypertension. It also had an estimated cost of $78.5 billion in medical expenditures in 1998 [24]. The Centers for Disease Control (CDC) believes that overweight and obesity are caused by what they call an energy imbalance: too many calories and too little physical activity. They believe the best areas for treatment and prevention are monitoring behavior and environment settings [24]. This epidemic needs to be stopped and reversed as soon as practicable.

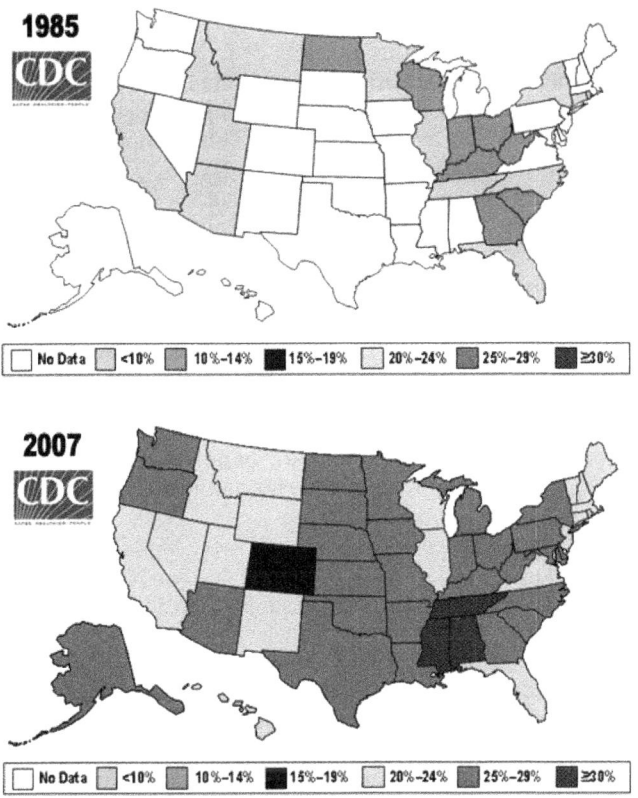

Fig. 1. Percent of obese (BMI ≥ 30) U.S. adults [24]

3 Related Work

Automatic monitoring has been found to be beneficial in a variety of fields. Berckmans [13] studied how automatic monitoring of livestock production processes help not only replace the farmer's eyes and ears but also provides a framework to help monitor variables such as infections and stress. Bhatia et al. [14] discuss a remote system that can monitor the blood pressure in a patient. This allows blood pressure to be taken even without the physician being present which is beneficial to both the patient and his/her doctor.

Diamond et al. [15] use current computing technologies to present an environmental nervous system that can automatically monitor certain factors such as water in lakes and at water treatment plants. However, wireless sensors are still very young technologies and can be applied in many different areas. S. Gupta et al. [17] help fill this need by studying the potential of smart sensors in biomedical applications. They stress that as new mobile sensing applications are developed, the knowledge learned from these sensors should be applied to even newer applications. An example of a unique application that uses remote sensing is presented by Lukowicz et al. [20]. They discuss how sensing can actually be used in meetings to automatically annotate meeting notes. The wearable sensors will allow for easy recording and user friendly retrieval of data.

Health monitoring is important in a variety of fields. For example, Linz et al. [19] study how remote monitoring actually helps patients manage their stress levels. They use textile contactless sensors integrated into a shirt, which measures their angular distance from certain muscles to measure stress. This illustrates another way in which automatic monitoring can help improve a person's health.

Wireless food intake systems have been attempted before and will be discussed in this section. One such system monitors real-time caloric balance [1]. Users interfaced with a food and activity database by selecting foods they have eaten and the amount of exercise they have engaged in. Thus the number of kilocalories eaten and burned during the course of a day is monitored. This system is similar to our work in the sense that its goal is one of our main objectives, caloric intake monitoring. However, the limitation of this system is that the selection of foods is manual; the user must browse a list to try to find the matching food. If the food item is not on the list, the user can enter an estimated number of kilocalories but this can affect accuracy. Also, it takes time to enter each entry and therefore compliance may be reduced. Our system attempts to bridge this gap by allowing for automatic and precise input of items by scanning the barcode of a consumed item.

Jovanov et al. [18] mention the limitations of using wires to connect different sensors to a person's body. They propose a Personal Area Network where a Wireless Intelligent Sensor can be integrated onto a single chip to eliminate the bulkiness of wires. They apply this to an environment where a patient's vital signs need to be measured consistently.

Brown et al. [2] present an application where users could take pictures of the food they eat and upload the pictures later using a desktop application. Users could later add kilocalorie amounts to the food images. This was found to be very successful when the users were tested about remembering what they ate. Again, this has the limitation that the user would have to look up each individual food item. Estrin et al. [12] discuss a system where a mobile phone is hung around the participant's neck and automatically takes pictures when it is on, in this case, during mealtimes. The pictures are then uploaded and analyzed to determine eating patterns and to attempt modeling nutritional intake.

Boissy et al. [6] present a study in which individuals over 60 used barcode scanners to self-report their health status. The study showed that users found the barcode scanner easy to learn and use and pleasant to use. This is an important reaction because the project in this paper utilizes a barcode scanner to provide automatic collection of foods.

Sensei.com [7] is a system that focuses more on customized recommendations. The program will send the user meal recommendations, weekly shopping lists, fitness tips and motivational messages. The motivation for the system is the same as ours: help users make healthier choices. Myca Nutrition [8] provides a service for nutritionists. Once the software is purchased, nutritionists can have clients send images of food to create a food journal. This improves nutritionists' understanding of their patient's eating habits and thereby helps them to devise more informed nutrition recommendations.

Toscos et al. [3] focuses on fitness monitoring, specifically targeting teenage girls. They created a mobile application that would allow teenage girls to create 'cliques' of 4 friends who could then compete against each other for how much exercise they did. This system was widely successful, showing that there is a latent demand for mobile applications that could help with developing healthier bodies. Similarly, [4] presents a system that would encourage activity by sharing step counts with friends. [5] has a mobile fitness demonstrator with different sensors to test user perceptions of mobile fitness systems.

As far as barcode scanners are concerned, many systems exist that scan 2D barcodes. However, these systems are not compatible with 1D barcodes, or UPC codes. I-Nigma [9] and iPhone 2D Barcode Reader [10] are examples of such systems. Other systems say they can scan 1D barcodes but they are all external devices. Microsoft was developing a system called AURA that would capture 1D barcodes with a Windows Mobile device camera and extract the number [11]. Unfortunately, development on this system has stopped. After extensive testing of the latest version on the HP iPAQ, we found that the system failed on processing every barcode captured.

4 Architecture

This food purchases and food intake monitoring system allows the entry of names of foods consumed through the use of manual keyed entry or by scanning UPC codes on packaged foods. The system connects with various databases to provide ingredient lists and nutritional information. The information is stored by the system for later examination on the web.

The system cues the user if food choices are inconsistent with the Dietary Guidelines for Americans, by subtle reminders such as "you are 50% over your daily energy density intake goal" or "you are consuming 'trans fats'." Additionally, the system allows users to obtain information about how well each scanned product could help the purchaser to adhere to the Dietary Guidelines, to help in choices made during grocery shopping.

The system helps users easily track what foods they have been eating with running totals of calories, fats, etc. We have developed three different versions of this application, on the Nokia N95 using the Symbian operating system, a Windows Mobile version on the HP iPAQ hw6945, and an Android version on the new G1 phone. The following sections discuss the architecture of each version.

4.1 Symbian Operating System

The first version of the application was developed on the Nokia N95, which uses the Symbian operating system. We chose this phone because it is one of the more popular operating systems and we had experience developing on the N95. It was programmed in Python and was under development for approximately nine weeks.

The software requires the user to first register an account on a public website and choose from a list of predefined diets or create a personalized one, consisting of the desired amount of calories, total fat, carbohydrates, sodium, and protein. This information is later used in the mobile phone application so the user can make an informed decision about the foods that they are about to eat or have eaten.

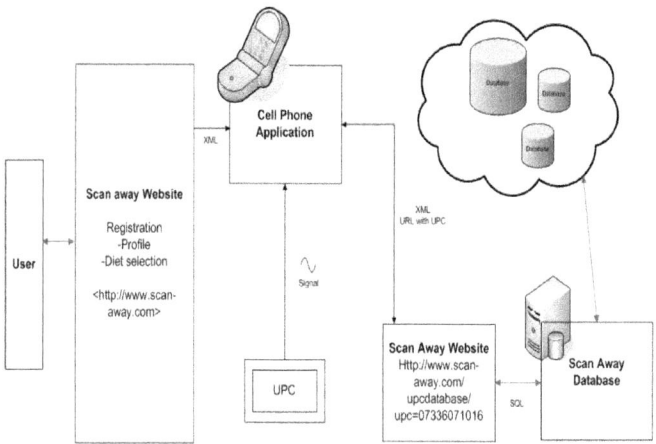

Fig. 2. Architecture of the Nokia application

The user can then choose whether to use an external Bluetooth scanner or manually input the barcode of an item. The application supports the LaserChamp Mobile Barcode Scanner [23]; this device scans a UPC barcode and transmits the code to the mobile device via Bluetooth. This barcode is then embedded in an XML query and sent to the web server. This server contains a list of barcodes and nutritional information, as this information is unavailable in a public database.

If the scanned item is not in the database, the query is sent to www.upcdatabase.com, a website that contains barcodes and their product descriptions. The product description is then sent to http://caloriecount.about.com/ that contains nutritional information about a large number of items. The HTML of that page is then parsed and the nutritional information is pulled out. The relevant data, such as kilocalories and fat, is then stored in the database on the application's server. In this manner, the server contains a self-learning database; with every item scanned, the database would "learn" the item and allow for processing the next time.

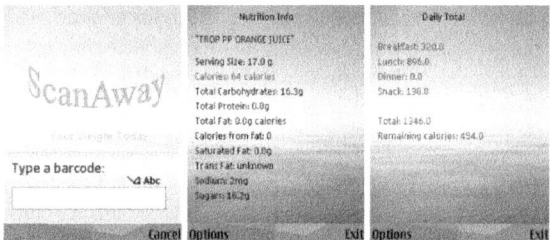

Fig. 3. Use case for the Nokia application

After scanning the barcode, the user then has a choice of whether to enter "grocery mode" or "meal mode". In Grocery Mode, the user can see detailed nutrition information about a single item and how this item would affect his/her overall daily diet. This mode is useful when a user is in a grocery store and would like to choose between two items. It is an important distinction whether an item, such as a frozen entrée, is 25% of a user's allotted daily kilocalories or 10%.

Meal Mode is used to keep track of a user's caloric intake throughout the day. The user can select to which meal an item is added: breakfast, lunch, dinner or a snack. S/he then scans the item and enters the serving size s/he is eating. The kilocalorie amount is added to his/her daily total so the user knows immediately whether or not s/he is sticking to his/her diet. This real-time knowledge allows informative decisions to be made about food consumed.

4.2 Windows Mobile Operating System

The second version of the application was developed on the HP iPAQ hw6945, which uses the Windows Mobile operating system. This is because there are many Smartphones that use this operating system and therefore this would help penetrate the market further. We used the Windows Mobile 5.0 SDK and programmed using Visual Studio 2007.

User procedures are similar to those required by the Symbian operating system. The user first enters his username and password into the mobile phone application. S/he then enters the barcode of a food product. In this system, however, the barcode can only be entered manually as there is a lack of research in embedded barcode scanner applications. The barcode is then sent directly to www.upcdatabase.com where a query is done to get the product description. That information is then sent to http://caloriecount.about.com/ where another query is done with that description. The returning HTML is then parsed and relevant information is displayed on the screen.

The user now has the option of adding this item to his/her total calorie count, or "consuming" this item. If s/he chooses to do that, the size of one standard serving is displayed and a query asks him/her how many serving sizes s/he is having. The user enters in a number and can then view how many calories have been added by that one meal.

Finally, the user can view how many kilocalories s/he has consumed for the day. This number is kept as a running total until the user quits the application. Windows

Mobile allows for many programs running in the background and therefore the user should keep this program open until s/he wants to reset his/her total calorie amount.

A few distinctions need to be emphasized between the first and second versions of the application. First, the "modes" were removed in the second version. Instead, the user is only able to see the nutritional information of a food product. Future work includes displaying the percentage of the total diet as well, on the same screen as the nutritional information.

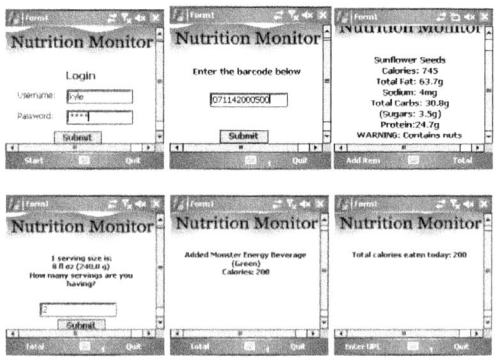

Fig. 4. Use case for the Windows Mobile application

The second variation is that the items are not separated into meals; we felt this was a step that would not be important to most users. Third, an allergy feature was added to the Windows Mobile version. That is, when an item is scanned, the nutritional information includes a warning if the product includes nuts. This is to help users who have a nut allergy; it provides an easy, immediate way to see whether or not a product would set off any allergic response.

4.3 Android Operating System

The software on the G-phone consists of three modes: eating, shopping and report mode. Each of these modes is described in detail below.

4.3.1 Eating Mode

In this mode, the user is presented with a text area where s/he can input the name or the UPC code of the item that s/he has eaten to generate a list of the consumed products. The menu consists of two other options: a product scan and the common list of foods.

The product scan provides the user with an interface that allows a photo to be taken of the UPC barcode. This is another way that the user can add an item to his/her list. The common list is a list of items that the user consumes frequently. This makes the management of a daily food list even easier; the system simply asks the user to select items from the common list.

4.3.2 Shopping Mode

When the user is in shopping mode, s/he is provided with the same options as the previous mode. However, after pushing the 'add' button, the user is directed to a view which shows him/her the sum of the nutrition facts of the shopping basket. The system also outputs warnings on the high-level amounts of some of the factors, such as calories and fat.

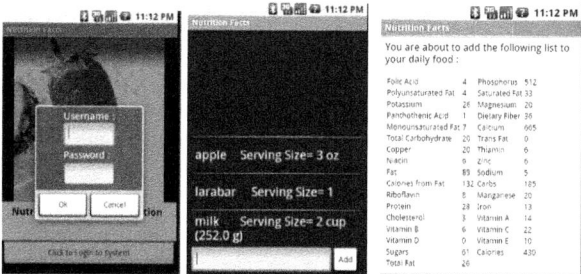

Fig. 5. Use case for the Android application

4.3.3 Report Mode

The information about the daily food intake or the shopping basket is recorded in a database upon the user's request. This information goes into related tables about food and shopping. Using the report mode, the user can input two dates to get a comprehensive report about the foods s/he consumed in that period of time.

4.3.4 User Information

The first page of the system prompts the user to login to the application by entering a user name and password. This account will link to his/her profile on the web page and will locate the relevant information in the database, located on a server.

The current database consists of three tables: the user's login information, data from shopping mode and data from eating mode. The eating mode table keeps a record of the nutritional facts of the items that the user consumes. Each row in this database represents a food that the user entered into the system. The shopping mode table records information about the shopping history of the user. Each row in this table represents the accumulated nutrition facts of the items inside a shopping basket on a certain date.

5 Conclusions

The obesity epidemic is real and threatening to affect the majority of the United State's population. With healthcare costs and the number of obese individuals increasing every year, new, low-cost measures are needed to turn this epidemic around, such as our proposed system. We propose a system that leverages mobile phones and provides the following:

1. A handheld system that is constantly with the user
2. A (mostly) automatic and accurate way to enter names of foods

3. A method to track user consumption and trends over time
4. A system that can be programmed to warn the user against unhealthy food choices

The future of this application is threefold. First, users will be able to access and view statistical data about their eating pattern history. That is, they will be able to see a daily history analysis of their food choices. Second, the application will support non-UPC foods, which is important because the eating of minimally processed fruits and vegetables, which typically have no UPC codes, are high-satiety foods. High-satiety foods fill people up with fewer calories and their consumption must therefore be encouraged. Third, users will be able to enter any dietary restrictions they have and, by scanning the barcode of an item, be able to see immediately whether or not their body can handle that food. This has been implemented with the nut detection on the Windows Mobile application but should be expanded to include other common allergies.

References

1. Tsai, C., Lee, G., Raab, F., Norman, G., Sohn, T., Griswold, W., Prick, K.: Usability and Feasibility of PmEB: A Mobile Phone Application for Monitoring Real Time Caloric Balance. Mobile Networks and Applications 12, 173–184 (2007)
2. Brown, B., Chetty, M., Grimes, A., Harmon, E.: Reflecting on health: a system for students to monitor diet and exercise. In: CHI 2006 extended abstracts on human factors in computing systems, pp. 1807–1812 (2006)
3. Toscos, T., Faber, A., An, S., Gandhi, M.P.: Chick clique: persuasive technology to motivate teenage girls to exercise. In: CHI 2006 extended abstracts on human factors in computing systems, pp. 1873–1878 (2006)
4. Consolvo, S., Everitt, K., Smith, I., Landay, J.A.: Design requirements for technologies that encourage physical activity. In: Proceedings of the SIGCHI conference on Human Factors in computing systems, pp. 457–466 (2006)
5. Ahtinen, A., Lehtiniemi, A., Häkkilä, J.: User Perceptions on Interacting with Mobile Fitness Devices. In: Pervasive and Mobile Interaction Devices (PERMID) workshop in Pervasive (2007)
6. Boissy, P., Jacobs, K., Roy, S.: Usability of a barcode scanning system as a means of data entry on a PDA for self-report health outcome questionnaires: a pilot study in individuals over 60 years of age. BMC Medical Informatics and Decision Making (2006)
7. Sensei, http://www.sensei.com
8. Myca, Nutrition, http://www.mycanutrition.com/
9. I-Nigma, http://www.i-nigma.com/personal/
10. Barcode 2D, http://sourceforge.net/projects/barcode2d
11. AURA, http://research.microsoft.com/research/downloads/Details/591ccd7c-708e-4155-a568-71ee7b979e80/Details.aspx
12. Estrin, D., Kim, D., Peterson, N., Rahimi, M., Burke, J.: Rewind: Leveraging Everyday Mobile Phones for Targeted Assisted Recall. UCLA Technical Report (2008)
13. Berckman, D.: Automatic on-line monitoring of animals by precision livestock farming. In: International congress, Saint-Malo, France, October 11-13, pp. 27–30 (2004)
14. Bhatia, D., Walker, W., Polk, T., Hande, A.: Remote Blood Pressure Monitoring Using a Wireless Sensor Network. In: Proceedings of the IEEE Sixth Annual Emerging Information Technology Conference, Dallas, Texas (2006)

15. Diamond, D., Sequeira, M., Bowden, M., Minogue, E.: Towards autonomous environmental monitoring systems. Talanta 56, 355–363 (2002)
16. Drawer, S., Fuller, C.: The application of risk management in sport. Sports Med. 34, 349–356 (2004)
17. Gupta, S., Schwiebert, L., Weinmann, J.: Research Challenges in Wireless Networks of Biomedical Sensors. In: Mobile Computing and Networking, pp. 151–165 (2001)
18. Jovanov, E.: Patient Monitoring using Personal Area Networks of Wireless Intelligent Sensors. Biomedical Sciences Instrumentation 37, 373–378 (2001)
19. Linz, T., Taelman, J., Adriaensen, T., van der Horst, C., Spaepen, A.: Textile Integrated Contactless EMG Sensing for Stress Analysis. Engineering in Medicine and Biology Society (2007)
20. Lukowicz, P., Kern, N., Schiele, B., Junker, H.: Wearable sensing to annotate meeting recordings. In: International Symposium on Wearable Computers (ISWC). IEEE Press, Los Alamitos (2002)
21. Weight Watchers, http://www.weightwatchers.com/plan/www/online_01.aspx?navid=onlineaag (accessed February 7, 2009)
22. Fogg, B.J., Eckles, D.: Mobile Persuasion: 20 Perspectives on the Future of Behavior Change. In: Stanford Captology Media, Stanford, California, USA (2007)
23. Serialio Products, http://serialio.com/products/scanner/LaserChampBT.php
24. Centers for Disease Control and Prevention, http://www.cdc.gov/nccdphp/dnpa/obesity/trend/maps/

A Personalised Body Motion Sensitive Training System Based on Auditive Feedback

Gerold Hoelzl

Johannes Kepler University, 4040 Linz, Austria
gerold.hoelzl@gmail.com

Abstract. In this paper the architecture and functionality of a personalized body motion sensitive training system based on auditive feedback is discussed. The system supports recognition of body motion using body worn sensors and gives the user feedback about his or her current status in adaptively selecting audio files accompanying the speed and path of exercise.

1 Introduction

1.1 Motivation

Being an enthusiastic sportsman, jogging (running) is one of my favourite sports. During the training sessions it is fun to listening to music, e.g. from an mp3 player, this also makes the training more amusing.

One problem is that the rhythm of the played audio file doesn't always fit to the running frequency of the exercising person. This can be very distracting.

The solution described in this paper is a system that adaptively selects audio files fitting to the rhythm of the runner. If the running rhythm changes during training (e.g. due to a power-up, exhaustion or change in terrain), the played music is automatically adapted to the new running rhythm.

Additionally the system can provide the user with status information like pulse, speed & distance or warn the user if predefined limits in those parameters are exceeded.

The aim of the project was to design a mobile, wearable system capable of fulfilling the above mentioned characteristics and to show its technical feasibility. The paper is structured as follows:

Part 2 identifies necessary system tasks needed to build such a system. Each task is described in detail and a framework is presented showing the integration of the different tasks into the whole system.

Part 3 focuses on developing a prototypical system. Requirements for hardware- and software components are defined and the prototype and its hardware components are presented.

Part 4 discusses and analysis the experimental results gathered from testing the prototype.

Part 5 concludes the paper giving a summery of the work and an outlook on future enhancements of the system.

1.2 Related Work

Being an enthusiastic sportsman and engineer as mentioned in 1.1, I'm always having a look at actual developments at the sports tool sector.

One interesting product from a cooperation of Nike and Apple is the so called "Nike+iPod" (http://www.apple.com/at/ipod/nike/). It supports the runner giving feedback via an iPod about the current running speed, distance and practicing time while the sportsman can listen to his favourite songs. It consists of special designed running shoes from Nike, an iPod from Apple and a sensor that is used to determine the running speed. The major disadvantages of the product are that on the one hand at least an iPod is needed for the audio feedback and on the other hand only a handful shoes from Nike support the needed sensor. (Runners mostly have their favourite shoes and don't want to change them).

Playlists and song playback have to be generated manually and don't adapt to changes of the context of the sportsman. There also is no support to monitor the heart rate (ECG) or the running route (GPS) and a MAC-computer is needed for further analysis of the data.

Especially for sportsmen it is important to get information about their biosatus during training. Actual projects like AMON, A Wearable Medical Computer for High Risk Patients [1] show that it's no problem to continuously monitor, analyse and log multi sensor biodata like heart rate, blood pressure, blood oxygen saturation and temperature even for high risk patients in a wearable watch like form.

This project didn't require to measure all physiological parameters and to use a device qualified for medical purpose. To get the most interesting biofeedback parameter for sportsmen, the heart rate, a Polar ecg-sensor was used (3.1).

A substantial amount of research has been performed in the area of wearable computing and context recognition. Many independent researchers have demonstrated the suitability and excellent further potential of body worn sensors for automatic context and activity recognition [2].

The available scientific literature reports about successful applications of such sensors to various types of activities, ranging from the analysis of the simple walking behaviour [3] to more complex tasks of everyday life like e.g. the recognition of Wing Chun movements [2] and even workshop assembly [4].

Many past works have demonstrated 85% to 95% recognition rates for ambulation, posture and other activities using acceleration data. Advances in miniaturization will permit accelerometers to be embedded withing adhesive patches, belts, wrist bands and bracelets and to wirelessly send data to a mobile computing device that can use the signals to recognize user activities [5]. This made the use of an acceleration sensor for analyzing the movement of a person the best option. An overview of projects using accelerometers to detect user activity is given in [5].

A more advanced, actual project dealing with sports activity is the "wearable trainer" for nordic walking mentioned in [6].

The aim of the project is to monitor user motions and ensure that the user gets the maximum benefit of the exercise while minimizing risk factors such as joint damage or overextension. It consists of unobtrusive body fixed sensors and correlates sensor signals, terrain data and user position. Inferring Motionpatterns from the collected data enables the possibility of teaching the practicing person doing the correct motion sequence and getting most out of its training.

So the "wearable trainer" tries not only to react on the behaviour of the user but rather to teach the user doing the right motion sequence and giving tips and analysis on how to improve the motion.

2 System Design

In order to adaptively select audio files corresponding to the actual running rhythm of the user the system has to handle three main tasks:

- analyzing the movement of the practicing person (2.1)
- analyzing the audio files for classifying them(2.2)
- mapping the movement-feature to the audio-feature to be able to select the correctly fitting music samples (2.3)

The development framework used for the system is described in section 2.4 showing how the above mentioned parts are working together.

2.1 Movement Analysis

The goal of the movement analysis is to calculate a feature corresponding to the running rhythm of the user.

According to Hay [7], a footstep can be split up into 13 sections (Figure 1). The important phases for analyzing the movement are phase 7, 8 and 9. During this sections defined as "front support phase (7)" and "rear support phase (8,9)" the ground-pressure of the foot leads to a peak.

Monitoring the footstep with an acceleration sensor [8] placed rear hip (on backbone origin) shows that the vertical acceleration during a footstep in a "support phase" passes the 1,8g gravitational acceleration. In all other phases ("non supported phases") it stays far beneath as seen in figure 2.

Fig. 1. Movement sections

The acceleration data shown in figure 2 were collected during a walk with moderate speed at around 4 km/h. The sensor was placed rear hip (on backbone origin). The faster the tempo is, the higher the acceleration value gets. This makes a detection of a step more easier because the spread of gravitational acceleration between a "support phase" and a "non support phase", indicating a footstep, is increasing.

The effect of different heights and weights of different people is minimal and only influences the spread of gravitational acceleration between a "support phase" and a "non support phase". The more lightweight a person is, the smaller the spread gets but easily passes the 1,8g acceleration value.

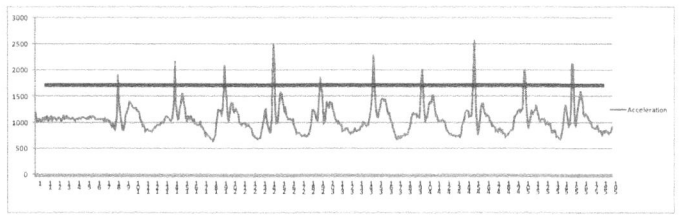

Fig. 2. Vertical acceleration

Based on knowing the characteristics of the acceleration values during a footstep, filtering the raw acceleration sensor data using a threshold filter with a hysteresis to avoid oscillations between states (as shown in figure 3) extracts the single footsteps.

Data analysis showed that a threshold value of 1,75g performs well in extracting the single footsteps. To avoid oscillations, the threshold values of the filter were set to 1,8g for high state (hH) and 1,7g for low state (hL). Figure 4 shows the raw acceleration sensor data overlaid with the extracted footsteps using a threshold filter as shown in figure 3, adjusted to the above mentioned threshold values.

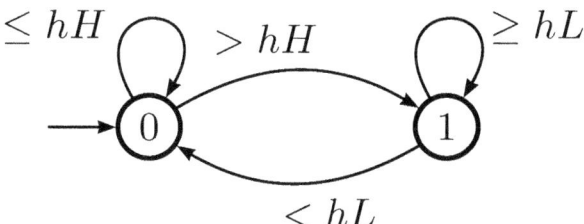

- **hH**... threshold for high state
- **hL**... threshold for low state

Fig. 3. Schema of the used threshold filter with hysteresis

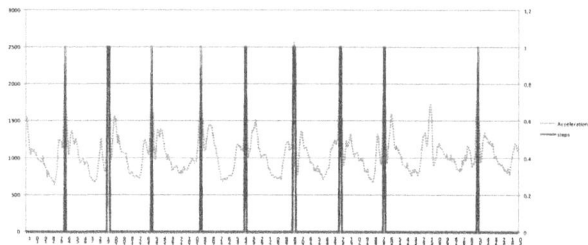

Fig. 4. Raw acceleration data overlaid with extracted footsteps

Having extracted the single footsteps out of the raw acceleration sensor data (figure 4) it is possible to define a feature that classifies the running rhythm over a specified time frame.

The feature defined as the *mean footsteptime mft[ms]* (1) is the arithmetic mean over the time between the single footsteps of the user over a specified time frame (illustrated in figure 5).

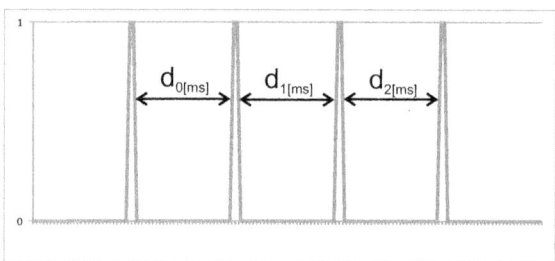

Fig. 5. Distances between footsteps to calculate mft

$$mft_{[ms]} = \frac{1}{n} \sum_{i=0}^{n-1} d(i)_{[ms]} \qquad (1)$$

where:
n ... *number of distances [ms] between the single footsteps*
$d(i)$ ... *distance between footsteps [ms]*.

2.2 Audio Analysis

To be able to map the running rhythm to music, a feature has to be extracted from the music that correspondences with its speed.

One feature that fulfills this requirement is the beats-per-minute feature [9]. It calculates how many beats per minute (bpm) occur in a given music sample. The beats-per-minute feature is typically used to measure the tempo of the music.

Humans perceive the beat as a binary regular pulse underlaying the music [9]. This qualifies the feature for mapping it to the running rhythm of the practicing person because the time between the beats can be multiple or part of the the time between the footsteps of the user.

Figure 6 shows schematically the detected beats out of an audio signal. An overview of techniques for beat tracking in music is given in [10].

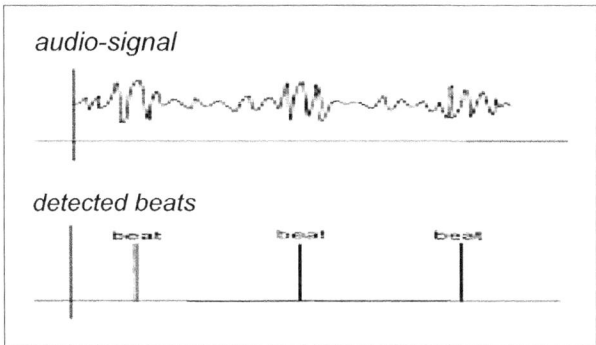

Fig. 6. Detected beats in audio sample

2.3 Mapping

Having a feature corresponding with the running-rhythm of the user (mean footsteptime (2.1)) and a feature corresponding with the tempo of the music (beats-per-minute (2.2)) it is possible to map one feature space into the other (mft↔bmp, bpm↔mft).

Therefore a mapping function has to be defined that maps the mft-feature to the bpm-feature. This enables selecting music fitting to the running rhythm of the user.

The mapping function F (2) is defined as:

$$\mathbf{O}_{[bpm]} = F(\mathbf{M}, f_{[mft]}) \qquad (2)$$

where:
$\mathbf{M} = \{m_1, \ldots, m_n\}$ ... set of music files
$f_{[mft]}$ ... mft-feature-value
$\mathbf{O}_{[bpm]} \subseteq \mathbf{M}$ ... set of corresponding music files.

Given a set **M** of music samples and a **mft-feature value** the mapping-function generates a set **O** of music samples with the corresponding **bpm-featue value**.

2.4 Development Framework

The concept framework for developing the system is shown in figure 7.

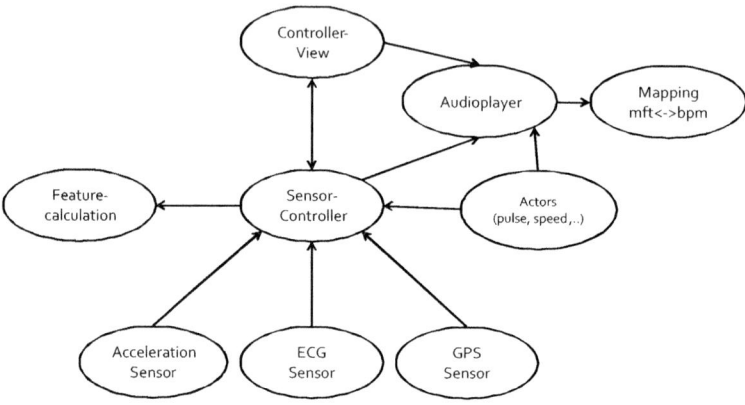

Fig. 7. Development framework

The key element of the framework is the *sensor controller*. It manages the attached sensor nodes (e.g acceleration sensor for movement analysis, GPS sensor for position, speed and direction, and an ECG-sensor). The sensor controller is expandable so new sensors can be attached easily to extend the system.

Further the sensor controller is responsible for managing the storing of the collected data from the attached sensor nodes and for periodically starting the calculation of the mft-feature value (2.1) from the acceleration-sensor data.

The mft-feature is sent from the sensor controller to the *audio-player* where the mapping (2.3) of the mft-feature to the fitting set of music files and the playing of the audio files itself takes place. The audio player is capable of playing audio files parallel and overlaid, so status information from the actors like exceeding pulse limits can be directly played into the current played music file.

The so called *actors* are bound to the data of a specific sensor (e.g. ecg) and perform checking of it's parameters in specified time intervals. If e.g. predefined limits in these parameters (e.g. heart rate) are exceeded the actors can raise warnings using the audio-player. This enables the system to inform or warn the user about different informations on different parameters of different sensors if necessary.

The *controller-view* enables the user to control the audio player (select audio files, change volume,..) and view status information from the sensors (heart rate, speed,..).

3 Prototype

To build a prototypical setup of the system, showing its technical realisation, components are needed that fulfill the special requirements of a wearable (mobile) system used during sports activity [11]. The components have to be selected considering especially the following aspects:

From the view of computing [12]:

- limited processing (computing) power and memory
- limited bandwidth
- limited user input, output
- limited power consumption

From the view of wearability [11]:

- placement (where on the body it should go)
- form (defining the shape)
- human movement (consider the dynamic structure)
- proxemics (human perception of space)
- attachment (fixing forms to the body)
- weight
- thermal (issue of heat next to the body)
- aesthetics (perceptual appropriateness)

Concluding the prototype system has to achieve two main requirements from the view of usability. It must be *unobtrusive* and *unrestrictive* meaning that it must not restrict the movement of the user and its wearing should be impalpable [11].

So it's necessary to use:

- as small and light sensors as possible with low power consumption (long operating time, avoiding heat)
- wireless communication between the components to not restrict the users movement with wires
- running on a device the user takes with him anyway (like mobile phone), minimizing the number of components the user has to take with him additionally

3.1 Hardware

During the past few years, devices like mobile phones or mp3 player became smaller and more light weight. They are frequently already integrated in one single device, equipped with communication technology like bluetooth [13], and their processing power was increased dramatically.

Many people take a mobile phone with them while doing outdoor sports to be on the one hand reachable and on the other hand able to make a call if something happens (like an injury). This and the ability of playing audio files and having more and more different sensors like GPS integrated in nowadays mobile phones make it the ideal hardware platform for hosting such a system.

Taking the requirements from 3 and the above mentioned circumstances in consideration the prototype focuses on being hosted on a mobile phone using its integrated components like the audio player or the gps-sensor being extended by components not offered by the host platform like an acceleration or heart-rate sensor using bluetooth [13] for interconnecting them.

To build up the prototypical system the following hardware components where selected:

- Processing unit (controller)
  For running the application a state of the art sony-ericsson mobile phone (www.sonyericsson.com), the C702 is used. Relevant features for the prototype are the built in gps sensor and the newest version of the javaME Platform (3.2) from sony-ericsson for mobile phones (Java Platform 8 (JP8)) (3.2). Users owning a mobile phone running this platform or higher don't have the need to take a separate processing unit with them as mentioned in 3.
- Sensors
  - acceleration sensor
    The *SparkFun WiTilt v3.0* sensor is used to measure the acceleration during a footstep. The sensor is set to 100hz sampling frequency and measures an acceleration up to 6g. This gives enough reserve for measuring the running movement when the sensor is placed rear hip (on backbone origin) [5].
  - ecg-sensor
    To enable the prototype of giving a feedback about the user's biostatus a Polar ecg sensor (http://www.polar.fi/en/) is used to measure the heart rate.
    For receiving and interpreting the signal of the polar ecg sensor the "polar heart rate monitor interface (HRMI)" from Active-Robots (www.active-robots.com) is used.
  - gps-sensor
    For getting location information like position, speed and direction the built in gps sensor of the processing unit is used.
- Component Interconnection
  For interconnecting the components of the prototype, the processing unit (controller) and the different sensors, a Bluetooth connection [13] is used.
  Bluetooth uses radio technology to establish wireless short-range communication between devices creating a "wireless personal area network" or an "ad hoc" connection [13].
  Bluetooth is de de facto standard to establish an instant wireless connection between locally distributed devices. Especially its high level of integration, very low power consumption and its reliability make it the ideal choice for connecting the components of the prototype [14].

Components of the prototype not equipped with Bluetooth capabilities (e.g. HRMI) were connected to a BlueNiceCom IV Bluetooth module from Amber Wireless (http://amber-wireless.de/) to be able to be connected to the prototype.

More and more mobile phones are equipped with built in acceleration sensors (e.g. iPhone). Use of them would be possible to analyse movement (2.1) and free users of taking a separate acceleration sensor with them. The disadvantage of this solution would be that it restricts the users of using the device for doing something else (e.g. show route). This is because for analysing the acceleration data, the sensor has to be placed and fixed in a defined position. If the position is changed, analysing is not possible any more. So a separate sensor is used to avoid this restriction.

Figure 8 shows the selected hardware components of the prototype.

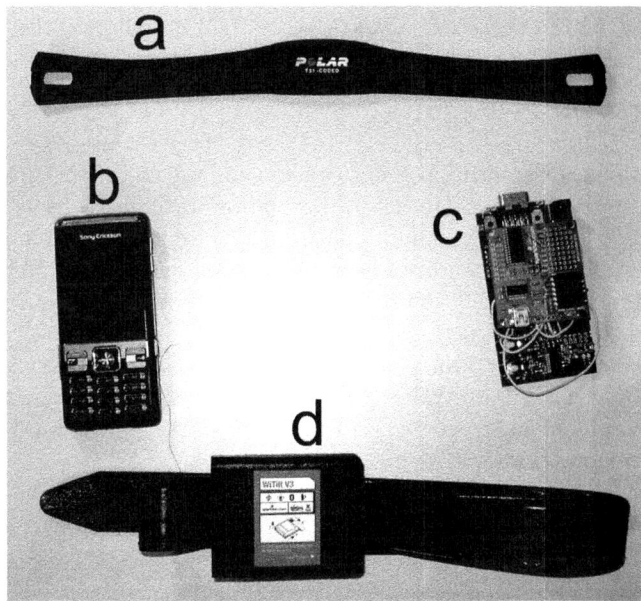

Fig. 8. Components of the prototype: a) polar-ecg-sensor, b) processing unit (Sony-Ericsson C702 JP8), c) HRMI connected to Bluetooth module (can be integrated and miniaturised), d) acceleration sensor fixed on a belt to be worn rear hip (on backbone origin)

3.2 Software

For implementing the prototype JavaME (http://java.sun.com/javame/) was selected. The reasons for selecting this application platform were the broad and free availability, being almost standard on nowadays mobile phones and the libraries especially supporting the needs of the system:

- JSR82: Java APIs for Bluetooth
  (http://jcp.org/en/jsr/detail?id=82)
- JSR179: Location API for J2ME
  (http://jcp.org/en/jsr/detail?id=179)
- JSR135: MobileMedia API
  (http://jcp.org/en/jsr/detail?id=135)

Sony-Ericsson, as other manufacturers as well, offers an implementation of JavaME for mobile phones called Java Platform in the actual version 8 (JP8) enabling it's mobile phones using JSR82 for connecting components using Bluetooth, JSR179 for accessing the built in GPS-Sensor and JSR135 for dealing with multimedia content. The schema of the prototype is shown in section 2.4.

Extracting the beat (2.2) out of the music is a computationally intensive work. For the prototypical setup the beat was extracted on a separate machine and stored as meta data to the music files.

For extracting the beat the software Mixxx [15] based on [9] was used.

4 Experimental Results

The prototype was tested in both naturalistic and laboratory environments. The laboratory tests where done to determine the performance and accuracy of the system. Testing the prototype under naturalistic circumstances should point out it's usability in ordinary conditions expected when using the system.

To determine the accuracy of the calculated mft-feature-value 2.1 a metronome was used to produce a defined, steady pulse to help the testing participant to get into the right running rhythm. The so collected data showed a very high accuracy of the calculated mft-feature-value (table 1). The time derivation in the measured values is explainable due to how exact the testing participants could adjust to the defined running rhythm.

Table 1. Accuracy of calculated mft-feature-value

expected [ms]	measured [ms]
400	380-420
500	475-530
700	660-730
900	870-935
1100	1050-1160

To get the performance of the system, the time needed to calculate the mft-feature was measured.

Therefore the possible options of connected sensors (acceleration, ecg and gps), the time frame of steps to calculate the mft-feature for, and the use of the audio player (ua→using audioplayer, no→no audioplayer) were varied (table 2).

Table 2. Duration of mft-feature calculation [ms]

	TIME FRAME FOR CALCULATING MFT-FEATURE					
	last 5 sec		last 10 sec		last 20 sec	
USED SENSORS	na	ua	na	ua	na	ua
acc	4-6	6-15	10-40	25-70	20-70	90-140
acc+ecg	4-7	7-17	10-50	20-75	20-80	95-150
acc+ecg+gps	GPS sensor causes crash of bluetooth connections					

Summarizing table 2: the system has a good response time (for this purpose) in calculating the mft-feature value even having a time frame of 20 sec

($\approx$ 2000 stored values) and the audio player is used. The time derivation in the calculation-time of the mft-feature-value is mainly explainable due to how many steps are recorded in the time frame, but also background tasks of the mobile phone affect the calculation time.

The use of the audio player significantly increases (doubles) the time needed for calculating the mft-feature, showing that for the parallel decoding of a mp3-file a lot of resources are needed, and it seems that the mp3-decoding is not being implemented in a separate hardware component.

The use of a second sensor (ecg) didn't increase the calculation time much. This can be explained because of the low sampling frequency the ecg-data is collected (750ms) and no further calculation intensive processing of the data is performed.

Unexpected behavior of the system was the crash of the bluetooth connections when the built in GPS-sensor was used. Using the GPS-sensor resulted in crashing both bluetooth connections (gps-sensor was still working) and the need to power off and restart the system (phone) to get bluetooth working again. This is difficult to explain but looks like a bug in the JavaME implementation.

After performing the tests, the mft-featue-calculation interval was set to 250ms to get a real time response of the system, and the time frame of steps to calculate the mft-feature for was set to 8 sec.

Test participants told that a longer time frame smooths the mft-feature too much when changing the running rhythm having a too long adjustment time to the new running rhythm. Contrariwise having a shorter time frame results in a too sensitive reaction of the system.

The calculation interval of 250ms was selected to get on the one hand a "feeled" realtime response and on the other hand not to fully load the system. A fully loaded system results in an instable gathering of the acceleration sensor data because the system is fully utilized doing the calculation of the mft-feature-value.

The calculation interval of 250ms and the time frame of 8 sec results in a processor utilization between 70-90% giving the system enough reserves for doing "mobile phone internal" stuff (e.g. managing bluetooth connection, network or broadband connections,make a call,..).

Testing the prototype under natural circumstances (running on the street and on loosely ground (crushed stone road)) showed the same good results as the testings in the laboratory environment.

5 Conclusion and Future Work

In this paper a novel approach for adaptively selecting music fitting to the running rhythm of a user was presented.

It solves the problem of listening to music having a distracting rhythm during training. Three main tasks, (1) analysing the movement of the practising person, (2) analysing the audio files for classifying them and (3) the mapping function between the movement-feature and the audio-feature were described.

A Framework on how to build such a system was shown and a prototype demonstrating the technical realisability on off the shelf hardware was developed.

Using and testing the developed prototype showed that its hard to develop a mapping function that fits to all users because the rhythm of the music is sensed very subjective by each person. Therefore an enhancement of the system would be to enable the learning of the mapping function on the fly during training using techniques from machine learning like decision trees [16] or nearest neighbour classification [17].

Another interesting focus for further research is not only to select music fitting to the running rhythm but rather build training programs forcing or retaining the sportsman in selecting pushing or assuasive music according to predefined parameters like e.g. pulse-limits.

Enabling the system of being tracked and monitored using a remote computer as shown in figure 9 enables a trainer or a third person to watch the users and their status in real time. Also competitions over e.g. continents are thinkable if the users run the "same" track (e.g. 400m in a stadium) and the systems can be tracked and synchronized comparing the position of one to each other.

Fig. 9. Tracking and monitoring the system using a remote computer

Acknowledgements

This work was done as a master project at JKU Linz[1]. My thanks go to Prof. Gabriele Kotsis[2] for her encouragement and helpful suggestions during this work.

---

[1] Johannes Kepler University Linz, Austria, Altenbergerstraße 69, 4040 Linz, http://www.jku.at

[2] Professor Gabriele Kotsis, Johannes Kepler University Linz, Austria, Altenbergerstraße 69, 4040 Linz, Email:gabriele.kotsis@jku.at

References

1. Lukowicz, P., Anliker, U., Ward, J., Troester, G., Hirt, E., Neufelt, C.: Amon: a wearable medical computer for high risk patients. In: Wearable Computers (ISWC 2002). Proceedings of the Sixth International Symposium on Wearable Computers, pp. 133–134 (2002), ISBN: 0-7695-1816-8
2. Heinz, E.A., Kunze, K., Gruber, M., Bannach, D., Lukowicz, P.: Using wearable sensors for real-time recognition tasks in games of martial arts. In: Proceedings of the 2nd IEEE Symposium on Computational Intelligence and Games (CIG), pp. 98–102. IEEE Press, Los Alamitos (2006)
3. Lee, S.-W., Mase, K.: Recognition of walking behaviors for pedestrian navigation. In: Control Applications (CCA 2001). Proceedings of the 2001 IEEE International Conference on Control Applications, pp. 1152–1155 (2001), ISBN: 0-7803-6733-2
4. Lukowicz, P., Ward, J.A., Junker, H., Stäger, M., Tröster, G., Atrash, A., Starner, T.: Recognizing workshop activity using body worn microphones and accelerometers. In: Ferscha, A., Mattern, F. (eds.) PERVASIVE 2004. LNCS, vol. 3001, pp. 18–32. Springer, Heidelberg (2004)
5. Bao, L., Intille, S.S.: Activity recognition from user-annotated acceleration data. In: Ferscha, A., Mattern, F. (eds.) PERVASIVE 2004. LNCS, vol. 3001, pp. 1–17. Springer, Heidelberg (2004)
6. Lukowicz, P., Hanser, F., Szubski, C., Schobersberger, W.: Detecting and interpreting muscle activity with wearable force sensors. In: Fishkin, K.P., Schiele, B., Nixon, P., Quigley, A. (eds.) PERVASIVE 2006. LNCS, vol. 3968, pp. 101–116. Springer, Heidelberg (2006)
7. Hay, J.G.: The biomechanics of sports techniques. Prentice-Hall, Englewood Cliffs (1978), ISBN: 0-13-077164-3
8. Beigl, M., Krohn, A., Zimmer, T., Decker, C.: Typical sensors needed in ubiquitous and pervasive computing. In: Proceedings of the First International Workshop on Networked Sensing Systems (INSS 2004), pp. 153–158 (2004)
9. Jensen, K., Andersen, T.: Beat estimation on the beat. In: Applications of Signal Processing to Audio and Acoustics, 2003 IEEE Workshop on Applications of Signal Processing to Audio and Acoustics, pp. 87–90 (2003), ISBN: 0-7803-7850-4
10. Hainsworth, S.W.: Techniques for the automated analysis of musical audio. Signal Proscessing Group, Department of Engineering, University of Cambridge, Tech. Rep. (December 2003)
11. Gemperle, F., Kasabach, C., Bauer, J.S.M., Martin, R.: Design for wearability. In: ISWC 1998: Proceedings of the 2nd IEEE International Symposium on Wearable Computers, pp. 116–122. IEEE Computer Society, Los Alamitos (1998)
12. Estrin, D., Culler, D., Pister, K., Sukhatme, G.: Connecting the physical world with pervasive networks. IEEE Pervasive Computing 1(1), 59–69 (2002)
13. Dideles, M.: Bluetooth: a technical overview. Crossroads 9(4), 11–18 (2003)
14. Krassi, B.A.: Reliability of bluetooth. In: Proceedings of the 12th Conference on Extreme Robotics, RTC, St. Petersburg (2001)
15. Andersen, T.H., Andersen, K.: "Mixxx" (2009), http://www.mixxx.org/
16. Bringmann, B., Zimmermann, A.: Tree 2 - decision trees for tree structured data. In: Jorge, A.M., Torgo, L., Brazdil, P.B., Camacho, R., Gama, J. (eds.) PKDD 2005. LNCS (LNAI), vol. 3721, pp. 46–58. Springer, Heidelberg (2005)
17. Cover, T., Hart, P.: Nearest neighbor pattern classification. IEEE Transactions on Information Theory 13(1), 21–27 (1967)

Context-Aware Authentication Framework

Diwakar Goel, Eisha Kher, Shriya Joag, Veda Mujumdar,
Martin Griss, and Anind K. Dey

Carnegie Mellon Silicon Valley,
NASA Research Park, Bldg. 23, Moffet Field, CA 94035
{diwakarg,ekher,sjoag,vmujumda}@sv.cmu.edu,
martin.griss@sv.cmu.edu, anind@cs.cmu.edu

Abstract. We present an extensible context-aware authentication framework which can adapt to the contextual information available in a smart environment. Having confidence in a user's identity and other contextual information is critical to the successful adoption of future mobile, context-aware services. This authentication framework can provide a standard base for the development of context-aware services, particularly while the user is mobile. Our implementation of the framework enhances usability during authentication by replacing the need for users to remember and enter their password with the act of a simple gesture. We discuss our architecture, implementation and policies and illustrate how they support usable authentication, using lightweight tagging and simple context from a smart environment.

General Terms: Barcode Management System, relational database, access policies.

Keywords: Barcode, context-aware computing, Wi-Fi signatures, location-based authentication, mobile computing, soft sensors, authentication.

1 Introduction

Context is the set of facts or circumstances that surround a situation or an event. [Dey 2001] The relative increase in the use of mobile systems as compared to desktop systems presents the opportunity to retrieve the dynamic context of mobile users and have it be ubiquitously available to various services. For example, in an increasing number of mobile applications, it is important to know the current location of a user. As context-aware systems become more pervasive, they have started incorporating a wide array of contextual cues pertaining to a user. Contextual cues help the computer system to better understand the state of the user and make informed decisions about any services requested by the user. The benefit of using context-aware systems is that a user does not have to explicitly provide information about her state, but this information can be sensed and used by the system nonetheless. This increase in the use of context-aware systems and smart environments suggests the potential value of a framework that would allow such systems to make a correct decision about a user's identity. Our implementation aims to address this need and presents an extensible framework for context-based authentication systems. The approach presented in this paper aims at circumventing the tedious task of remembering complicated passwords or carrying any additional tags at the user's end for frequent accesses to a resource; rather it enables the user to act naturally in an environment, which is unique to him, and have that unique, natural interaction authenticate him.

Through this work we aimed at striking a balance between relying completely on automated actions taken by the system on one hand, and those performed explicitly by the user for authentication on the other. We selected a lightweight mechanism for capturing user actions, QR codes (a two-dimensional barcode), owing to the several advantages in deployment they have over other access control systems. They are easier to generate and can be read by nearly all camera-equipped phones (which are reaching ubiquity). Moreover, this technique is robust against any kind of sniffing attacks which other radio based tags like RFID are susceptible to. We have further combined role- and location-based access control to leverage context, thus allowing the creation of rules that are based both on roles and location, *e.g.*, a *student* can access the environmental control system in the *room* where he is attending his class or attending a meeting in a conference room. The set of policies written for this system facilitates its decision-making capability, based on the user's role and the location she attempts to access.

The remainder of the paper proceeds as follows: Section 2 acknowledges the related work performed in this field and elucidates how our proposed implementation extends this work. Section 3 introduces the basic context-based authentication scenario used to explain the working of the proposed framework. Sections 4 and 5 explain the implementation of the architecture and the design for our context-aware based authentication framework. Section 6 describes the various experimental scenarios which help us demonstrate the strength of our framework under different circumstances. It also introduces the policies used for defining the access control mechanism in our implementation. Section 7 discusses the attacks and threats against the system and how they can be handled. In the final Sections 8 and 9, we discuss the future work which can be performed to improve this framework and make it more scalable, and provide a conclusion.

2 Related Work

We leverage the concept of context-aware computing from the many papers related to Context Models and frameworks, Context and Devices, and Context and Mobile Professionals [Dey & Abowd, 2000]. In one such paper, 'Seeing is Believing' [McCune et al., 2009], the authors propose to use the camera on a mobile phone as a new visual channel to achieve security properties formerly attainable with techniques as passwords, but now in a more intuitive manner. This collection of approaches is termed: Seeing is Believing (SiB). The paper discusses the several possible configurations considering the presence of a camera, a display or both. We have used this as a basis for our approach here. SiB can be used to establish a mutual security context between the devices, without a trusted authority. The protocol discussed in the paper involves a pre-authentication phase where the user captures the digest of the other user's public key (with whom they want to authenticate). The paper also describes Unidirectional Authentication using SiB in which one device is display-less and the other one does possess a display.

In a similar fashion, our project focuses on unidirectional authentication of the user with the help of a smart phone equipped with a camera. The uniqueness of our project stems from the fact that we aggregate contextual cues with the QR codes to authenticate the user. In our protocol we present three levels of authentication depending upon the risk associated with the location the user wants to access and the role of the user. In our scenario, access to the foyer of our facility requires the lowest level of authentication, while the conference room and server room have the intermediate and highest levels of authentication, respectively.

We propose to combine both passive and active authentication in order to achieve the aforementioned authentication levels. Cerberus [Al-Muhtadi et al., 2003] is a system which supports multilevel authentication, where principals are associated with confidence values. Its context infrastructure captures rapidly changing context information and incorporates it into the knowledge base. Rather than relying solely on expensive cryptographic measures like public key exchanges and certificates, we too use contextual information from the user's device such as calendar hints, GPS location, and Wi-Fi location to authenticate after the QR code has been scanned at the location where the user intends to gain access. In Cerberus, context-aware security policies are described in an expressive language that supports binary operators, quantification and complex inferring while our access policies are based on the XACML standard (eXtensible Access Control Markup Language). Unlike Cerberus, we use a combination of light weight tagging using QR codes and contextual information for authentication.

In the paper, Using camera-equipped Mobile phones for interacting with real world [Rohs & Gfeller, 2004], the authors demonstrate the feasibility of recognizing 2-dimensional visual codes with resource-constrained mobile phone devices. This has been achieved using low-quality images obtained from integrated CCD cameras and even QR codes. A light-weight recognition algorithm has been designed such that it minimizes the use of floating point operations in order to achieve reasonable recognition rates of the code. Since the codes can only encode a limited amount of information, they normally serve as a key that is resolved to the actual data of interest, which is the basic design concept of our framework.

Our solution uses both role- and location-based access control to leverage context, and thus allows creation of rules that are based both on roles and location. Covington et al. [2002] present a Context-Aware Security Architecture (CASA) for providing authorization services in context-aware environments and applications. This framework supports the collection of contextual information from resources, the environment and the users who interact in that environment. In addition, they have explored, through the context-aware security architecture, an implementation of the Generalized Role-Based Access Control mode. Like our work, XML is used to specify access policies, role definitions and relationships and is also used as a common representation to share data between the various services in the architecture. The architecture is based on the context toolkit and a complex algorithm is used to authenticate users on the basis of the context and environmental data. In our work, the architecture contains an access management module which derives data from the store/database and the user is authenticated on the basis of the context and roles they belong to. In future work, we will propose an algorithm which identifies the user through his unique profile along with the environmental cues.

3 Understanding Context-Aware Authentication Scenario

To better explain the implementation of our context-aware authentication framework, we present a general scenario of a student attending school on a regular day.

The student's personal information is registered with the server managed by the school. His personal information includes name, email ID, username, academic track as well as context such as his default location (study room), meeting times and usual arrival time on campus. He is also assigned the role "student" and his access permissions are determined accordingly.

The QR codes are displayed at screens at various locations in the school. On scanning a QR code with his smart phone, the user receives a URL on his phone. When he accesses this URL using Wi-Fi or some other network, information such as the id of the application and history about his previous access (encrypted on the phone) and current location is posted in the response. This information is verified against his personal data and location coordinates on a relational database on the server. Upon success, her presence is registered and she is permitted access to the interior premises. These stages can be seen in Fig. 1.

Fig. 1.a. QR code displayed on a screen

Fig. 1.b. User receives URL on his phone after scanning the QR code

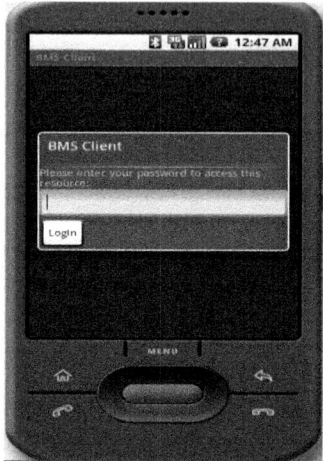

Fig. 1.c. User prompted to enter password for high level authentication

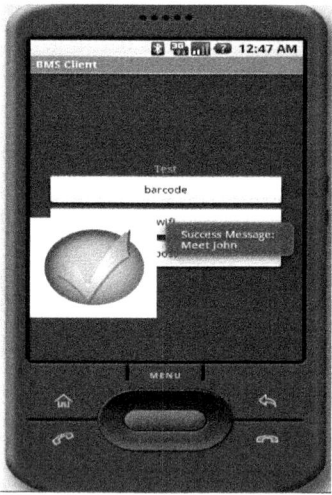

Fig. 1.d. User granted access after successful authentication

4 Architecture

Our context-aware authentication system has the following major components, as seen in Fig. 2.

1 Core Access Management Module(CAMM)

This is the access management module which decides what challenges are presented to the user at the authentication sites and maintains usage patterns for every user. This module has two main entities:

a. QR Code Generation and Management

This system maintains specific information for each QR code being displayed across all authentication sites. The characteristics that are stored are:
- The unique code being displayed
- Expiry time
- Message (optional)
- Owner
- Display location

b. Usage Patterns

The system keeps logs of all authentication attempts, their times, their results and other context information that was provided or sensed in relation to every attempt. This provides valuable data to interpret system usage, learning user behavior, checking malicious usage and providing feedback on efficiency of deployment sites by giving information on which sites are popularly used and if a better distribution of sites could be achieved.

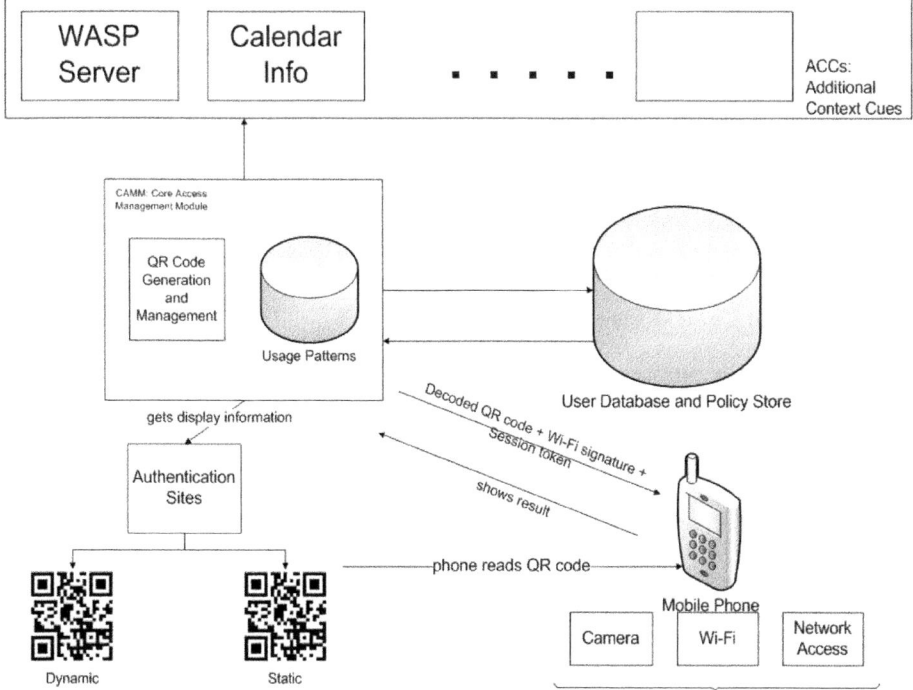

Fig. 2. Architecture of Context Aware Authentication Framework

2 User Database and Policy Store

This database stores user-specific information and policies to make decisions on whether to authenticate a request or not. Two essential pieces of information that are stored here are:

- Session token: The server maintains a list of active session tokens for each user, which is checked on each request. This token is freshly generated upon the completion of every successful authentication attempt and stored on the user's device to be used with the next attempt.
- Calendar id: To access the user's calendar information retrieving the relevant details to be used in access policy rules.

3 Client Mobile Device[1]

Our implementation requires the following capabilities on the mobile device:
- a. Camera equipped: To be able to scan the QR code displayed.
- b. Wi-Fi enabled: To utilize the WASP [Lin et al., 2009] interior positioning system's implementation on our campus.
- c. Network Access: To make authentication requests to the server.

4 Authentication Site

The system can have different kinds of authentication sites based on security and cost requirements.
- a. QR code dynamic display: These are linked to the server and display a dynamically generated code. These may be displayed on kiosks, tablet phones or computer screens.
- b. bQR code static display: These are static codes which may be printed on paper or any other material and displayed at necessary locations which may require authentication.

5 Additional Context Cues (ACCs)

These are the other pieces of contextual information that may be available in a smart environment. Policy rules can be easily added to accommodate such ACCs. These are very important for learning a user's behavior and to eventually make our system more intuitive. Also, the availability of multiple ACCs provides a fallback mechanism in cases where some context information is not available. Some of the ACCs that were considered in our implementation were:
- a. Weighted Access Point Similarity Positioning System. (WASP)[2] [Lin et al., 2009]: This system was implemented in our test environment; it provides a probabilistic location of a mobile handset in a building.

---

[1] The specific client application is not necessary and any bar code reader that can make a web request may be used. Although, using our client application enables a rich set of additional features which would be not available on other devices.

[2] The system can be used even in the absence of an implementation of the WASP system. Though, this would lead to the absence of the context cue available from indoor locationing.

b. Calendar information: User's time-schedule information from his online calendar. The attendees for an event are also considered to aid in the decision-making process.

These additional context cues can be used in situations where the system fails to authenticate a legitimate user in a location requiring a low or intermediate level of authentication. If, for example, a user is scheduled to attend an event to which he is unable to gain access, and the calendar information of a person authenticated for that event shows that the person stranded outside is also an attendee, the stranded person can be granted access.

The contextual cues used are Wi-Fi locationing, calendar, IMEI, user's role (such as student or faculty in a school environment). The cues can be extended to user alarm, to-do list, battery usage etc. and history can collectively be used to create a unique user profile. This unique profile can be used for authentication in absence of certain contextual cues.

We have defined a scalable system where the components CAMM, Client Mobile Device and ACCs can be easily expanded or replaced by other technologies. To illustrate this: the Access Management Techniques can be extended with NFC[3], Bluetooth security systems [4] or other biometric systems; the client mobile device can be extended with other capabilities through usage of NFC or Bluetooth; and the additional context cues can include the user's movement history or social context. We implemented this system using QR codes to first illustrate the basic concepts and secondly to create a system with low cost and certain definite advantages provided over other alternatives such as:

- QR codes support fast reading, using only the image recognition available across camera phones.
- Screens for displaying codes are everywhere.
- Codes can be printed and deployed anywhere.

5 Design

The user scans the QR code, which may be statically or dynamically generated depending on the security needs, at the reception area of the school building or any of the internal locations like the common study room, classrooms, library, and cafeteria or conference room. The QR code system provides a unique code to the user application, which along with other context from the phone, such as the session token and Wi-Fi signatures are sent back to the server as part of a request for further access. The relational database on the server maintains specific details about a user's roles, personal information, and alarms set in the user's calendar, and a pattern such as regular weekday routines, which may include the details of meetings, conferences or classes that the user is scheduled to attend. Information about the user's social context such as

---

[3] FeliCa-Contactless IC card technology by Sony,
  http://www.sony.net/Products/felica/abt/index.html
[4] BlueAccess BAL-100-BL, a wireless access control system.
  http://www.bluelon.com/index.php?id=248

whom else is in same location or neighborhood, gesture information determined from an accelerometer or from time history of location may also be recorded.

The access policies are based on XACML which is an OASIS standard that describes both a policy language and an access control decision request/response language. Using XACML simplifies the implementation and standardizes the policy set specification. Depending on the response collected from the user and the access policies defined, the user may be authenticated and granted access rights. The access rights can be role-based or location-based or a combination of both.

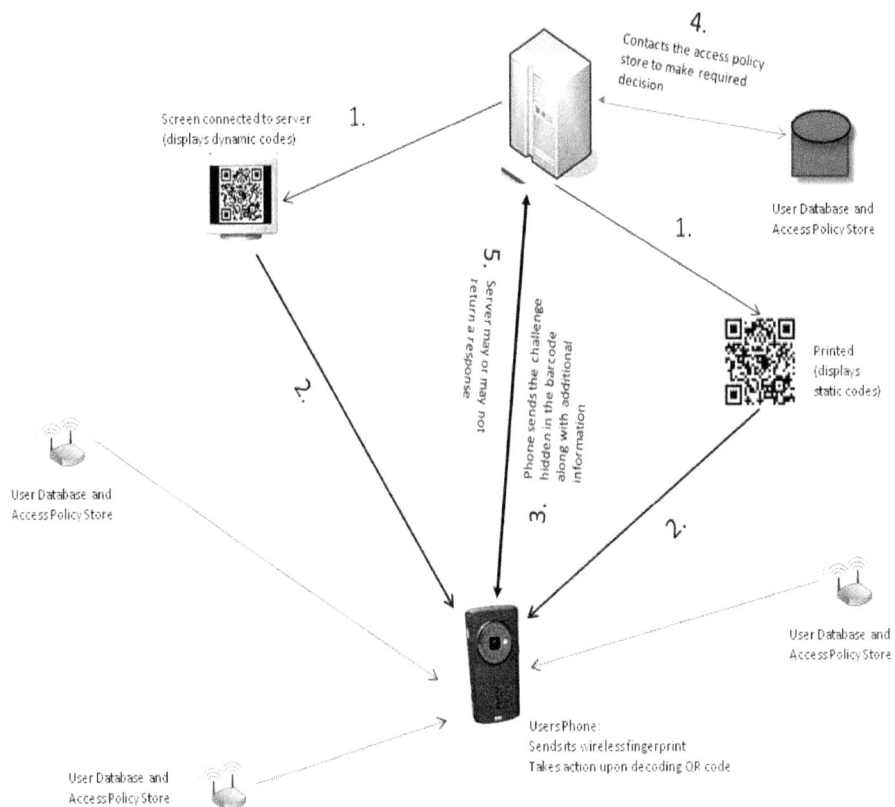

Fig. 3. Usage Scenario for the Framework

An example illustration of the working of our system is shown in Fig 3. It is comprised of the following steps
1. The user sees the deployed screen and touches it to see a QR code.
2. The user uses his phone to scan the code, and the phone automatically takes action sending the decoded code and its valid session token.
3. The server communicates with the user database and policy store to make a decision on whether to authenticate the user or not.

4. The server returns the response, which may be a success and a personalized message or a failure, or depending upon the access rules may prompt the user for further information.
5. The phone screen displays immediate feedback to the user about his action.

6 Experimental Scenarios

For the initial experimentation and testing purpose, we have developed several scenarios for location-based authentication on our campus. We establish role-based access control by using parameters such as Wi-Fi location and soft cues like meeting schedule from our campus' calendar and users' usual arrival time.

The users need not carry additional tags like RFID for authentication nor do they need to provide manual information about their schedule. The QR codes are displayed dynamically on screens as well as static printouts at any place for which access needs to be controlled. The QR codes represent the location as well as provide a link to access the server to request access. Also, by using dynamic screens we can generate new QR codes at different time intervals which are safer tokens owing to the randomness. Using static printouts, on the other hand brings predictability and they again need to be changed manually and repeatedly to ensure that the same token is not used for a longer period of time. The interaction between the smart phone and the Barcode Management System is the crux of the application.

6.1 Scenario Based Roles

The users have been assigned different roles per their occupation. These roles determine what access permissions are to be given to the user for various locations. The four current roles include:
- Faculty
- Students
- Admin
- Visitor

Example: A user with role "Student" has access to the classroom but not to the computer server room. Only the user with role "Admin" has access to the server room. "Faculty" can always access their offices but "Student" needs prior permission to access the faculty offices, for example, for a scheduled faculty-student meeting.

The user ID consists of any of the parameters like phone IMEI number, user name and password. Wi-Fi locationing helps verify the presence of the user at a particular location using Wi-Fi co-ordinates. The calendar schedules are extracted from our campus' calendar.

6.2 Scenario 1

Goal: To recognize the presence of the user for registration/monitoring purpose and provide him with a session key

- User enters the school building
- User receives a QR code on a screen at the foyer (A new QR code is generated at regular intervals)
- User scans the QR code and gets a URL
- After confirming his User ID (with the personal information stored in the database), arrival time and Wi-Fi locationing, the data are appended to the URL
- The user now accesses that URL and thus sends context to the server using HTTPS
- The server checks the request parameters and ascertains authenticity of the user depending on the entries in the relational database and access polices. Depending on this, the server makes decisions on the access rights that should be made available to the user
- Moreover, further services can be made accessible via the person's phone.

6.3 Scenario 2

Goal: Access to specific locations like classroom, conference rooms, faculty offices per calendar hints

- There are QR code screens in front of classrooms, faculty offices, and conference rooms
- User receives a QR code on a screen at the entrance of the desired room
- User scans the QR code and gets a URL
- After confirming his IMEI number, calendar settings and Wi-Fi locationing, the data are appended to the URL
- The user now accesses that URL and thus sends context to the server
- Depending on the request parameters, the server ascertains authenticity of the user for access to specific locations by checking the meeting schedule (in the campus calendar) and permissions. The server makes decisions about the access permission to be given to the particular user. In case of access to the server room an additional authentication is required by the user by entering a password valid only for 'Admin'. On validating the password, the server permits access to the 'Admin'. On an unsuccessful attempt (invalid password), the permission is denied.

6.4 Access Rules and Policies

The access rules and policies defined for this project are role-based as well as location-based. These access policies work as the guidelines for the server to allow or deny access to a particular location such as classroom, conference room, faculty room or computer server room for the user. Again, per the role, the user is granted access to the specific location.

We have three levels of authentication associated with different locations depending upon the need for access control and associated risk at that location. The risk is minimal at places like classrooms because the access to location and data is not highly confidential, whereas server rooms are most critical in terms of security of data. The following table maps levels of authentication and risk levels with their associated locations in our scenario:

Table 1. Authentication vs. Location Map

Level of Authentication	Location associated with it
Low	Classrooms
Intermediate	Conference , Faculty room
Highest	Server room

These levels help in identifying the different levels of authentication required for various areas in the building. For example: A student gets access to his usual study area, but a classroom would be accessible only when the class on his schedule is conducted. A student is never allowed to get access to the server room which is associated with the highest level authentication in our scenario. For any other location access, the dynamically fetched values from the student's calendar are matched against the time at which the code is scanned and the other details such as whether there is a scheduled meeting at that time, and finally, according to the policies he/she is granted access for a certain period of time. In the future, we will explore the use of multiple levels of authentication for services within a particular location.

An example of the policy corresponding to the low level of authentication, where a user gets an access to the classroom is:

Rule 1

If ((Room type = Classroom) && (User ID belongs to Role = Student)) – According to Table 1
{
    Permission = allowed
}

An example of the policy corresponding to the intermediate level of authentication, where a user gets an access to the conference room is:

Rule 2

If ((Room type = conference room && User ID = belongs to the list of User ID allowed))
  {
    if (the start time of the event from the CMUwest_calendar > = [within limit of +- 15 mins] scan time of the code)
    {
      Permission = allowed
    }
  }
else
    Permission = Deny.

An example of the policy corresponding to the highest level of authentication, where a user is denied of the access to the server room is:

Rule 3

If ((Room type = Server Room)
&& (User ID belongs to Role = Student))
{
  Permission = Deny
}

For visitors wanting to authenticate, a smart phone with the appropriate application must be given to them (likely in the foyer or reception area), if they do not already have one. Policies would have already been defined for the role Visitor, providing access to different locations in the building. In the future, to keep this project more scalable and the policies easy enough to be used in a distributed environment, we will design policies based on the concept of XACML.

7 Threats and Attacks

In this section we discuss the possible threats and attacks possible specific to our implementation considering the QR Code Generation and Management System to be the CAMM (Core Access Management Module). Here, we present several scenarios that could arise with such a system in place and evaluate its robustness in light of the same.

7.1 Replication of a Displayed Code for Reuse

Each QR code displayed at any authentication site has an expiry time and a binding with the location it has been deployed at. The validity periods can be calibrated to the security needs at the site, but are usually set to short intervals of a few minutes. This prevents a user from replicating a displayed QR code and using it later. Also, in the event a user tries to use a QR code from a location different from the authentication site, the request would be legitimate if made within the validity period because it would mean that the user was actually able to read the code from the correct location. In case of static codes, this is not considered a issue as they are aimed at a lower level of security and for services that do not need physical proximity for usage, for example: printers or Wi-Fi access.

7.2 Cloning or Theft of a User's Device

In the event that a user's phone is covertly cloned the framework is able to significantly limit the attack window. The attack window is limited to the time from which the phone was cloned to the next time the legitimate user makes an authentication attempt. If the legitimate user is able to successfully authenticate himself on his attempt after the cloning took place, it automatically makes the cloned copy defunct because now a new token sent by the server is stored on the phone which must be used for any subsequent attempts. While, in the other case where the legitimate user is not able to authenticate himself in his next attempt, he knows that there has been a breach and can take actions to deactivate his account.

On the other hand, the event of device theft is more difficult to contain but several properties of our implementation offer interesting solutions. Firstly, the system relies on physical presence of the device in front of the authentication site. It may not always be possible for the thief to appear publicly in front of a secured area risking suspicious activity and identification by others in the vicinity. Moreover the system can easily adapt to sense of insecurity or possible breach as discussed in 7.4.

7.3 Brute Force or Guessing Attacks

Each authentication attempt requires a user-requested short-lived code (through interaction with the touch-screen) and also a secure token received by the user from his past successful request. The generation of code on a site is highly random owing to the fact that they are generated when a user is at the site. This lends a unique randomness and limits the possibilities of finding a pattern in code generation. Moreover, codes expire in a couple of minutes which make it extremely difficult for an attacker to be able to brute force a 200 character value (as of the current implementation).

Moreover, to check brute force attempts, the CAMM can automatically disable a user account on receiving more than n invalid requests.

7.4 Sense of Insecurity or of Possible Breach

If a sense of insecurity or of a possible breach is felt in the deployment space, the system can be easily adapted to guard against both. If a general sense of insecurity is felt at a site, the level of security can be elevated temporarily to add another level, which would request the user to enter his password upon a valid initial request.

Also, if specific devices or user accounts are suspected of being breached, those accounts can be de-activated or security levels could be elevated for just that user.

7.5 Attempts at Faking and Manipulation of Context Information

Attempts could be made by malicious users to try and manipulate context information to trick the system into granting extra privileges. For example, a malicious user may fake calendar entries in order to access a specific resource. To thwart such attacks, we plan to assign weights to various context cues considering their susceptibility to such attacks and also try and achieve peer verification where possible. For example, calendar entries would have a lower weight as compared to past history information or group membership. Also, an attempt would be made to see if there were other participants involved and if they have actually confirmed this calendar edit.

7.6 Other Attacks

The system relies on the strength of random number generation algorithms and security of over the air transmissions (on WLAN or available telecommunication networks) and weaknesses in these could directly affect this framework. But, securing over the air transmissions is outside the scope of this paper, and for all purposes it is assumed secure, though HTTPS is used for web requests.

Moreover, there arises an opportunity for an attacker to disrupt the system if he is able to gain access to any of the user's accounts like email or calendar due to the user's carelessness or system vulnerabilities. Prevention against such attacks is outside the scope of this paper.

8 Challenges

Our suggested system is able to overcome critical shortcomings in currently used access control systems like physical keys or RFID tags like reducing the extent of breach on loss, doing away with the inconvenience of carrying an additional object just for access privileges alone and other attacks like passive sniffing in the case of RFID.

But, there are other challenges that crop up with our system like, support across the wide range of handsets, lighting conditions, ensuring network availability on the phone. The current prototype is built using the android platform. In case of poor lighting conditions or unavailability of network, the system will have to revert to password mechanism for authentication. Moreover, as with any new system, convincing people of the reliability and trustworthiness of this mechanism is a challenge with access control to physical resources.

9 Conclusion

In this paper we have described a new model for authentication in context-aware environments. We have used a combination of a user's context, authentication policies and light weight tagging to support role-based and location-based access control. This framework can be extended to support other contextual information from available resources, the environment and the users who interact with that environment.

QR codes have been presented as a means of authentication owing to their unique advantages like simple and inexpensive deployment, increased centralized control over authentication sites, rising presence of such capabilities in today's mobile phones and the novel 'line-of-sight' property they provide which is non-existent in other radio-based options.

10 Future Work

We will continue to iterate on our design and implementation of the Mobile Context Management System and demonstrate that we can apply it to assist in authenticating users and controlling access to data and services. We will prototype and evaluate several multi-user mobile applications incorporating this framework, such as a mobile transportation advisor and arranger, mobile eldercare concierge and monitor, mobile people finder, mobile environment controller, mobile meeting planner/scheduler and mobile advertising/shopping [Griss et al., 2002].

Furthermore, we will explore if this framework actually improves a user's experience while mobile, by providing appropriate services and information in appropriate situations, all in a secure manner. We would also like to extend our system to provide a robust platform for significant groups of users, adding multi-user context sharing, additional context for critical levels of authentication and management to support mobile social applications.

Our future focus will also be on *fingerprinting* context attributes and using cues from user's history which would help the system to learn and define unique user profiles to authenticate users. The system will hence learn to give access to a user depending on the user profile rather than the role assigned to the user. Example: If a student has been attending meeting at a particular time every Monday for a month or more, he should be allowed access in the future depending on his previous registrations at the location.

Acknowledgements

This research was supported by grants from Nokia Research Center and by CyLab at Carnegie Mellon under grant DAAD19-02-1-0389 from the Army Research Office. The views and conclusions contained here are those of the authors and should not be interpreted as necessarily representing the official policies or endorsements, either expressed or implied, of ARO, CMU, or the U.S. Government or any of its agencies. We also acknowledge the efforts of Patricia Collins who reviewed multiple versions of the paper.

References

1. Al-Muhtadi, J., Ranganathan, A., Campbell, R., Mickunas, M.D.: Cerberus: A Context-Aware Security Scheme for Smart Spaces. In: IEEE PerCom 2003 (2003)
2. Covington, M.J., Fogla, P., Zhan, Z., Ahamad, M.: A context-aware security architecture for emerging applications. In: Proceedings of 18th Annual Computer Security Applications Conference (2002)
3. Rohs, M., Gfeller, B.: Using Camera-Equipped Mobile phones for interacting with real World. In: Published in Advances in Pervasive Computing, Austrian Computing Soc (OCG), pp. 265–271 (2004)
4. Dey, A.K., Abowd, G.D.: The Context Toolkit: Aiding the Development of Context-Enabled Applications. Presented in the Workshop on Software Engineering for Wearable and Pervasive Computing, Limerick, Ireland, June 6 (2000)
5. Dey, A.K., Abowd, G.D., Salber, D.: A Conceptual Framework and a Toolkit for Supporting the Rapid Prototyping of Context-Aware Applications. Published in Human-Computer Interaction Journal 16(2-4), 97–166 (2001)
6. Griss, M., Letsinger, R., Cowan, D., Sayers, C., VanHilst, M., Kessle, R.: CoolAgent: Intelligent Digital Assistants for Mobile Professionals - Phase 1, Retrospective HP Laboratories report HPL-2002-55(R1) (July 2002)
7. Lin, T., Zhang, J., Griss, M.: Enhancement of Wi-Fi Indoor Locationing, Carnegie Mellon Silicon Valley, CyLab Mobility Research Center technical report MRC-TR-2009-04 (March 2009)
8. McCune, J.M., Perrig, A., Reiter, M.K.: Seeing-is-Believing: Using Camera Phones for Human-Verifiable Authentiction. International Journal of Security and Networks Special Issue on Secure Spontaneous Interaction 4(1-2), 43–56 (2009)
9. Meng, J., Yang, Y.: Application of Mobile 2D Barcode in China, Published in Wireless Communications, Networking and Mobile Computing. In: Presented in WiCOM 2008, 4th International Conference, October 12-14, pp. 1–4 (2008)

The Tradeoff between Energy Efficiency and User State Estimation Accuracy in Mobile Sensing

Yi Wang[1], Bhaskar Krishnamachari[1], Qing Zhao[2], and Murali Annavaram[1]

[1] Ming Hsieh Department of Electrical Engineering, University of Southern California, Los Angeles, USA
{wangyi,bkrishna,annavara}@usc.edu
[2] Department of Electrical and Computer Engineering, University of California, Davis, USA
qzhao@ece.ucdavis.edu

Abstract. People-centric sensing and user state recognition can provide rich contextual information for various mobile applications and services. However, continuously capturing this contextual information on mobile devices drains device battery very quickly. In this paper, we study the tradeoff between device energy consumption and user state recognition accuracy from a novel perspective. We assume the user state evolves as a hidden discrete time Markov chain (DTMC) and an embedded sensor on mobile device discovers user state by performing a sensing observation. We investigate a stationary deterministic sensor sampling policy which assigns different sensor duty cycles based on different user states, and propose two state estimation mechanisms providing the best "guess" of user state sequence when observations are missing. We analyze the effect of varying sensor duty cycles on (a) device energy consumption and (b) user state estimation error, and visualize the tradeoff between the two numerically for a two-state setting.

Keywords: mobile sensing, energy efficiency, user state estimation accuracy, tradeoff.

1 Introduction

Mobile smart phones these days contain a wide range of features including sensing abilities using GPS, audio, WiFi, accelerometer, and so on. By utilizing all the sensing units and extracting more meaningful characteristics of users and surroundings in real time, applications can be more adaptive to the changing environment and user preferences. For instance, it would be convenient if the phones can automatically adjust the ring tone profile to appropriate volume and mode according to the surroundings and the events in which the user is participating. Or imagine an "automated twittering" application (Twitter: [10]), where the user's activity can be automatically recognized and updated online in real-time by mobile device instead of manual input from the user. As these kinds of applications require that the user context be automatically and correctly recognized, in this paper we use "user state" as an important way to represent human context information, similar to the work in [12]. A set of user states can be detected by embedded sensor on mobile device with high accuracy. For example, a single accelerometer is able to discriminate a set of human motions such as walking, running, riding a vehicle and so on with high confidence.

Although we believe that user's contextual information brings application personalization to new levels of sophistication, a big problem in context detection is that the limited battery capacity of mobile devices has restricted the continuous functioning of sensors without recharging the battery, because embedded sensors on mobile devices are major sources of energy consumption. For example, our experiments show that a fully charged battery on Nokia N95 device would be completely drained within six hours if its GPS is operated continuously. In this paper, we address this problem and study how to assign different duty cycles to a sensor at different states such that the energy consumption can be significantly reduced while maintaining high user state recognition accuracy.

In general, some user states once entered may stay unchanged for a long while; therefore, if the sensor is sampled too frequently within that state duration, an excessive amount of energy will be consumed. For example, if the user is identified to be sitting in a class, sensors such as accelerometer and microphone do not need to be monitored every minute if we know that on average the user is going to spend more than an hour in the class. On the other hand, if the user is in a mode of frequently transiting among different states, e.g., frequently traveling in an area with multiple tasks, we prefer to sample the corresponding sensor more frequently to capture his/her context, as otherwise it becomes harder to figure out "what happened" between two sensor observations.

In this work, we assume that time is discretized into slots. Ideally, if the sensor can be sampled in each time slot, the true user state sequence will be obtained perfectly. However, due to energy constraints the sensor have to adopt duty cycles in order to extend mobile device lifetime. We use a discrete time Markov chain (DTMC) to model the user states, with different state transition probabilities indicating different amounts of average duration a user spent in each state. We consider a stationary deterministic sensor sampling policy which assigns different duty cycles to the sensor based on different user states. In other words, at each user state the sensor will adopt a fixed sampling duty cycle. As a consequence, given the fact that the sensor does not make observation in every time slot, there exist periods of time when user state information is not observed and has to be estimated using only available data and knowledge of the transition probabilities of the user state Markov chain.

We propose two methods based on the Forward-Backward Algorithm and the Viterbi Algorithm [11,8] in order to perform state estimation for those time slots with missing observation. In the first method, for each time slot we pick the most likely user state at that time slot and compute the overall expected per-slot estimation error. In the second method, we choose the most likely state sequence with the highest probability, and compute the probability that the estimated sequence is incorrect.

One can easily tell from intuition that if the sensor makes frequent observations, the user state estimation error should be small. In this paper, we derive the expression for both the expected energy consumption and the expected state estimation error in terms of the user state transition probability matrix and sensor duty cycle parameters, and visualize the tradeoff between them.

The rest of this paper is organized as follows. In section 2, we present relevant prior work. In section 3, we introduce our mathematical formulation of sensor duty cycle selection, propose two methods that estimate user state sequence with missing observation data, and for each method derive its expected energy consumption and the corresponding expected state estimation error. In section 4, we enable the visualization of the tradeoff between energy consumption and state estimation error numerically for a two-state DTMC and discuss our findings. Optimal sensor duty cycles are found for different state transition probability matrices and energy budgets. Finally, we conclude and present directions for future work in section 5.

2 Related Work

Embedded sensors on mobile phone such as GPS, Bluetooth, WiFi detector, accelerometer, and so on have been well studied and explored in order to conduct user activity recognition [7,5,1,12]. For example, Annavaram *et al.* [1] show that by using data from multiple sensors and applying multi-model signal processing, seemingly similar states such as sitting and lying down can be accurately discriminated.

Due to the fact that low battery capacity on mobile device limits the application lifetime, the problem of energy management on mobile devices has been extensively studied in the literature such as [9,6,12]. Shih *et al.* [9] study event-driven power saving method and focused on reducing the *idle power*, the power a device consumes in a "standby" mode, such that a device turns off the wireless network adaptor to avoid energy waste while not actively used. In terms of sensor management, Wang *et al.* [12] propose a framework for energy efficient mobile sensing system (EEMSS), and show that by only operating necessary sensors and manage sensors hierarchically based on user state the device lifetime can be extended by more than 75% than existing similar systems while maintaining high user state recognition accuracy. Krause *et al.* [6] investigate the topic of trading off prediction accuracy and power consumption in mobile computing, and the authors showed that even very low sampling rate of accelerometer can lead to competitive classification results while device energy consumption can be significantly reduced.

In this paper, however, we study the energy management problem from a different perspective: we assume that each sensor sampling takes a unit of time to complete with a corresponding energy cost. Whenever the sensor makes an observation the user state will be discovered; on the other hand, when the sensor stays idle the user state is unknown and needs to be estimated using existing information. Instead of focusing on how a single sensor sample can be optimized like the work in [6], we study how the sensor can select its duty cycles in order to achieve the best tradeoff between the expected energy consumption and expected user state estimation error along the observation process.

Several previous works have investigated the problem of using Markov models to model user states and their transitions in the context of mobile networks and applications. For example, Bhattacharya and Das [4] use Markov chain to model user movement profile as he or she transits between cell towers, and the model could be updated as users move between cells or stay in a cell for a long period of time. Ashbrook et al. [2] incorporate user location traces into a Markov model that can be consulted for use with a variety of applications.

As introduced in the previous section, we model the background user state as a hidden discrete time Markov chain and as the chain evolves a sensor makes observations to detect user state in real-time. The Hidden Markov Model (HMM) technique is one of the most well-known techniques in the field of estimation and recognition [3,8]. Both the Forward-Backward Algorithm and the Viterbi Algorithm are well known algorithms that derive the most likely underlying state sequence of HMM given an state observation sequence [11,8]. Yu and Kobayashi [13] propose a new Forward-Backward algorithm with linear running time for Hidden Semi-Markov Model (HSMM) which can be applied to missing observations and multiple observation sequences. In this paper we simplify this process by assuming that sensor makes perfect observations, i.e., the observed state always equals the true user state. Recall that in order to achieve energy efficiency, the sensor would not be sampled in each time slot, and therefore the unknown state sequence only contains all the time slots without an associated sensor observation.

3 Model and Analysis

In this section, we present our mathematical model of sampling a DTMC based user state transition process. A stationary deterministic sensor sampling policy is proposed and two different mechanisms based on the Forward-Backward Algorithm and the Viterbi Algorithm are investigated which estimate user state sequence for all periods with missing observations, and finally derive the expression for expected sensor energy cost as well as expected state user estimation error, in terms of sensor duty cycle parameters and the DTMC transition probabilities.

3.1 Preliminaries

We assume that there are N user states, and the state transition follows a N-state discrete time Markov chain with transition probability $p_{ij}(i, j = 1, ...N)$ from state i to state j. We denote the discrete time line as $T = \{t : t = 0, 1, 2, ...\}$. As the Markov chain evolves, a sensor is operated based on the following deterministic stationary sensing policy: If the sensor detects user state i at some time

t, it will stay idle for I_i time slots (I_i belongs to the set of positive integers), and is re-sampled at time slot $t + I_i$. In other words, I_i is the length of the sensor idle period when state i is detected[1]. If the sensor makes an observation, there is a corresponding unit energy cost incurred in that time slot, whereas if the sensor stays idle the energy cost in that time slot is 0. We define O as the set of time slots when the sensor makes observation and O is thus a subset of T, as shown by figure 1. Consider the case where $I_1 = I_2 = ... = I_N = 1$, which indicates that the sensor performs state detection in every time slot. In this scenario $O = T$, and the user state can be correctly obtained in each time slot. However, due to energy consumption constraint, the sensor may have to stay idle sometimes in order to increase the device lifetime. As a result, there may exist time slots where no sensing is performed and the user state has to be estimated using available information, which is also illustrated by Figure 1. In particular, the unknown user state (state sequence) between two subsequent observations is going to be estimated using only the information from those two observations. We will present two mechanisms that estimate the missing state information between two subsequent observations in Section 3.3.

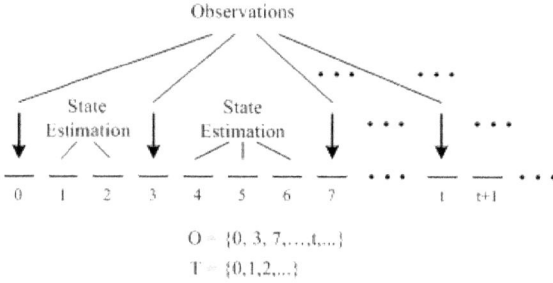

Fig. 1. Illustration of a discrete time Markov chain which is sampled at some time slots

Our goal is to identify the tradeoff between the energy consumption and state detection (estimation) accuracy by quantifying both metrics as we vary the values of the idle periods I_i for each state i. The longer the sensor waits until making the next observation, the less energy the device consumes. However, a longer waiting time also leads to more possibilities of state sequence taking place in that idle interval and thus more uncertainty about what "happened" between two observations.

---

[1] Note that the stationary deterministic sensor sampling policy studied in this paper is not claimed to be always optimal. In fact, a set of stationary randomized policies which select sensor duty cycle randomly may achieve higher state estimation accuracy than deterministic ones while preserving the same energy consumption constraint; however, in this paper we only focus on the analysis towards stationary deterministic policies and leave the study of randomized policies to future work.

Let s_t and s'_t denote the true user state and the estimated state at time slot t, respectively. The true state represents the user's real condition, whereas the estimated state is the "best guess" of user state which may be incorrect with certain probability. We assume that the sensor makes perfect observations, therefore if $t \in O$ we have $s'_t = s_t$ with probability 1.

We first derive the steady state probability of detecting a particular state i in an observation, under our deterministic sampling policy. We denote this probability as p_i. Since the state transition follows the Markovian rule, the current state observation result only depends on the previous observation status, thus:

$$p_i = P[observing\ state\ i] \tag{1}$$

$$= \sum_{j \in \{1,2,...,N\}} P[observing\ state\ j\ I_j\ steps\ ago] \cdot P_{ji}^{(I_j)} \tag{2}$$

$$= \sum_j p_j \cdot P_{ji}^{(I_j)} \tag{3}$$

where $P_{ji}^{(I_j)}$ denotes the probability of state transition from j to i in exactly I_j time steps. Equation (3) accounts for all possible cases of last observation that may lead to state i in the current observation.

It is obvious to see that

$$\sum_i p_i = 1 \tag{4}$$

The values of $p_1, p_2, ..., $ and p_N can be obtained by solving the $N+1$ simultaneous equations in (3) and (4).

3.2 Expected Sampling Interval and Expected Energy Cost

Given the probability of "seeing" each state when making an observation and the length of waiting time interval until next observation, the overall expected sampling interval $E[I]$ (idle time between two observations) can thus be expressed as:

$$E[I] = \sum_i p_i \cdot I_i \tag{5}$$

Let $E[C]$ be the overall expected energy cost whose value is defined as

$$E[C] = 1/E[I] \tag{6}$$

It can be seen that $E[C]$ characterizes the average per-slot energy cost. Note that since $I_i \geq 1$ and $\sum_i p_i = 1$, it is easy to conclude that $E[I] \geq 1$ and hence $E[C] \leq 1$. When $E[I] = E[C] = 1$, the Markov chain is observed in every time slot.

3.3 Estimation of Missing State Information and Expected Error

The problem of estimating missing state information between two subsequent observations is formulated as follows: given the fact that one observation detects state i at some time t_m ($t_m \in O$) and the next observation detects state j at time $t_m + I_i$ ($t_m + I_i \in O$), what is the most likely state sequence that connects them?

In our study, we address this problem by two different methods:

3.3.1 Method 1: Pick the Most Likely State for Each Time Slot and Compute the Expected Per-Slot Error (Utilizing the Forward-Backward Algorithm)

Given the sensor detects state i at time t_m, and detects state j at time $t_m + I_i$, the probability of being at state k at time t ($t_m < t < t_m + I_i$), which happens between the two observations, can be given by:

$$p[s_t = k | s_{t_m} = i, s_{t_m + I_i} = j] \quad (7)$$

$$= \frac{p[s_t = k, s_{t_m + I_i} = j | s_{t_m} = i]}{p[s_{t_m + I_i} = j | s_{t_m} = i]} \quad (8)$$

$$= \frac{p[s_t = k | s_{t_m} = i] \cdot p[s_{t_m + I_i} = j | s_t = k]}{p[s_{t_m + I_i} = j | s_{t_m} = i]} \quad (9)$$

$$= \frac{P_{ik}^{(t - t_m)} \cdot P_{kj}^{(t_m + I_i - t)}}{P_{ij}^{(I_i)}} \quad (10)$$

In order to obtain the "most likely" state at time t, the quantity given in equation 7 needs to be maximized. Thus, the estimated state s'_t can be chosen as:

$$s'_t = \operatorname*{argmax}_{k} \left\{ p[s_t = k | s_{t_m} = i, s_{t_m + I_i} = j] \right\} \quad (11)$$

$$= \operatorname*{argmax}_{k} \left\{ \frac{P_{ik}^{(t - t_m)} \cdot P_{kj}^{(t_m + I_i - t)}}{P_{ij}^{(I_i)}} \right\} \quad (12)$$

and the probability of correct estimation is given by:

$$p[s_t = s'_t | s_{t_m} = i, s_{t_m + I_i} = j] \quad (13)$$

$$= \max_{k} \left\{ p[s_t = k | s_{t_m} = i, s_{t_m + I_i} = j] \right\} \quad (14)$$

$$= \max_{k} \left\{ \frac{P_{ik}^{(t - t_m)} \cdot P_{kj}^{(t_m + I_i - t)}}{P_{ij}^{(I_i)}} \right\} \quad (15)$$

Having chosen the mostly likely state at time t, the corresponding expected estimation error for time slot t, denoted by e_t, is simply given by the probability that the true state is different from the estimation result:

$$e_t = 1 - p[s_t = s'_t | s_{t_m} = i, s_{t_m+I_i} = j] \tag{16}$$

$$= 1 - \max_k \left\{ \frac{P_{ik}^{(t-t_m)} \cdot P_{kj}^{(t_m+I_i-t)}}{P_{ij}^{(I_i)}} \right\} \tag{17}$$

Let e_{ij} be the aggregated expected estimation error between two consecutive samples that observe state i and j respectively. This e_{ij} accounts for the aggregated estimation error for all time slots between the two observations. Again, suppose the first observation is made at time t_m, e_{ij} can be expressed as

$$e_{ij} = \sum_{t=t_m+1}^{t_m+I_i-1} e_t \tag{18}$$

Now we derive the overall expected per-slot estimation error, denoted by $e_{perslot}$, which can be calculated by dividing the total number of time slots where incorrect estimation is made by the total number of time slots, i.e.:

$$e_{perslot} = \lim_{t \to \infty} \frac{\sum_{k=1}^{o(t)} W_k}{\sum_{k=1}^{o(t)} N_k} \tag{19}$$

$$= \lim_{t \to \infty} \frac{\frac{1}{o(t)} \cdot \sum_{k=1}^{o(t)} W_k}{\frac{1}{o(t)} \cdot \sum_{k=1}^{o(t)} N_k} \tag{20}$$

where $o(t)$, W_k, and N_k denote the total number of observation intervals till time t, the number of incorrect estimations in the kth observation interval, and the number of time slots in the kth observation interval, respectively. Considering the Law of Large Numbers, equation (20) can be further written as

$$e_{perslot} = \frac{E[W]}{E[N]} \tag{21}$$

which indicates that the overall expected per-slot estimation error is given by the ratio of expected number of estimation errors per observation interval and expected observation interval length. Considering all possible cases of a pair of subsequent observations and their corresponding aggregated estimation error, the overall expected per-slot estimation error is thus given by:

$$e_{perslot} = \frac{\sum_{ij}(e_{ij} \cdot p_i \cdot P_{ij}^{(I_i)})}{\sum_{ij}(p_i \cdot P_{ij}^{(I_i)} \cdot I_i)} \tag{22}$$

$$= \frac{\sum_{ij}(e_{ij} \cdot p_i \cdot P_{ij}^{(I_i)})}{\sum_i p_i \cdot I_i} \tag{23}$$

$$= \frac{\sum_{ij}(e_{ij} \cdot p_i \cdot P_{ij}^{(I_i)})}{E[I]} \tag{24}$$

in which the component $p_i \cdot P_{ij}^{(I_i)}$ accounts for the probability that observing state j in I_i time slots after observing state i.

3.3.2 Method 2: Estimate the Most Likely State Sequence and Compute the Expected Sequence Estimation Error (Utilizing the Viterbi Algorithm)

Instead of selecting the most likely state for individual time slots, *Method 2* returns the most likely state sequence between two subsequent observations. We first construct a trellis which contains all possible paths from state i to state j in I_i time steps. Without loss of generality, again, let t_m and $t_m + I_i$ be the first and second observation time slots. Define the quantity

$$q_k^t = \max_{s_{t_m}=i, s_{t_m+1}, \ldots, s_{t-1}} P[s_{t_m} = i, s_{t_m+1}, \ldots, s_{t-1}, s_t = k] \qquad (25)$$

where $k \in \{1, \ldots, N\}$ and $t_m < t$, i.e., q_k^t is the highest probability along a single path starting from state i at time t_m, and ends at state k at time t, which accounts for the first t observations after t_m.

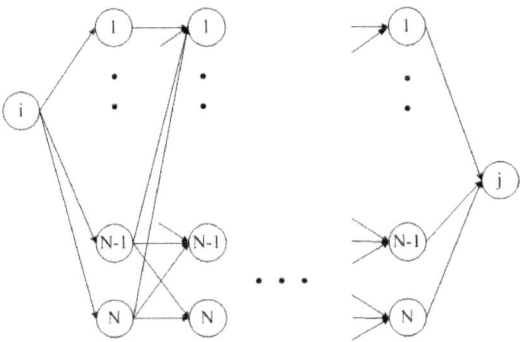

Fig. 2. All possible paths starting from state i at time 0 and ending at state j at time I_i

In order to retrieve the most likely state sequence, we need to keep track of all the states that maximize the quantity given by equation (25), for each t and i. Thus, the complete process for finding the mostly likely state sequence and expected sequence error probability between two observations can be listed as follows:

– Initialization:
$$q_i^{t_m} = 1. \qquad (26)$$

– Induction and state estimation:
$$q_k^t = \max_m(q_m^{t-1} \cdot p_{mk}), t_m < t \leq t_m + I_i - 1 \qquad (27)$$
$$s_t' = \underset{k}{\operatorname{argmax}}(q_k^t), t_m < t \leq t_m + I_i - 1 \qquad (28)$$

– Construct the most likely estimation sequence:
$$S' = \{s'_{t_m+1}, s'_{t_m+2}, \ldots, s'_{t_m+I_i-1}\} \qquad (29)$$

– Compute expected sequence error e'_{ij}:

$$e'_{ij} = 1 - \max_k(q_k^{t_m+I_i-1}) \qquad (30)$$

It should be noted that q_k^t denotes a "path probability" and this method computes the most likely state sequence instead of estimating the most possible state for each individual time slots. The estimated sequence is considered to be incorrect even if it contains a single bit error. Therefore the expected sequence error probability can be computed based on equation (30), which is the complement of the maximum probability over all possible paths starting at state i at time t_m and ending at time $t_m + I_i - 1$. Finally, recall the fact that estimation may be performed within all possible subsequent observation pairs; hence the overall expected sequence error e_{seq}, which is the average probability that a sequence estimation is in error, can be thus expressed as:

$$e_{seq} = \sum_{ij}(e'_{ij} \cdot p_i \cdot P_{ij}^{(I_i)}) \qquad (31)$$

It is worth noting that the expected energy consumption $E[C]$, the expected per-slot error $e_{perslot}$ (computed in *Method 1*), and the expected sequence estimation error e_{seq} (computed in *Method 2*) all depend on the choices of I_i ($i \in \{1, ..., N\}$) as well as the Markov chain transition probabilities $p_{ij}(i, j = 1, ...N)$.

3.3.3 Discussion of the Two Methods

As can be seen, the key difference between the two methods lies in the fact that *Method 1* requires two subsequent observation results and estimate user state for each time slot and finally construct the state sequence, whereas *Method 2* only requires a single observation and returns the most likely state sequence directly. The choice of algorithm would depend on which metric is most important to a given application. It may also be possible to consider the combination of the two objectives, but we have not studied this in our current work.

4   A Case Study: Two-State DTMC

In order to visualize the tradeoff between energy consumption and estimation error, we have conducted a case study where for simplicity, we consider a two-state discrete time Markov chain with states "1" and "2" and transition probability matrix

$$P = \begin{bmatrix} p_{11} & p_{12} \\ p_{21} & p_{22} \end{bmatrix} \qquad (32)$$

According to the deterministic sampling policy, if the sensor detects state "1", it will be re-sampled in I_1 time slots, and if the sensor detects state "2" it is re-sampled in I_2 time slots. As the potential possibilities of transition matrix is infinite, we select six representative ones, namely, $\begin{bmatrix} 0.9 & 0.1 \\ 0.1 & 0.9 \end{bmatrix}$, $\begin{bmatrix} 0.9 & 0.1 \\ 0.5 & 0.5 \end{bmatrix}$, $\begin{bmatrix} 0.9 & 0.1 \\ 0.9 & 0.1 \end{bmatrix}$,

$\begin{bmatrix} 0.5 & 0.5 \\ 0.5 & 0.5 \end{bmatrix}$, $\begin{bmatrix} 0.5 & 0.5 \\ 0.9 & 0.1 \end{bmatrix}$, and $\begin{bmatrix} 0.1 & 0.9 \\ 0.9 & 0.1 \end{bmatrix}$. Recall that our goal is to identify the effect of selecting different sampling intervals on the performance including energy consumption and state estimation error. We examine all possible combinations of I_1 and I_2 with constraints $1 \leq I_1 \leq 30$ and $1 \leq I_2 \leq 30$.

4.1 Analytical Results

The corresponding energy cost and expected errors are computed based on equation (6), (24), and (31). figure 3 shows the relationship between the expected per-slot estimation error $e_{perslot}$ and energy $E[C]$ based on the *Method 1*, whereas figure 4 shows the incorrect sequence estimation probability e_{seq} with respect to energy cost $E[C]$, computed based on *Method 2*. Note that these figures present a scatter plot of the tradeoffs obtained for all values of I_1 and I_2 considered. Of most interest is the lower envelope of these figures, which represent the optimal tradeoff between energy consumption and state estimation error.

It can be seen from figure 3 that the average per-slot state estimation error is within the range $[0, 1 - \max\{\pi_1, \pi_2\}]$ (π_1 and π_2 are the steady state probability for state 1 and state 2 respectively) as the expected energy consumption changes within the range $[0, 1]$. This result is quite intuitive because when $E[C] = 1$, the sensor makes observation in each time slot therefore no estimation is needed and the observed user state sequence matches the true state sequence perfectly. On the other hand, if the sensor does not make any observation where $E[C] = 0$, the estimated state in each time slot is always going to be the one with larger

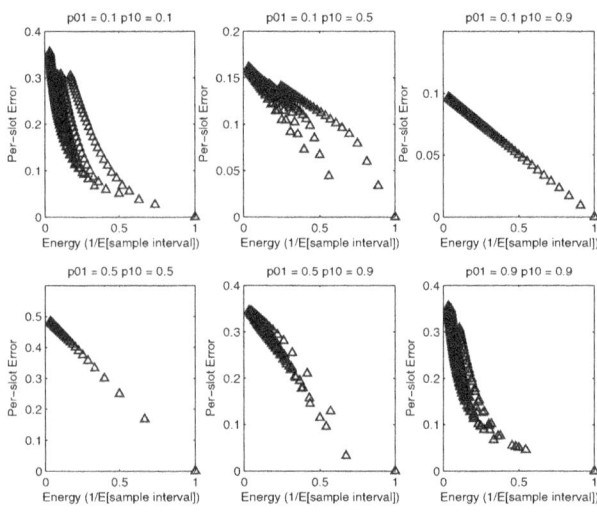

Fig. 3. Analysis result of expected per-slot estimation error with respect to energy usage based on *Method 1*

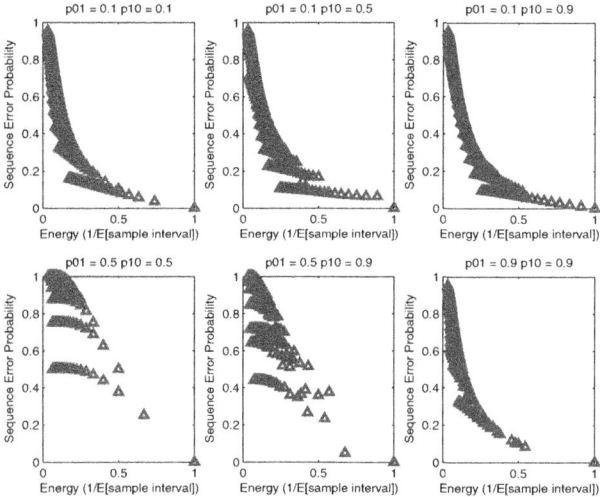

Fig. 4. Analysis result of expected sequence estimation error with respect to energy usage based on *Method 2*

steady state probability. Therefore the upper bound of expected per-slot state estimation error can be expressed as $1 - \max\{\pi_1, \pi_2\}$. For N-state DTMC, the expected per-slot state estimation error can be generalized as:

$$0 \leq e_{perslot} \leq 1 - \max\{\pi_1, \pi_2, ..., \pi_N\} \tag{33}$$

A similar explanation from above can be applied to figure 4, where the expected sequence estimation error $e_{seq} = 0$ whenever $E = 1$. However it can be seen that $e_{seq} = 1$ if no sensor observation is make. This is because in *Method 2* we examine the expected sequence error probability, and an estimated sequence is considered incorrect as long as it contains a single bit error.

It can be easily concluded from figure 3 that the energy consumption have a large impact on user state estimation accuracy. The results verify the intuition that the more energy utilized, the more accurate state estimation result will be, and vice versa. It can be also found out that given different user state transition probability matrices, the tradeoff curves between energy and error appear to have different characteristics. For all the tradeoff curves, it may be desirable to keep the system running at the knee of the curve which provides the best tradeoff between the two metrics. For example, if $p_{12} = p_{21} = 0.1$, if 10% error is allowed, the system energy consumption can be reduced by as much as 80% compared to a fully operating device.

4.2 Simulation Results

In order to verify the analytical results presented in Section 4.1, we have conducted simulations on a 2-state discrete time Markov chain with same settings as discussed in Section 4.1. For each I_1 and I_2 combination, a duration of 5000 time slots is simulated. At the beginning of the simulation run the initial state of the Markov chain is arbitrarily set to "1" and the first sensor observation takes place at time slot 0. State estimation is performed right after a single observation is made and all the missing state information from last observation point are estimated using both methods. The system records the estimated state sequence as well as all the observation points in order to evaluate both the state estimation error and energy cost.

The expected energy consumption is measured by calculating the ratio between the total number of observations made and the length of simulation time (5000 time slots). For *Method 1*, we have found out the expected per-slot estimation error by dividing the number of incorrect estimations (as represented by the number of time slots where the estimation result does not match the true state) by the total number of time slots where the state needs to be estimated (i.e.: observation is missing). The results can be found in figure 5.

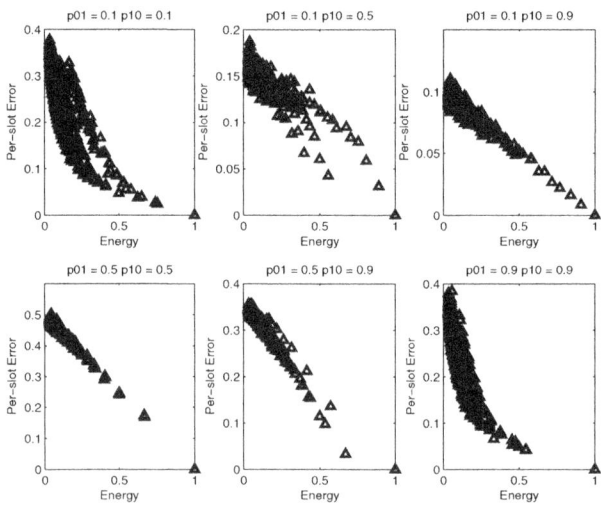

Fig. 5. Simulation result of per-slot error versus energy cost based on *Method 1*

For *Method 2*, we examine the percentage of sequence estimations that are incorrect. Recall that the estimated state sequence is considered incorrect even if a single bit of estimation does not match the ground truth. The plot of percentage of incorrect sequence estimations with respect to energy cost can be found in figure 6. Moreover, since the system is able to record the estimated state sequence which can be further compared to the ground truth, we have also identified the per-slot error based on *Method 2*, and the results can be found in figure 7. Interestingly, we find that the per-slot error of *Method 2* appears to be comparable to that obtained by *Method 1*.

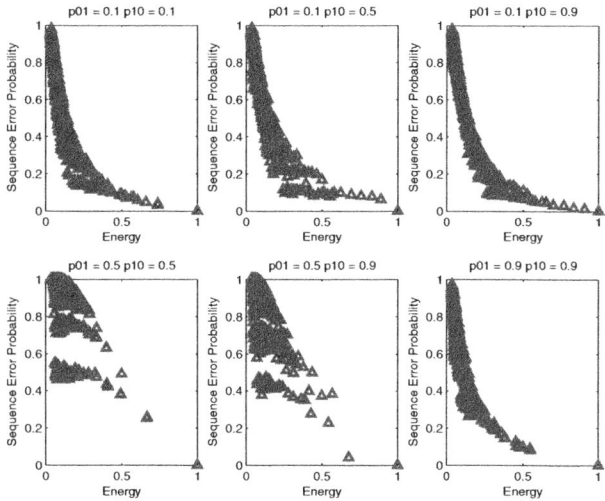

Fig. 6. Simulation result of expected sequence estimation error versus energy cost based on *Method 2*

It can be seen that the energy-error tradeoff plot in figure 5 and 6 demonstrate a very close match to the numerical analysis result shown in figure 3 and 4, and this validates our analytical process and derivations of expected energy and state estimation error in section 3.

4.3 Optimal Sensor Idle Intervals for Different Energy Budget

In practical system designs, a device energy consumption budget is usually specified in order to ensure long enough device operating duration. We define ξ (where $0 \leq \xi \leq 1$) as the energy budget which is the maximum average energy cost allowed. We investigate the following problem: given an energy budget ξ, and user state transition probability matrix P, what is the optimal length of sensor idle interval at each state such that the state estimation error can be minimized?

All integer combinations of I_1 and I_2 have been tested and the one that leads to the minimum estimation error while satisfying the energy constraint is picked, and figure 8 shows the optimal I_1 and I_2 combinations to utilize in order to achieve the least per-slot estimation error, given a particular energy budget, i.e., the sensor idle intervals must be such that the average energy cost is less than the budget value[2].

From figure 8 it can be seen that for symmetric user state transitions, i.e., when $p_{12} = p_{21}$, the optimal I_1, I_2 values are close to each other. This is due to

---

[2] A more general approach is to model the optimal sensor idle interval selection with energy constraint problem as a Constrained Markov Decision Process (CMDP) such that an optimal sensor sampling policy can be deduced. We leave this to future work.

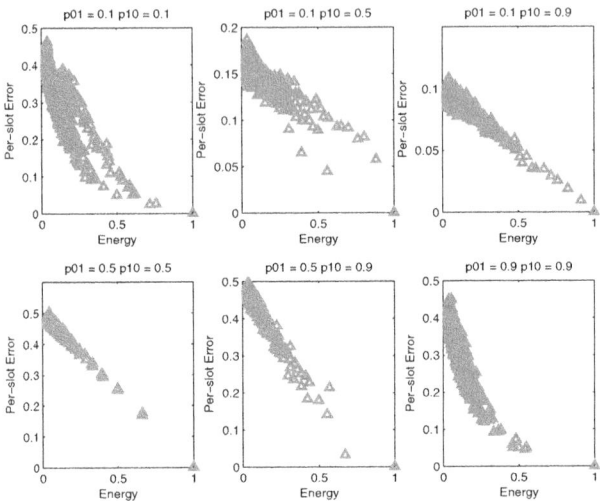

Fig. 7. Simulation result of per-slot error versus energy cost based on *Method 2*

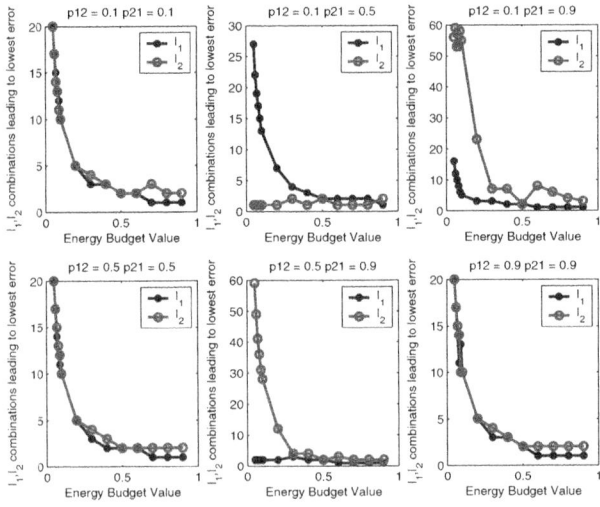

Fig. 8. The best I_1, I_2 combinations that lead to the lowest error given a particular energy budget

the fact that the average duration spent in each state are the same. Therefore, the sensor should be assigned same duty cycles in each state as well. An interesting observation is the following: when $p_{12} = 0.1$ and $p_{21} = 0.5$, sampling the sensor more frequently when the user is at state 2 leads to the minimum state estimation error. Intuitively, if the average duration spent in a state is low (in this case, state 2), we prefer to sample the sensor more frequently in order to capture state transition quickly. However, an inverse observation is made for both cases when $p_{12} = 0.1, p_{21} = 0.9$ and $p_{12} = 0.5, p_{21} = 0.9$. In these cases sampling more often in user state 1 leads to the minimum state estimation error. This illustrates that the optimal sensing policy can be quite sensitive to the transition probabilities.

5 Conclusion and Future Work Directions

Mobile device based sensing of human user states is able to provide rich contextual information to applications such as social networking and health monitoring. However, energy management remains a critical problem due to limited battery capacity on mobile devices.

In this paper we have studied the effect of applying different sensor duty cycles on the expected energy consumption as well as the expected user state estimation error, and have identified the tradeoff between the two metrics. In particular, we have modeled the user state as a hidden DTMC, and have proposed a stationary deterministic sensor sampling policy which assigns different sensor duty cycles for different user states. We introduced two state estimation methods in order to compute the most likely state sequence whenever sensors are idle. The tradeoff between the expected energy consumption and the expected user state estimation error have been visualized numerically.

In future work, we plan on modeling and studying the sensor duty cycle selection with energy constraint problem as a CMDP with stationary randomized sensor sampling policies and comparing them to our current stationary deterministic policies. We also plan to apply our sensor sampling policy and state estimation algorithms to real user data and identify the relationship between energy consumption and state estimation error.

References

1. Annavaram, M., Medvidovic, N., Mitra, U., Narayanan, S., Spruijt-Metz, D., Sukhatme, G., Meng, Z., Qiu, S., Kumar, R., Thatte, G.: Multimodal sensing for pediatric obesity applications. In: UrbanSense 2008 Workshop at SenSys, Raleigh, NC, USA (November 2008)
2. Ashbrook, D., Starner, T.: Learning significant locations and predicting user movement with GPS. In: IEEE International Symposium on Wearable Computers (2002)
3. Bahl, L.R., Jelinek, F., Mercer, R.L.: A maximum likelihood approach to continuous speech recognition. IEEE Transactions on Pattern Analysis and Machine Intelligence (1983)
4. Bhattacharya, A., Das, S.K.: Lezi-update: An information-theoretic approach to track mobile users in PCS networks. In: Proceedings of the 5th annual ACM/IEEE international conference on Mobile computing and networking (1999)
5. Biswas, S., Quwaider, M.: Body posture identification using hidden markov model with wearable sensor networks. In: BodyNets Workshop, Tempe, AZ, USA (March 2008)

6. Krause, A., Ihmig, M., Rankin, E., Gupta, S., Leong, D., Siewiorek, D.P., Smailagic, A., Deisher, M., Sengupta, U.: Trading off prediction accuracy and power consumption for context-aware wearable computing. In: IEEE International Symposium on Wearable Computers (2005)
7. Lester, J., Choudhury, T., Borriello, G., Consolvo, S., Landay, J., Everitt, K., Smith, I.: Sensing and modeling activities to support physical fitness. In: Proceedings of UbiComp, Tokyo, Japan (2005)
8. Rabiner, L.R.: A tutorial on hidden markov models and selected applications in speech recognition. In: Proceedings of the IEEE (1989)
9. Shih, E., Bahl, P., Sinclair, M.J.: Wake on wireless: an event driven energy saving strategy for battery operated devices. In: Proceedings of MobiCom, Atlanta, Georgia, USA (2002)
10. Twitter, http://www.twitter.com
11. Viterbi, A.J.: Error bounds for convolutional codes and an asymptotically optimum decoding algorithm. IEEE Transactions on Information Theory (1967)
12. Wang, Y., Lin, J., Annavaram, M., Jacobson, Q.A., Hong, J., krishnamachari, B., Sadeh, N.: A framework of energy efficient mobile sensing for automatic user state recognition. In: Proceedings of MobiSys, Krakow, Poland (June 2009)
13. Yu, S., Kobayashi, H.: A hidden semi-markov model with missing data and multiple observation sequences for mobility tracking. In: Signal Processing (2003)

Dynamic Migration of Computation through Virtualization of the Mobile Platform

Shivani Sud, Roy Want, Trevor Pering, Kent Lyons, Barbara Rosario, and Michelle X. Gong

Future Technology Research, Intel Inc
2200 Mission College Blvd
Santa Clara, CA 95024
{shivani.a.sud,roy.want,trevor.pering,kent.lyons,
barbara.rosario,michelle.x.gong}@intel.com

Abstract. Virtualization and live migration techniques have long been used in the enterprise server space and have been tuned to address data center usages. These capabilities are now expanding to personal computers including desktops and laptops and more recently into smaller mobile devices such as Netbooks and Mobile Internet Devices (MID). Hardware support for virtualization in these platforms, such as that offered by Intel® Atom™ processor, enables the use of existing operating systems and virtualization software. Our experiments demonstrate that live migration can be used to dynamically offload computation to a nearby desktop computer from a Netbook, taking only 25 seconds over a 100Mbps Ethernet network and approximately 100 seconds over an 802.11n interface with a measured throughput of 70Mbps. Additionally, these experiments highlight the limitations of existing virtualization solutions for migrating computation between small form-factor mobile devices and desktop computers which have widely varying resources and processing capabilities. Finally, we discuss the challenges observed in these early experiments and raise key questions that need to be resolved to enable the design of effective systems that support this use model.

Keywords: Virtualization, MIDs, Intel® Atom™ processor, mobile, Virtual Machine (VM), live migration.

1 Introduction

In this paper, we discuss the use of virtualization on small mobile computers such as smart phones, Mobile Internet Devices (MIDs) and Netbooks which are increasingly enabling low-power high-performance computing in a small portable device form factor. As an example of such systems, we focus specifically on the Intel's Atom class of mobile processors and how virtualization can be used to migrate a virtual machine between these types of mobile devices, and other IA systems such as a laptop. The overall goal of such live migration is to allow a user to move their computation from an ultra mobile device such as a MID to a more capable machine such as a laptop or desktop when it becomes available.

Mobile Internet Devices

Users are increasingly choosing smart phones as their 24/7 connection to the world. These devices are not only primarily communication devices, but personal portals to the internet and are used to access data as frequently as they are for voice. Further, mobile devices are increasingly being used as media generation devices with users capturing videos and photos, text messaging, twittering, micro blogging etc. in addition to more traditional media consumption such as listening to MP3s or watching videos, or using the device as an e-reader. Moreover, social networking applications now allow mobile users another level of instant notification when an event occurs in their social network. Statistics indicate that the number of internet users in the developing world will increase dramatically as many users who never had access to a wired Internet connection start using their smart phones to access the internet for the first time [7]. The processing power available in smart phones is already able to host operating systems such as embedded Linux and Windows Mobile 6.0. The new low power x86 based Intel's Atom processors can host full desktop operating systems such as Windows XP or Linux desktop editions. These processors are being used in an emerging class of devices called Netbooks and Mobile Internet Devices (MIDs). Since desktop class operating systems are now available on these small mobile devices, new usage models can be enabled through the increased available compute power in ways that enrich the mobile user experience. These trends indicate that in the future smart phones are likely to become the primary computing platforms for mobile users [1]. Research is also on-going to allow these devices to take advantage of nearby computers and wirelessly share peripherals to extend their capabilities with more peripherals, such as full-sized displays and keyboards, and components not available on the mobile platform [15].

Opportunities Enabled by the Atom Family of Processors

Atom is a Low Power Intel Architecture based SoC that support Intel x86 ISA. The Atom Z550 processor has a clock speed of 2GHz with hyper-threading and supports up to 2GB of memory and can be packaged into a handheld device consuming 220mW average power and 100mW idle power. In addition, Moblin [6] is a Linux software platform optimized for Atom that has drawn in many manufacturers, vendors and dev elopers. With processing capabilities that far outstrip many desktop computers sold only a few years ago, these processors are likely to change the landscape of smart phones, MIDs, Netbooks and other mobile computers. We compare the various smartphone and MID processors in the market today in Figure 1.

The Atom processor has hardware support for virtualization (Intel VT extensions) and allows hosting hardware accelerated VMs that provide better performance compared to software emulation based solutions. Hardware extensions to support virtualization are a capability not seen in the handheld space before [3]. This widens the playing field by allowing readily available open source and proprietary hypervisors designed for x86 platforms to be run on devices such as Netbooks and MIDs. This capability allows for the leveraging of experience and existing knowledge base in this area.

Processor-> Feature	Intel Atom Z550	ARM 11 core	Qualcomm MSM7201A RISC Chipset (based on ARM 11 core)	Texas Instruments OMAP 2430 RISC Chipset (based on ARM 11 core)	Nvidia Tegra APX 2500
Frequency	2.00 GHz	620 MHz	528MHz	450 MHz	750MHz
Cache	512K	16K	-	32K(data)/ 32K(instruction)	-
Power	2.4W	0.45 mW/MHz	-	-	2.5-4W
Instruction Set	x86	ARMv6	ARMv6	ARMv6	ARMv6
CPU Core	SoC codename Lincroft	ARM11	ARM1136EJ-S	ARM1136	ARM11MP

Fig. 1. Smart phone and MID processors

In this paper, we demonstrate the use of this virtualization as a technology for creating a container for a set of applications (like office productivity applications in the case of a mobile enterprise user) on a mobile device. Using a high bandwidth radio, we treat the VM as a unit to migrate the computation state onto a PC for better utilization of opportunistically available resources.

In Section 2 we present uses of virtualization in the mobile device space and introduce a new usage model, for dynamic computation offload based on live migration and virtualization built from off-the-shelf components. In section 3, we demonstrate how we can achieve the migration of VM state in approximately 100 seconds over existing wireless interfaces. We present results that characterize the offload time using fast local area networking and discuss their viability. As this is done using existing technologies it demonstrates the concept is viable, although technology improvements can clearly benefit the approach. In Section 4, we present the research questions brought to the fore that need to be addressed to make this usage model attractive to users through seamless and transparent operation, and compare our research with related work in Section 5. We conclude with key findings that support the processing offload vision.

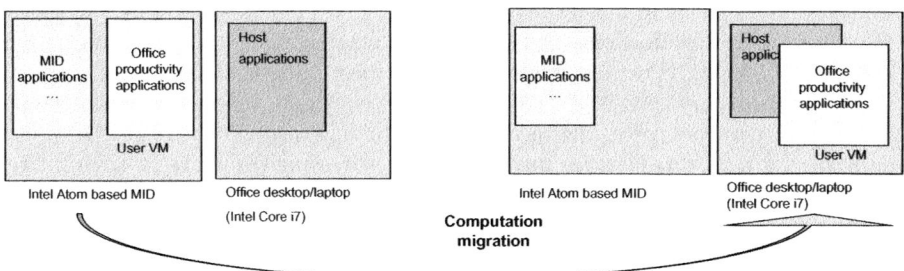

Fig. 2. Virtualization and live migration provide a mechanism to migrate the dynamic computational state

2 Virtualization in the Mobile Space

Virtualization technology has long been used in the IT industry to increase reliability and availability, and optimize performance of server class machines. This technique is employed in data centers and is typically transparent to the end user, who is unaware how it is benefiting server applications. Virtualization has its roots in the mainframe era, and has since made its way to personal computers but its use is still relatively limited, and development of the broader market is still at a nascent stage.

As discussed above, the advent of MIDs that provide accelerated performance in a mobile platform, combined with the increased desire of users for full on-the-go information access, means that users can now run their favorite PC operating systems and applications on their handheld devices.

Computation Migration for Mobile Devices

We explore additional uses of virtualization in the MID space from the perspective of enhancing the user experience through better performance and interactivity with more capable processors and peripherals found on desktop machines. Our usage model is based on the dynamic computation offload from a mobile device to a PC, to take advantage of a richer computation environment. To illustrate the problems and benefits of such a solution we first describe a scenario involving a mobile enterprise user who needs constant access to office productivity tools, along with voice and data connectivity, irrespective of their location, and the optimizations that could be achieved using this approach:

> *Jane uses her home computer to check her email and reviews a presentation she needs to deliver later that morning. As the day progresses, she seamlessly migrates her work environment from her home PC to her mobile device before leaving home. While traveling she continues reviewing the presentation, adding notes as she rides the subway to work. Soon it is time for her to dial into a teleconference. On reaching her desk, her work environment seamlessly migrates from her mobile device to the office PC, and she can now use the office PC to continue reviewing the presentation, while she continues her teleconference from her mobile device.*

The processing power of a MID device is now large enough to support the full suite of Microsoft Office applications, and most of the other applications users run on their laptops or desktops. However, while their small form-factor and low power consumption make them very suitable for mobility, it also constrains long term user interaction because of the poor user interface, and the limitations of the battery. It is thus desirable for a user to leverage any additional processing resources to augment performance and overcome power constraints on the device and nearby peripherals to overcome the user interface limitations, while retaining the complete independence of functionality when environmental support is not available. Using virtualization and VM migration, we can migrate our office applications to nearby PCs and utilize their full hardware capabilities e.g. processor, display, keyboard, network as shown in Figure 2.

Consider the usage scenario described in the section above, in which a seamless transition for the user can be achieved by hosting the desired office applications in another guest VM on the MID. This ensures that all applications are available to the user when (s)he is on the go. And when the user reaches a resource rich environment, such as their office, the VM hosting the office productivity applications can be migrated to a more powerful PC, while the user continues using their MID as an independent standalone device.

At a later time when the user becomes mobile again they can migrate the context back again to their MID retaining the current state of their applications and they can thus seamlessly transition computation among the most suitable computers in their environment and adapt to their situational constraints.

Live Migration

Some of the technologies to enable the capability described above are currently available for experimentation. Both proprietary and open-source virtualization software are readily available for the desktop environments, such as VMWare, Xen desktop, VirtualBox and KVM. Some of these virtualization software solutions support *live VM migration* that allows a user to migrate a live VM session from one system to another.

Of the above mentioned products, KVM is an open-source kernel virtual-machine infrastructure based on hardware virtualization support available in most of today's processors (Intel's VT and AMD's SVM). KVM adds virtualization capabilities to the Linux kernel while benefiting from its memory management and scheduling algorithms, such that VMs exist as regular Linux processes. KVM leverages processor extensions for hardware virtualization and has a performance advantage over software emulation based solutions. KVM *live migration* uses an NFS mounted shared disk space to store the guest VM image which needs to be accessible from both source and target migration end-points. The KVM live migration algorithm makes use of a cyclic process of copying dirty memory pages from the source to the destination. A write access to a page that has already been copied causes it to be re-marked as dirty and it needs to be copied again in the next cycle. When the number of dirty pages falls below a certain threshold, the cycle is stopped and the rest of the dirty pages are copied to the destination; at which point the VM has been completely migrated.

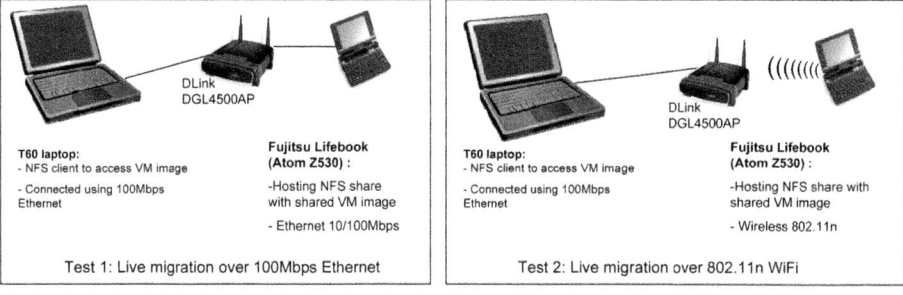

Fig. 3. Two test environments used to capture live migration timings

3 Experiment

Wireless Docking is a term used to extend the notion of a wired docking station to wireless access to peripherals. Our usage with virtualization and live migration extends this model to the processing resources available on other computers, and we call this usage *Wireless Dock+*. So, Wireless Dock+ is the extension of the wireless docking model that allows for enhanced computation through VM migration, in addition to enabling access to functionally better peripherals on the target device.

Test Environment Setup for Wireless Dock+

As the name indicates, Wireless Dock+ should offer a cable free experience for the user. In terms of our mission for live migration this implies that when a mobile user comes into the vicinity of their office PC, VM migration can be triggered (either manually or automatically) over a short-range high-bandwidth radio channel.

To experiment with the scenario described in section 2 we employ an Atom based Netbook (Fujitsu Lifebook U820) as the mobile device. It is powered by an Intel Z530 Atom 1.6GHz processor with 512K cache and 1GB of memory. For the office computer we used a ThinkPad T60 notebook computer based on an Intel T2400 Core Duo 1.83GHz processor with 2MB cache and 2GB of memory. Both devices are connected to a DLink DGL 4500 802.11 Draft N WLAN AP. The host OS on both the T60 and Lifebook is the Ubuntu 8.10 desktop edition. The guest VM is also created using the Ubuntu 8.10 desktop edition on the Lifebook. We set up the NFS server for hosting the VM image on the Lifebook, because the mobile device should be able to operate the office applications VM independently, without requiring access to a remote server. The mobile device can then share the VM image, and support ad hoc VM migration to nearby PCs over a local wireless link. The T60 has an NFS client installed and can access the VM image from the NFS share on the Lifebook.

In our effort to use off-the-shelf and open-source components, for this experiment, we use KVM (kvm-72) as the open source hypervisor for supporting the guest VMs on both the Lifebook and T60, and makes use of Intel hardware virtualization extensions on both devices. The guest VMs are configured with a single CPU, 512 MB of memory space, and the allocated a disk image of 10GB. The network is configured with a user-space network-stack and a Cirrus graphics driver provides the graphics emulation.

We performed two tests, setups shown in Figure 3, in which we investigate the feasibility of computation handoff between a Netbook and a PC by measuring the migration time. First, we measured it with an Ethernet connection between the devices to ensure the viability of our proposal and provide a baseline. And in the second test we employ a Wifi connection. Details are presented below:

Test 1: The T60 was connected to the access point with 100Mbps Ethernet, and the Fujitsu Lifebook was also connected over a Realtek RT8139 100Mbps Ethernet interface.

Test 2: The T60 setup was the same, while the Fujitsu Lifebook connected wirelessly using the Atheros 928 chipset, providing 802.11agn Wifi with maximum theoretical physical layer data rates of 600Mpbs. There was no other traffic through the AP during the test.

For each test, we ran three scenarios:
- o Basic Computation: a couple of terminals and browser windows active,
- o Media Browsing: viewing a YouTube video,
- o Interactive Collaboration: an openGL collaborative application.

We used the qemu[1] console window to trigger the migration process and made coarse grained measurements of the timings from when the migration was initiated to the time the VM became active at the destination computer. The measured results, summarized in Figure 4, were averaged over 3 iterations for each scenario.

App Load-> NW i/f	Terminal and browsers windows	YouTube video	OpenGL Cobalt collaboration
100Mpbs	20 sec	25 sec	25 sec
802.11 Draft N	110 sec	150 sec	130 sec

Fig. 4. Timings of live migration from Lifebook to T60

Performance Statistics

Our results show that for VM migration over a 100Mbps Ethernet link, typically 25 sec of migration time is needed; but with an 802.11 Draft N link, the average migration times are on the order of 2 minutes.

Using *iperf*, we determined the 802.11 Draft N throughput was about 70 Mbps. The average time for the migration was proportional to the size of the memory allocated to the VM. In the case of applications that dirty memory pages more frequently, multiple iterations are needed to transfer the dirtied pages to the destination, and add to the migration time. This explains why the YouTube test took the longest out of our three test examples. Due to the lower throughput and collision prone nature of the wireless link, the migration of same amount of data takes longer compared to an Ethernet link. This additional delay contributes to more dirty data being generated during the migration process and thus the average migration time over the 802.11 Draft N link is much longer than that over the Ethernet link. The results of our experiments show the viability and current limitations for seamless wireless *live migration* of a mobile users compute environment to infrastructural PCs (2 minutes is a long time when you are waiting).

With improvements in WLAN network technologies we can envision scenarios in which a fast WLAN or WPAN (such as UWB) can significantly reduce migration time, thus requiring less waiting on behalf of the user and therefore provide a more seamless user experience. In addition, developments in storage technologies such as Solid State Disks (SSDs) allow faster disk access than spinning media, Together these technologies provide additive performance improvements for accessing files across the local network and hence result in a better user experience.

Discussion of Results

Our usage requires a high-speed short-range radio technology for live migration and dynamic computation offload; and does not depend on high-speed broadband access to the Internet or another server, so it is not affected by the latency considerations of

---

[1] KVM has a console, which provides command line interface used to interact with the VM. This command line provides commands to set the migration speed, initiate the migration process and check the status of an active migration.

WAN technologies. WLAN technologies are however rapidly evolving towards much higher speeds than WAN technologies and the gap continues to grow. With 802.11 Draft N interface, we saw only 70 Mbps throughput that is a fraction of the maximum theoretical physical layer data rate of 600Mbps. And with the upcoming standards for 60GHz, we can only expect the WLAN throughput rates to go higher (with a projected theoretical raw bit rate of 5 Gbps [5]). Based on our measurements and the expected new standards, we can project that a wireless migration time of around 120 seconds for 802.11 Draft N will become a couple of seconds using a 60GHz radio.

Additional optimizations for our experimental set-up can be achieved by providing faster access to the remote VM disk image on the MID using an additional radio technology such as WUSB, to augment the WLAN radio bandwidth in the migration process. Thus, we can leverage multiple high-bandwidth WLAN/WPAN technologies to achieve an optimal user experience. High speed Ethernet links (e.g. 1Gbps) are the norm on PCs nowadays however they are not typically available on MID or mobile devices, so we were not able to compare the performance over the best of class wired and wireless interfaces.

4 Mobile Virtualization and Migration – Some Research Questions

Our experiments demonstrate the viability of the usage scenario proposed using existing off-the-shelf and open source components available for x86 platforms; and underlines the benefits of employing an Atom processor to support VMs and migration, on a mobile device. During the experiment described in section 3, we observed some issues that arose out of the nature of client side virtualization, and from the platform differences of mobile and desktop computers. Some of these have been identified by the research community in the recent past but there are still no off-the-shelf solutions to compare against. We summarize all our findings in the remainder of this section.

Effective Platform Resource Utilization vs. Generalization

Virtualization is typically implemented by using VMs which simulate a set of generic device interfaces to the guest OS. This works well for migration when VMs are compute bound, as long as the CPU instruction set is supported and the needed optimizations to support these typical uses of VMs in data centers that provide faster access to storage and networks, are available.

In the end-user/consumer space, when using a VM to access all the resources of the host device through generic abstractions of the hardware that are presented, undermines the opportunities users might benefit from by migrating to a more powerful host. For example, consider a game that migrates between a high performance desktop computer and a mobile device. If a user has a discrete graphics processor on their device they can run Open GL based games and other applications that are tuned for enhanced graphics, but they will be deprived of the enhanced experience when running inside a VM. There are solutions available, such as VMGL [8] that counter this issue to some extent, but they are not standard and are not guaranteed to work for all applications. For a user, the use of the complete working set of applications hosted inside a VM will not be transparent if it involves the loss of performance because hardware accelerators, normally available on the base platform are no longer accessible. This holds for the myriad of special peripherals that are available on mobile devices and are an important part of the user experience, but which may not be handled by the generic VM abstractions. Some virtualization solutions improve on this by offering para-virtualized drivers that enable the host and guest VM's to cooperate and provide a better approximation of the actual hardware to the guest VM. Such drivers will need to be made available for all of the special hardware and for each host OS and guest VM that can be hosted. An important take-away is that this issue is accentuated in the mobile market because the variety of peripherals is likely to be quite different on mobile devices and PCs.

Migrating across Platforms with Heterogeneous Resources

If we consider the usage scenario described in section 2, using live migration of the system state between a MID, and a desktop computer, the disparity in platform resources between the two becomes apparent. Live migration features have been realized for the personal computer as an evolution from the needs of servers, where VMs are migrated among servers with similar set of abstracted resources to attain highly available services in the presence of actual or anticipated failures. However in our mobile model employing out-of-the-box live migration features, the user experience will be limited to the minimum common denominator of the resources that are available among the target devices. This will limit the user experience on resource rich devices, and make it less transparent, which could deter user adoption.

In the light of our discussion in section 3 and the use of VMs for the migration of dynamic system state, interesting research questions arise. First, if and how to make it possible to provide a user with the appropriate level of abstraction of resources in the guest OS, while allowing the guest to be able to detect and dynamically adapt to resources on the target host, while maintaining the resource availability and continuity of its use after the process of migration. This would avoid the abstraction of resources down to the lowest common denominator, and instead match the capabilities of available resources on the target host. For example, it may be a scaling from a modest display resolution and screen area on a MID, to a high-definition display using a 3D accelerated graphics adapter on a desktop PC. The reverse question is equally likely, how can we migrate and adapt from a resource rich environment to a more modestly resourced environment in an effective way?

In summary, when using live migration the resulting client side experience should not deter the user by dumbing down the platform, instead they should have access to the full potential of all the hardware resources on the target. This includes both platform capabilities, and the processor performance; whether it be on a MID or a desktop computer. Live migration should have the capability to dynamically scale the user experience to take advantage of all the available resources.

One dimension of scaling of the migration process is to adapt to the available peripherals – video (display, camera), audio (speakers, Bluetooth headset), I/O (joystick, keyboard, track point, mouse), peripheral devices connected to the platform (printer, scanner), or network connectivity (broadband, cellular, WLAN, WPAN). Some of these peripherals do not have state information that needs to be migrated (e.g. keyboard, and mouse etc.) as the human operator ensures the continuity. However, some peripherals require state changes transition across the migration process. For example, in a network interface, the connections established need to be transferred to the destination host to provide seamless operation. Also a display needs to convey the same information that the user sees on the source host as it transitions to the target host. When using para-virtualized drivers to provide a closed approximation of the hardware resources, the transmission of peripheral state during the migration process will be key to making the experience seamless for the user. There is research in the area of capability adaptation for interface virtualization [13] that discusses some of these issues.

Another dimension in scaling the migration process is adapting to new computation resources - the system memory, the number of processor cores, available, and any special purpose co-processors that are being used (e.g. TPMs). In our experiment we used an Atom processor that has 1 core and 1GB memory, where as the laptop processor T60 had 2 cores and 2GB of memory. As technology evolves the variety of options will increase for both mobile and static computers making migration more complicated, though desktop computers will typically outpace the capabilities of MIDs. When a user employs live migration to move state between a MID and desktop PC, based on today's virtualization solutions the desktop computer will be under-utilized. This observation inspires the research question: how do we dynamically scale the OS to recognize and utilize the increased processing resources (multiple cores), and memory available on the desktop PC when migrating from the mobile device. And when the migration moves state from a desktop to the MID, how does an OS recognize that it has fewer cores and less memory available and adapt accordingly. Gracefully handling these transitions at the OS level could make these transitions less visible to applications, and hence allow legacy applications to be used without modification and achieve the best possible result.

User Interface Adaptation

The disparity of resources between computers is most keenly seen by changes in resources that affect the user interface. The modalities of interacting with a mobile device with significant processing power are themselves topics of active research. Even though very high-resolution displays are increasingly becoming available on the mobiles, interactions tend to be brief when compared to the use of a desktop PC. Today the interactions with mobile devices, mostly a legacy form the time when they primarily supported voice communication devices, are still evolving. Using legacy PC applications also has problems as they are honed for the keyboard-video-mouse model of interaction and it is difficult to use the same mode of interaction on a MID. There are numerous research initiatives that are addressing these issues, some employing sensor-based solutions to improve user interaction. However, this is potentially complicating VM migration as not all devices will have the same set of peripherals and sensors for interaction.

This raises the research question: how do we dynamically change the mode of interaction and user interface during migration to suit the target device, and how do we dynamically adapt the state of applications to these changing interfaces. User interface adaptation is also a hot topic in the HCI research community and also applies to scenarios such as designing web pages that display well on both mobile and desktop browsers.

Virtual Wireless I/O

One of the enabling platform features to support mobility is ubiquitous wireless access. As we described in our experiment, we used an 802.11 Draft N interface on the mobile device, and a user-space network-stack to provide network access inside the VM. The network stack limits the network performance seen by the guest, and furthermore the guest is not visible as a host by other devices on the LAN, or other guest VMs on the host, it is only visible to the hosting OS. Bridging the network interfaces makes the guest visible as a host on the LAN and lets other systems access this guest VM. This allows for constant and uninterrupted visibility (the guest maybe hosting services) as the guest VM migrates from one host to another. Host network bridging is available only for wired network interfaces and is not available for most of today's wireless interfaces because it involves manipulation of the MAC address for the interface it bridges. This function is not currently available in the firmware API of most wireless chipsets. Another option is to use IP routing to achieve a bridging function, but this has a detrimental impact on the VM network performance. A key learning is that wireless access is a key requirement for driving mobile usage, and the lack of support for bridging between wireless interfaces is a major hurdle for effective live migration.

Security

Data security is an important requirement for most computers, and it is even more critical for mobile devices that may be exposed to a wider variety of malware while connecting to networks in un-trusted locations. Secure partitioning, or sandboxing, mechanisms, can be based on TPM to establish trust for components that provide partitioning and protection among the partitions. In that case, related research questions arise for secure migration: how do the credentials that are rooted in the trusted hardware on one platform migrate or map on to the hardware credentials of the target platform.

When a user works with a set of known host computers for wireless Dock+ there can be a shared and preconfigured set of trust credentials among the computers. But, when the user tries to use a public kiosk for the Wireless Dock+ additional security concerns will arise for protection of both the kiosk computer and the mobile device. For example, malicious software could be left on the kiosk computer by another rogue user or rogue kiosk operator. This requires establishing a trust mechanism where the source and destination computer can ensure hardened boundaries of interaction between the devices. When trust is not established, interaction can be curtailed or established in a sandbox which limits the potential damage that could result from malware.

5 Related Work

ISR [12] and SoulPad [2] are research projects that have the goal of migrating VM state to maintain the continuity of a user session between computers. ISR enables a mobile user to access their remote session stored on a network server from any other ISR instrumented computer connected to the network. This involves checking out a *parcel* from the server and then checking it back in to allow access by the next computer that needs to continue the session. High-speed connectivity to the server is required for the best operating experience. There are various optimizations available to improve this process like VM overlays [16] and Opportunistic replay [14]. SoulPad provides a passive solution for transferring VM state in a USB flash drive. State can be moved from a computer to the USB drive and then restored at another computer by physically transporting the drive and plugging it into the target machine. There are additional process migration techniques that use sandboxing for migrating groups of applications, or process domains that involve check pointing and restart of the target PC at a later time [11]. Goyal & Carter [4] also use virtualization as a means of off-loading specific tasks to a server in the infrastructure for remote execution, and return the results to the source device, thus exploiting the potential performance benefit of more capable remote computers. Most other research and products around mobile virtualization solutions are proprietary in nature [9][10].

6 Conclusion

In this paper we propose the use of virtualization to enhance the utilization of processing capabilities available on MIDs, an emerging class of mobile device. Through Wireless Dock+ a computer can be configured as a communication device and at the same time serve as the mobile user's primary personal computer. Live migration techniques can be used to extend the MID using nearby high-performance and resource-rich computers, while retaining independent processing capability in their absence. Further, as our usage is based on high-throughput WLAN or WPAN radios to achieve this transition, it places no additional strain on a WAN or MAN to effect migration, and no centralized server is required for hosting the VM images. Wireless Dock+ therefore decouples WAN connectivity from the process of migration, and the migration process can be ad hoc and completely opportunistic.

As we propose using the MID as a mobile user's primary compute platform, there is no requirement to carry any additional devices or computer hardware. Our experiments show that a seamlessness user experience is possible along with modest migration times of ~100 seconds using currently available technologies: In the future this has the potential to decrease to a few seconds with the development of 60GHz radio standards (already underway). In addition, more research questions are brought highlighted by our migration experiments, arising from the heterogeneity of the devices involved, and the requirements for making the migration process transparent to users, particularly across platforms with a diverse set of system resources and user interface technologies. These questions need to be addressed to use live VM migration as a useful tool for computation transfer from a MID to a PC to right-size the user experience to the PC's enhanced processing capabilities, and then back to the MID capabilities when migrating the state back to it for enhanced mobility.

References

1. Barton, J., Zhai, S., Cousins, S.: Mobile phones will become the primary personal computing devices. In: WMCSA (2006)
2. Caceres, R., Carter, C., Narayanswami, C., Raghunath, M.: Reincarnating PCs with Portable SoulPads. In: Mobisys 2005. IBM Research Center, Watson (2005)

3. Cox, L., Chen, P.: Pocket Hypervisors: Opportunities and challenges. In: HotMobile 2007 (2007)
4. Goyal, S., Carter, J.: A lightweight Secure Cyber Foraging Infrastructure for resource constrained device. In: WMCSA 2004 (2004)
5. 60GHz: the gigabit wireless we are waiting for, http://apcmag.com/60ghz_the_gigabit_wireless_we_are_waiting_for.htm
6. Moblin Home Page, http://moblin.org
7. High Growth Forecasted for the Mobile Internet (2010), http://www.cellular-news.com/story/35529.php
8. VMGL, H Andres Lagar-Cavilla, http://www.cs.toronto.edu/~andreslc/xen-gl/
9. OK Labs Enables World's First Virtualized Smartphone, http://www.ok-labs.com/releases/release/ok-labs-enables-worlds-first-virtualized-smartphone-with-mobile-virtualizat
10. Secure Architecture and implementation of Xen on ARM for mobile devices, http://www.xen.org/files/xensummit_4/Secure_Xen_ARM_xen-summit-04_07_Suh.pdf
11. Osman, S., Shubhraveti, D., Su, G., Nieh, J.: The design and implementation of Zap: A system for migrating Computing environments. In: OSDI 2002 (2002)
12. Satyanarayanan, M., et al.: Internet Suspend Resume. CMU, http://www.isr.cmu.edu
13. Suh, S., Song, X., Jumar, J., Mohapatra, D., Ramachandran, U., Yoo, J., Park, I.: Chameleon: A capability Adaptation System for Interface Virtualization. In: Mobivirt 2008 (2008)
14. Surie, A., Cavilla, H., Lara, E., Satyanarayanan, M.: Low Bandwidth VM Migration via Opportunistic Replay. In: HotMobile 2008 (2008)
15. Want, R., Pering, T., Sud, S., Rosario, B.: Dynamic Composable Computing. In: ACM HotMobile 2008 (2008)
16. Wolbach, A., Harkes, J., Chellappa, S., Satyanarayanan, M.: Transient Customization of Mobile Computing Infrastructure. In: MobiVirt Workshop 2008. Carnegie Mellon University (2008)

Intelligent Telemetry for Freight Trains

Johnathan M. Reason[1], Han Chen[1], Riccardo Crepaldi[2], and Sastry Duri[1]

[1] IBM T. J. Watson Research Center
19 Skyline Drive, Hawthorne, NY 10532
{reason,chenhan,sastry}@us.ibm.com
[2] University of Illinois, Urbana-Champaign
201 N. Goodwin Avenue - M/C 258, Urbana, IL 61801
rcrepal2@illinois.edu

Abstract. Within the North American freight railroad industry, there is currently an effort to enable more intelligent telemetry for freight trains. By enabling greater visibility of their rolling stock, including locomotives and railroad cars, railroad companies hope to improve their asset utilization, operational safety, and business profitability. Different communication and sensing technologies are being explored and one candidate technology is wireless sensor networks (WSN). In this article, we present Sensor-Enabled Ambient-Intelligent Telemetry for Trains (SEAIT), which is a WSN-based approach to supporting sensing and communications for advanced freight transportation scenarios. As part of a proof-of-technology exploration, SEAIT was designed to address key requirements of industry proposed applications. We introduce several of these applications and highlight the challenges, which include high end-to-end reliability over many hops, low-latency delivery of emergency alerts, and accurate identification of train composition. We present the architecture of SEAIT and evaluate it against these requirements using an experimental deployment.

Keywords: freight trains, sensor networks, on-board telemetry, outlier detection.

1 Introduction

Train accidents are large sources of lost revenue for the railroad industry. An undetected mechanical issue may lead to a critical failure, such as a derailed train, which requires expensive on-site repairs. In addition, when a train becomes immobile, part of the railroad system becomes inaccessible, thus creating delays for other trains. While the lost revenue due to delays can be substantial, the loss of optimum train speed and balance throughout the train network also affects equipment utilization and customer confidence levels. In addition to revenue loss because of delays, freight trains carry hazardous materials of various kinds, which are at risk of spilling into the environment during an accident, potentially causing detrimental damage to the environment and further economic loss. Train accidents also have a third dramatic cost: loss of human lives. Each year dozens of lives are lost, and hundreds are injured, because of train accidents, which include crossing collisions, equipment failures, and other operating incidents.

Some of the equipment manufacturers for the railroad industry have conducted studies of advanced technologies, such as remote-controlled locomotives and electronically controlled brakes. These prior works have focused almost exclusively on upgrading specific subsystems with modern technology. This is the conventional approach to improving operating efficiencies. On a parallel track, the railroad industry is also searching for a disruptive technology that will enable more advanced applications that currently are not supported with their existing technology. The conventional wisdom is that greater visibility of the health and status of trains would enable business transformation in numerous areas, including predictive maintenance, schedule optimization, and asset utilization.

In this paper, we present Sensor-Enabled Ambient-Intelligent Telemetry for Trains (SEAIT), which is a WSN-based approach to support sensing and communications for novel and advanced freight railroad applications. Freight trains – which are comprised of un-powered and unwired railroad cars – can exploit the ad hoc wireless networking capabilities of WSN technology to provide timely data about the identity and condition of the cars composing a train. SEAIT (pronounced "see it") was designed and realized as a part of an ongoing feasibility study to discern whether WSN is a viable approach to provide the next-generation sensing, computing, and communications infrastructure for the freight railroad industry.

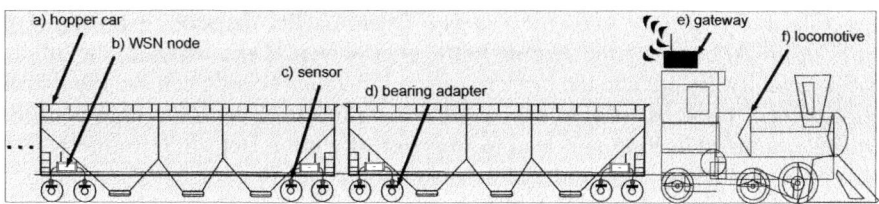

Fig. 1. A freight train hauling hopper cars (a) with WSN nodes (b) attached to the car body and sensors (c) attached to the bearing adapters (d). A gateway (e) is attached to the locomotive (f).

By adding local sensing, computation, and communication at the critical points of each car, SEAIT provides the infrastructure for enabling advanced railroad applications. Fig. 1 represents a possible deployment view of SEAIT on a freight train. On each car, WSN nodes (two per car) are monitoring the temperature of the wheel bearings using a thermocouple connected to the bearing adapter. The WSN nodes form a multi-hop network to communicate data towards the locomotive. The locomotive houses a special node, or gateway, that performs real-time analysis on data and events; thereby yielding faster response time to any outlier conditions (e.g., bearing is too hot). With this in mind, this paper makes the following contributions.

1. Proposes a WSN architecture that offers some novel features and optimizations, including semantics-based wakeups.
2. Proposes novel WSN solutions to key exemplary railroad applications and realizes these solutions using a reference implementation of the architecture.
3. Studies the feasibility of using WSN technology for railroad applications by comparing performance of the realization to key application requirements.

Considering the themes above, the paper is organized as follows. Section 2 describes the more challenging application requirements as exemplified by three advanced railroad applications. Additionally, this section discusses the key characteristics of an onboard WSN for trains. Section 3 describes the architecture of SEAIT and motivates the design choices. This section also cites related work when appropriate. Lastly, Section 4 describes the realization of the architecture and discusses how the experimental results compare to the application requirements.

2 Background

Through collaborative discussion with the railroads, we were able to collect a comprehensive list of desirable application scenarios that their next-generation, on-board sensing and communications infrastructure should support. While this list is too long to describe in detail here, we do describe a few choice scenarios to help highlight the more challenging requirements.

2.1 Application Requirements

On a railroad car, the hub where wheel meets axle encloses lubricated ball bearings that are hermetically sealed. The ball bearings help reduce friction when rolling and are collectively referred to as the *bearing*. Occasionally, the seal on a bearing will crack or break, exposing the bearing to the environment. Once exposed, the lubricant will eventually dry out and the bearing will fail. A failed bearing can cause a wheel to seize, which in the worst case can result in derailment of the train. However, more typically, a failed bearing will lead to stoppage of a train. Detecting (or more desirably predicting) a bearing failure can substantially reduce or eliminate derailments and stoppage delays caused by such faults.

The temperature of a bearing can be a strong indication that a bearing has failed or is about to fail. As the lubricant inside the bearing begins to wane, the temperature of the bearing increases because of increased friction between wheel and axle. It is common for the surface temperature of a failing bearing to exceed 300° F above the ambient temperature. The time between an overheating event and an actual failure may only be tens of seconds to a few minutes. Thus, timely reporting of an overheating (or hot bearing) event is paramount.

The current approach to detecting hot bearings employs trackside hot bearing detectors at sparse locations, on average about every 30 miles. Given the sparse deployment, coupled with slow average train speed (about 20 miles per hour), the current railroad technology does not support timely detecting or reporting of hot bearing events. On average, the current technology offers visibility every 45 minutes. Moreover, the hot bearing analysis takes place at the enterprise level – not in the field, adding to the overall delay of event notification. In fact, event notification is often performed manually via two-way radio between the train's engineer and operations.

While there is certainly more need for algorithm study to detect hot bearings in situ, the low-latency reporting of hot bearing events is the more challenging requirement. Critical alert messages must traverse a potentially long distance and over many communication hops. Additionally, for energy consumption consideration, nodes will likely employ a sleep schedule with a low duty cycle. Thus, the end-to-end latency will include the time it takes to wake up enough of the network to provide a forwarding path to the gateway.

Hot bearing detection is just one of several fault detection scenarios envisioned by the railroads, each having the requirement of low-latency reporting of alert messages. Other scenarios include cracked wheel detection, flat wheel detection, low/high brake pressure detection, car intrusion detection, high temperature detection on refrigerated cargo, and biological or chemical hazard detection. In addition to outlier detection, the railroads would also like to have periodic reporting of these key operational measurements, where the target reporting interval is every 10 minutes or better. In this paper we will evaluate SEAIT's effectiveness in delivering alert messages with low latency and regular synchronous reports, both over the characteristic distances and number of hops encountered on a real train.

In North America there are more than 1.5 million freight railroad cars and more than 17000 freight locomotives either owned or operated by the nine major Class I Railroads. Locomotives and railroad cars represent a significant portion of the capital assets invested by the freight railroad industry and their customers. A freight train may comprise up to 150 freight cars of different types from different customers to be delivered to different destinations. At every stop of a route, cars may be dropped off or added to the train. Maintaining the accurate and up-to-date information about the train consist (listing of the railroad cars and their order with respect to the lead locomotive) is an essential requirement for railroad operations. Such information may be used for correcting and preventing operational mistakes, logistic planning and optimization, and reconciling billing and other financial settlements. The information about the length and weight distribution of the train can also be derived from the consist, and these measures, along with speed, brake pressure, location, and knowledge of terrain can be used to help operate the train safely at the optimal speed for maximum fuel efficiency and train network throughput. Additionally, consist information is needed to fully utilize applications that perform outlier detection. Knowing the ordering of cars allows service personnel to be directed to the exact car and component that generated the outlier condition.

In the early 1990's, the North American Railroad industry adopted an Automatic Equipment Identification (AEI) system to identify and track railroad equipment while en route. Today, over 95% of the railcars in operation have been tagged with an industry standard-based UHF AEI transponder, one on each side of a car, and more than 3000 AEI readers have been deployed across North America. As a train passes by a trackside reader, the unique identification numbers of the locomotives and railcars are automatically captured by the reader and transmitted to a central server to construct a train's consist.

While useful, this approach suffers from the same drawback as the trackside hot bearing detectors. Namely, sparse deployment leads to gaps in coverage and untimely information. AEI readers are not available at all possible drop-off and pick-up locations. In fact, many waysides, where cars are often dropped off for customers to load/unload, are cut off from communications, including AEI. Alternatively, the railroads would like to identify a train consist in real-time, where real-time is measured in minutes versus the tens of minutes achieved using AEI. However, there are some challenges to indentifying consists in real-time. First, to keep the total solution costs down, it is desirable that consist identification be performed using radio frequency (RF) measurements only. Second, the communications infrastructure must be able to deliver data reliably over many hops. Lastly, assuming 100% reliability cannot be achieved, any proposed consist identification algorithm must be robust to some packet loss and noisy RF measurements.

The on-board, WSN-based infrastructure may also help the railroads to address another problem: detecting "dark" cars. Railroad cars are routinely left at wayside locations for customers to load or unload. Occasionally the information system would lose track of the locations of some cars after they are detached from a train. There is no existing solution for this problem. AEI is not suitable because the range is inadequate (up to 30 feet). Using WSN-based consist identification, each train network will be able to detect when and where cars are dropped off at waysides, thus reducing the occurrence of dark cars. Additionally, any WSN-equipped train that passes a wayside can potentially query the wayside for the presence of any dark cars.

This scenario is an example of an on-demand query/response communication paradigm, where there is a mobile gateway and one or more nodes at a stationary wayside. This paradigm presents an interesting challenge. Namely, the mobile gateway potentially passes many sensor nodes at the wayside simultaneously and each sensor node may only be in range of the wayside gateway for a few seconds (depending on the train speed). Thus, robust delivery and low latency are also challenges for dark car detection.

In anticipation of a progressive roll-out of WSN, a train might have some railroad cars equipped with WSN technology while others are not. In such a mixed-mode environment, a train might be without a gateway or the network might become partitioned. For example, sensor nodes on a train without a gateway can store pertinent data until they are in proximity of a wayside gateway, which then can query the train for data as it passes. This scenario is similar to the familiar data mule scenario, where the train is the mule. It is also another example of an important mobile railroad scenario that relies on the on-demand query/response paradigm.

2.2 Preliminaries

Conceptually, an onboard WSN can be viewed as a logical representation of the train composition. Recall Fig. 1, WSN nodes are attached to railroad cars, creating a physical association between the node and the car. Thus, tracking a WSN node is logically equivalent to tracking a railroad car. To make this binding concrete, a WSN node must be provisioned with a unique identifier for itself and for the railroad car before deployment. For this purpose, SEAIT uses unique 64-bit identifiers, which are referred to as the *nodeId* and *carId*, respectively. The resolution of each identifier is more than adequate to support uniqueness for the number of railroad cars and locomotives in North America. This binding must be persistent in a database at the enterprise, and gateways must have access to it. A gateway should be able to learn everything it needs to know about a WSN node/railroad car by querying the database with the nodeId/carId. We use node and car interchangeably throughout the discussion. Additionally, the location where each node is attached to the railroad car is needed to help determine the ordering of cars and to identify the location of faulty components (e.g., a bad wheel bearing). In SEAIT, location refers to a designated end, and side. In particular, railroads designate the end with the hand brake as the *B end* and the other as the *A end*. Similarly, each side of the car is designated as either the *left side* or the *right side* as viewed from the B end.

In many sensor network applications, the WSN is considered to be a single flat network controlled by one or more gateways, where the WSN nodes are free to choose any available gateway to sink their packets [1]-[5]. This approach to network discovery does not apply well to a WSN onboard a freight train because most railroad applications imply each node should belong to a distinct network. For example, let's consider the case when two trains, Train A and Train B, are within radio communications range of each other (as is the case in railroad yards and when trains pass each other). Operational data, such as a hot bearing alert, from a WSN node on Train A has no meaning to Train B, and vice versa. Therefore, nodes on Train A should never choose Train B's gateway as a sink, even if the route to Train B's gateway is the best route. This observation implies that there should be a logical binding between train and network. SEAIT accomplishes this by assigning a unique 16-bit identifier to every train WSN (TWSN) and binding the TWSN ID to a unique train identifier that already exists for every train. The resolution of a TWSN ID is sufficient to statically assign a unique ID to all the locomotives in North America. To support low power operation for motes not associated to a TWSN, there is one common system-wide wakeup channel. All motes not associated to a TWSN operate in a low power sleep mode on the wakeup channel. During association, the TWSN gateway assigns a channel distinct from the wakeup channel for TWSN traffic.

3 SEAIT Architecture

Fig. 2 depicts an overview of the SEAIT software architecture for a sensor node. Message flow is designated by the grey arrows, while the dotted black arrows represent cross-layer configuration. Railroad applications and services send and receive messages through the interface to the communication stack, which provides networking, link delivery, and an IEEE 802.15.4 medium access control (MAC) and physical layer (PHY). Applications also interact with node services, which have global scope. The information and reporting services realize the execution of a novel and uniform information and messaging model (Section 3.1 and 3.2, respectively). Because the messaging and information model is unified, most applications are realized in SEAIT without any application-specific messages; dark car detection is one example (Section 4). The synchronization service realizes robust management of a real-time clock (RTC), which provides application-level alarms (Section 3.3).

The network layer includes the message manager, time-scheduled transmit and receive queues, a list of neighbors, and a router (Section 3.6). The message manager supervises message flow and coordinates interaction between other components in the network layer. The router uses cross-layer configuration to provide a novel approach to optimizing the routing protocol based on the operating mode of the node. A time-scheduled queue in SEAIT is simply one where each message in the queue is assigned a specific time-to-live (TTL) in the queue. The queues use the following service policy: the entry with the lowest TTL gets served first. To break ties, a FIFO policy is employed. The time-scheduled queues provide a mechanism to randomize responses

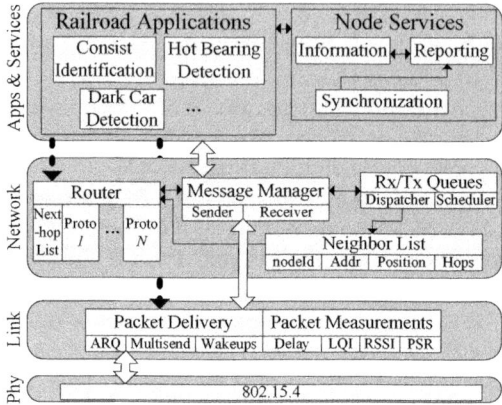

Fig. 2. Applications and services send/receive messages through the communication stack (grey arrows). Cross-layer configuration is used to optimize routing and link message delivery (dotted black arrows). Components within a layer interact with each other through their interfaces (solid black arrows).

to one-shot queries, provide a means to support sub-second timing for multi-phase protocols and applications (Section 3.7), and it support priority-based queuing. The neighbor list stores a node's one-hop neighbors' attributes, including nodeId, network address, position in consist, and hop count to the gateway.

The link layer provides packet delivery and packet measurements. All packets that flow through the communications stack are annotated with the packet measurements. The delivery mechanisms, automatic repeat request (ARQ) and multisend of broadcast messages, are well known, so we do not describe them here. However, we do highlight our approach to measuring delay and wakeup signaling (Section 3.3 and 3.4), which differ from prior works. We use 802.15.4 for the MAC and PHY; however, there is no real dependence on 802.15.4-specific MAC functions. The architecture can easily be adapted to any comparable multi-channel radio that uses carrier-sense multiple-access with collision avoidance.

3.1 Information Service

The goal of the information service is to provide a common information model for managing and reporting sensor, configuration, and application information. Most of the architectures we surveyed did not present an information model. One related work, Tiny Web Services (TWS), uses XML to support web services on resource constrained nodes [6]. However, this approach requires an underlying TCP/IP stack and about a third of the ROM resources on the device (15.8KB of 48KB). In contrast, the information service in SEAIT uses about 300 bytes of ROM using the same device as TWS. Because of this costly resource overhead and the dependence on a specific communications stack, we opted to design a very light weight information model for our study.

The information service manages a collection of information blocks. An information block is an object that describes the attributes and value[s] of some physical measure or a logical abstraction derived from some physical measure. The latter provides a means to track complex events of interest. From a gateway's perspective, attributes are read/write and values are read only. Each information block has a unique one-byte identifier, allowing for the definition of up to 256 unique information blocks. An attribute is a distinct characteristic of an information block, such as the unit of measure for a block's value. Attributes provide the information necessary to allow a sensor node to properly encode/decode the value of a sensor. For example, to model any analog temperature sensor we can use the following information block, expressed as the tuple *{T, Vref, ADCMax, unit, offset, resolution}*, where *T* is the temperature value, *Vref* is the reference voltage of the analog-to-digital converter (ADC) on the sensor node, *ADCMax* is the maximum range of the ADC, *unit* is a one-byte code representing the unit of measure for *T*, *offset* is the zero calibration offset of the sensor, and *resolution* is the sensor resolution (usually in volts/°C). The temperature value for the analog sensor can be determined as follows

$$V = \frac{ADCValue}{ADCMax} \times Vref \quad (1)$$

$$T = \frac{(V - offset)}{resolution} \quad (2)$$

where *V* is the voltage representation of the digital output (*ADCValue*) from the ADC.

The example above describes how an information block can be used as a generic model for a particular sensor type. That is, if the analog temperature sensor is changed to one having a different specification, the node does not need to be reprogrammed or taken offline. The new sensor can be connected and the analog temperature information block can be dynamically configured with new attributes. Attributes also provide a means to configure the application logic associated with a particular information block. To describe this aspect, we consider a threshold-based hot bearing detection application. We can model a hot bearing detection block using the following tuple *{hasAlert, avgBearingTemp, unit, threshold, window}*, where *hasAlert* is the state of the detector, *avgBearingTemp* is the average temperature of the wheel bearing, unit is a code representing the unit of measure for *avgBearingTemp*, *threshold* is a temperature threshold, and *window* is the time interval to average the bearing temperature over. The attributes *threshold* and *window* are used to configure the threshold detection logic. In particular, *avgBearingTemp* is derived by averaging the wheel bearing temperature over the last *window* milliseconds. For each new sensor sample, *avgBearingTemp* is computed and then compared to *threshold*. If *avgBearingTemp* is greater than *threshold*, then the bearing temperature is considered to be outside normal operating range and the value *hasAlert* is set to true. If *avgBearingTemp* falls back below *threshold*, then *hasAlert* is set to false. This simple example motivates using attributes to configure application logic. The same approach can be used to configure more sophisticated logic, such as trend analysis.

SEAIT provides a uniform structure for encoding and decoding of information blocks; Table 1 depicts this structure. Each information block contains two parts: 1) a small three byte header and 2) a number of data fields that store a block's values and attributes. The first byte of the header is always the block id and the second byte is always the length of the block in fields. The third byte represents the size of each data field in bytes. When the field size is greater than zero, then the information block is said to have fixed-length encoding, where each field has the same size. For example, if the field size is two, then each data field will be two bytes wide. When the field size is equal to zero, then the information block is said to have variable-length encoding, where each field can vary in size. In variable-length encoding, the first byte of each data field represents the length of the field in bytes. The first M data fields always contain the read only values, where the size of M depends on the block definition. The remaining fields contain read/write attributes.

Table 1. General structure of an information block

Header Byte	Meaning
0	block id
1	length of block
2	field size
Data Field	**Meaning**
0	read-only value
...	...
$M-1$	read-only value
M	read/write attribute
...	...
up to packet length	read/write attribute

To configure information blocks, gateway applications issue *Configure Command (ConfCmd)*. Configure Command enables configuration of all attributes for any information block supported by a sensor node, and multiple blocks can be configured with a single message. Using a nodeId or carId and a flag indicating which type of identifier is present in ConfCmd, the message can be targeted towards all nodes, to a specific node, or all nodes on a specific railroad car. End-to-end acknowledgments can optionally be requested for ConfCmd.

3.2 Reporting Service

The reporting service provides the messaging means by which a node's information blocks are sent upstream to a gateway. It keeps track of what information blocks require service and the type of reporting paradigm required by each active block. It also manages the execution of each reporting paradigm (see below). In a typical situation, a gateway will request one or more information blocks by issuing a *Report Request (RepReq)* and nodes respond using *Report Response (RepResp)*. Report Request supports retrieval of just the values in an information block or the entire information block (values and attributes). As with ConfCmd, RepReq can be targeted towards all nodes, to a specific node, or all nodes on a specific railroad car, and multiple blocks can be requested with a single message. RepResp is the unified message format that carries information blocks from a sensor node to the gateway.

SEAIT supports two types of synchronous reporting paradigms: periodic and N-times, where N-times reporting is a special case of periodic reporting that has a preconfigured finite duration. In both paradigms, a node will send all active information blocks periodically in its car's designated time slot, which is assigned when a car associates to a train (Section 3.5). A synchronous paradigm is invoked when the report service receives a periodic alarm event from the synchronization service

(Section 3.3). Because the specific use and granularity of intra-frame slot timing can depend on application semantics, applications can override the default slot timing. Consist identification is one application that employs this option to support robust intra-slot messaging needed to measure neighbor closeness (Section 3.7). Gateways can also modify a node's default duty cycle through an information block.

SEAIT supports asynchronous reports using on-demand and event-trigger paradigms. In the on-demand paradigm a gateway will request one or more information blocks using RepReq and all nodes will respond immediately. In contrast, the event-trigger paradigm is initiated by node applications and is used to communicate alert events. The hot bearing detection block described in the previous section is an example that employs the event-triggered paradigm. When the hot bearing detection logic detects an overheated wheel bearing, it updates the hot bearing detection block and notifies the report service that the block requires service. Since alert events will likely occur outside a synchronous duty cycle (when nodes are sleeping), SEAIT accomplishes fast delivery of asynchronous reports by using wakeup messages (Section 3.4). Any node that has an alert message to report will just send it if the node is in a synchronous duty cycle, otherwise it will send a wakeup message first and then send the alert. An intermediate node that receives an alert first tries to forward the alert to a next hop. At any intermediate hop where a forwarding node fails to find a next hop, the forwarding node will send a subsequent wakeup message and then send the alert message. Each intermediate node follows this procedure until the alert message is delivered to the final destination. This approach allows us to delivery alerts within tens of seconds over a long train (Section 4).

3.3 Synchronization Service

The synchronization service manages and provides an interface to a real-time clock (RTC). The RTC can be realized in software or hardware; SEAIT currently uses a software implementation. Applications can register alarms with the synchronization service, where one alarm is always dedicated for the report service. Alarms can be configured to provide second, minute, or hour granularity. A pre-computed schedule is set up by configuring an alarm to alert at a discrete epoch that evenly divides into one cycle of the next highest granularity. For example, for an alarm configured to a granularity of seconds, the valid periods are every 2, 3, 4, 5, 6, 10, 12 15, 20, and 30 seconds. There is also a notion of a starting epoch for each alarm. The starting epoch represents a shift of the alarm's epoch zero, and it can be any discrete value in one cycle of the clock. For example, an alarm configured to go off every 15 seconds with a starting epoch of 2 will alarm 4 times each minute of each hour at XX:XX:02, XX:XX:17, XX:XX:32, and XX:XX:47. The starting epoch provides a means by which two or more gateways operating distinct networks in close proximity on the same channel can coordinate their synchronous schedules. Using discrete alarm times allows for a simple and robust implementation because the alarm update procedure is memoryless; compute once and the schedule repeats each cycle of the clock. In contrast, when using a continuum of alarm times, the schedule does not repeat each cycle; requiring a node to maintain history and perform computations at each update.

To maintain its clock, our prototype gateway uses NTPv4, which provides a tolerance of 1ms or less over a LAN [7]. On a real locomotive, the gateway would have access to GPS time, which can provide microsecond tolerance. A synchronization update for each WSN node is achieved by broadcasting a reference clock from the TWSN gateway and measuring the accumulated delay at each hop, where the accumulated delay is used to adjust the clock offset. To combat clock drift, SEAIT uses a two-pronged approach. First, the gateway periodically sends out new reference broadcasts, with a frequency determined by application requirements. Second, each node proactively polls its neighbors for a fresh clock, whenever a node determines its synchronization state is stale. Our measured delay approach is similar in concept to DMTS or Delay Measurement Time Synchronization [8] in that it seeks to accurately measure the clock offset by measuring the delay between transmitter and receiver using a reference broadcast. Our approach differs from DMTS in how the measured delay is propagated throughout the network. DMTS requires the network to organize itself in a parent-child hierarchy, where each parent node updates its clock first, then generates and sends a new reference broadcast to its child nodes. In contrast, our approach does not require such organization. Instead, the network packet header includes a two-byte field that represents the accumulated forwarding delay, as measured from the reference source to each sink. Thus, at any given sink, the clock offset is simply the forwarding delay of the received reference broadcast message plus the queuing delay incurred between hops. Directly measuring the accumulated forwarding delay provides other benefits, such as measuring end-to-end packet latency and providing an accurate temporal notion of packet freshness.

Using a similar measured delay approach, the authors in [8] and also in [9] show that, when packet reception/transmission timestamps are taken as close to the actual event as possible, the worst case offset error can be limited to two clock ticks per hop, where clock tick is measured in the local time base of the sink node. Alternative approaches try to algorithmically estimate the clock offset and drift by using multiple broadcast references during synchronization update [10][11]. In our implementation, the local time base and the forwarding delay field have the resolution of a 32KHz clock, which gives a worst case error per hop of ~61µs. Assuming a liberal number of hops (about 30) to cover a maximum length train (150 cars with average length 60 feet), the worst case offset error would be ~1.8ms.

3.4 Semantics-Based Wakeups

Preamble sampling is a common low power, asynchronous approach used to wake up low duty-cycling (sleeping) nodes on-demand using a long preamble that is continuous or pulsed. Sending long preambles as a communication wakeup was first proposed in [12] and has since been refined in numerous works [13]-[16]. SEAIT's approach to wakeup signaling differs from these prior works in two respects: 1) a node's decision to wakeup is based on application semantics and 2) a novel packet structure to embed the semantics and wakeup signaling. In the common approach, wakeups would normally be obscured in the MAC layer, not accessible to applications. All the prior works cited above wakeup a node's radio from the MAC layer, which, from a layering perspective, seems like the appropriate layer to exercise control over the radio. However, our experience in designing a WSN for railroad applications shows that wakeup signaling can have greater utility if the decision to wakeup is realized at the application layer. In prior works, wakeup signaling is used exclusively as a mechanism to facilitate asynchronous, multi-hop transfer during upstream communications from WSN nodes to a gateway. In addition, SEAIT uses wakeups for a swath of applications and network operations, including the dissemination of network commands (e.g., reset), fast delivery of alert events from outlier applications (e.g., hot bearing detection), and as an on-demand wakeup for a group of nodes that share the same mode of operation (e.g., all dark cars).

Octets:	1	7	1	variable	1	1	7	1	2	1	1	variable	2	2
Fields:	Outer Length	Outer MAC Header	Pkt Type	Preamble	SFD	Inner Length	Inner MAC Header	Pkt Type	Wakeup Delay	Wakeup Type	Timeout	Max Hops	Options	Inner FCS	Outer FCS

Inner packet headers | Inner packet payload
Outer packet payload = Inner packet

Fig. 3. The inner packet is preceded by a long preamble and a full MAC header. The inner packet payload contains the application semantics.

A wakeup message in SEAIT is a specially formatted data message, whose structure is depicted in Fig. 3. The structure shown in the figure was designed for use with IEEE 802.15.4, but the same approach can be generally applied to other radios. Each packet is a maximum length message that contains a complete nested packet within a packet. The nested packet is called the inner packet and it is said to be contained in the outer packet. The outer packet has a normal PHY and MAC header and footer. The inner packet is the payload of the outer packet and is self contained. It includes its own PHY and MAC headers followed by the inner payload and ending with the standard two-byte frame control sequence (FCS).

In SEAIT, the first byte after the MAC header is always the packet type field, which indicates the type of packet. The packet type field provides a flat classification for all packets, leaving a designer free to decide which layer in the stack will decode a packet. To accommodate variable-sized inner packet payloads (and optionally variable-sized MAC addressing), the inner packet's preamble is variable (up to 100 bytes). To be complete, the inner PHY header must also contain the start of frame delimiter (SFD) and the length of the inner MAC frame. For our application space, we found 802.15.4 destination MAC addressing with short addresses sufficient for most applications. The inner packet payload has four other fixed fields and one variable length field. The wakeup delay field of milliseconds remaining in the wakeup burst. This field allows a receiver to schedule when it should wakeup and listen for data. The authors in [17] use a similar approach to reduce receiver idle listening time while waiting for completion of the wakeup burst. The wakeup type field indicates the type of command/application/protocol that is associated with the wakeup. The timeout field allows each wakeup to have deterministic duration of arbitrary length. The max hops field indicates the maximum number of hops for forwarding the wakeup. This field is decremented before forwarding to the next hop. When the field's value is zero, the wakeup is not forwarded. The options field carries optional application data, which is wakeup type dependent.

Because full MAC addressing is used in the wakeup signaling, a wakeup can target a specific node or an entire TWSN. Additionally, a wakeup can target only those nodes that have interest in the specific wakeup type, which is an effective approach to reduce idle listening that is not supported by prior works. Our realization of consist identification and dark car detection (Section 3.5 and 4) use this approach to target only nodes that share the same mode of operation. Forwarding of wakeups over a finite number of hops is another useful feature not supported by prior works. Wakeup forwarding can be used to ensure that the entire TWSN is awake before the gateway issues an important command. Applications can override the default parameters for the number of packets in a wakeup burst, the channel, and the destination addressing fields – on a per packet basis. When waking up a single node, the authors in [16] propose a pulsed approach to transmitting long preambles to allow time for the target node to acknowledge the wakeup, thereby shortening wakeup time and energy consumption. Alternatively, SEAIT also supports this approach by using the time-scheduled queues in the network layer to schedule multiple wakeup bursts that can be separated by gaps of arbitrary length, based on application requirements.

3.5 Associating Cars to a Train

The process of automatically associating cars with a train is a vital first step in the important application of consist identification. This process depends on the train's situational awareness: Is the train moving? How fast is the train moving? Did the train just start/stop moving? Does a manifest exist for the expected consist? Has the train received a clear-to-go message from yard operations? These business and operational constructs are most effectively assessed at the application level. However, because of the binding between WSN node and car and between train and network, network discovery is inherently coupled to the process of associating cars to a train. In particular, SEAIT employs a coordinated approach, where the network layer tracks networks (via gateways), but the application layer determines the association to a specific train/network. Thus, the network layer's default gateway is set by the application layer. Throughout the discussion, we use association and discovery interchangeably.

For obvious safety reasons, cars can be physically attached to (or removed from) a train only when the train is stationary. The TWSN gateway in the locomotive is best equipped to discern a train's situational awareness because it has access to GPS speed and location, as well as connectivity to the enterprise; therefore, association is initially directed by a TWSN gateway. There are numerous scenarios under which a gateway might start associating cars to a train; however, the basic approach is the same. The TWSN gateway detects a start condition (e.g., clear-to-go message, manual push-button, train begins moving, etc.). The TWSN gateway optionally sends a wakeup message on the wakeup channel, followed by one or more Car Association (CarAsc) messages (described below). The wakeup has type WAKE_CARASC, which indicates that CarAsc will follow. Any nodes not requiring association can ignore the wakeup and go back to sleep to conserve energy. Car Association contains the TWSN ID, synchronization data, the gateway's carId, the gateway's position in the consist, and an optional list of carIds. The synchronization data includes the clock reference and optionally the alarm time and duty cycle for the synchronous reporting paradigm. The list of carIds specifies which cars should associate to the train. This list is only available when a manifest exists. When no manifest exists, the list is empty, and the message is referred to as the all join CarAsc. When receiving an all join CarAsc a node will optionally join a network based on its car's motion. A SEAIT node uses motion detection (via an accelerometer) to determine its status. If an unassociated node is experiencing persistent motion upon receiving an all join CarAsc, then the node will join the network. To extend the range of CarAsc, associated nodes periodically take turns at rebroadcasting their association data on the wakeup channel, with an updated timestamp and optionally preceding it with a wakeup.

3.6 Routing

A railroad car experiences three distinct modes of operation: 1) it is not associated to a train (dark car), 2) it is associated to a train with a known consist (known linear topology), and 3) it is associated to a train with unknown consist (unknown linear topology). This observation led to a modal design approach to routing messages in SEAIT. Specifically, SEAIT supports multiple protocols for selecting the next-hop and the appropriate routing strategy is selected based on the operating mode of the TWSN. When the consist is unknown, SEAIT employs a hop-based routing strategy, where the next-hop always has a hop count one less than the current hop. During consist identification, this simple protocol is used because it has low complexity and it can quickly adapt to a changing topology. Once the consist is known, very efficient geography-based routing is used to multi-hop the ordered linear topology. A car's position in the consist represents it's one-dimensional coordinate along a line. The next-hop is selected by choosing a node that is P positions closer to the gateway. When a car is dark, it employs single-hop communications only because the typical distance from a wayside to the track does not require multi-hop (40 feet or less).

Through the cross-layer configuration interface (Fig. 2), the application layer configures the operating mode of the network, along with any operational parameters such as the car's position in the consist. While we found existing hop- and geography-based routing protocols adequate for our evaluation, SEAIT's modal approach does not preclude using other techniques.

3.7 Consist Identification Application

Consist identification is an iterative application that has four major phases: 1) associating cars to the train (Section 3.5), 2) measuring closeness between each pair of nodes, 3) reporting closeness measurements, and 4) determining the consist composition and order. The gateway sets up the second and third phase by sending a RepReq for the closest neighbor information block using the periodic reporting paradigm. This request tells each node to perform a series of RF closeness measurements and report the results. An RF measurement is a broadcasted request for neighbor information followed by unicast responses from each neighbor in range; each node takes a total of four measurements. During the measurements, sender and responder use the same power level, which is set to cover a range of about three cars. A closest neighbor information block contains a ranked list of each node's closest neighboring nodes. Each list entry contains the pair *{nodeId, measurement}*, where the measurement is the bi-directional link quality. We define bi-directional link quality as the average of the sender and responder link quality indicator, which is produced by the 802.15.4 radio. In the final phase of the consist identification application, an ordering algorithm (described below) at the gateway inputs the closest neighbor information blocks and produces the consist as its output.

Consider n cars in a train $\mathbf{N} = \{N_i \mid i = 1,...,n\}$. The ordering algorithm operates in three steps: 1) compute a car closeness metric $\{d_{ij}\}$ from the node measurements, 2) refine the car closeness metric using a correlation based operator, and 3) construct a weighted digraph, $G = (\mathbf{N}, \mathbf{E})$, where edge $e_{ij} \in \mathbf{E}$ has a weight of d_{ij}. The car closeness metric reflects the closeness between a pair of cars N_i and N_j. The closer the two cars are, the greater the value for d_{ij}. Forming a consist from $\mathbf{N}$ is then equivalent to finding the maximum Hamiltonian path for graph G.

Step 1: Each car N_i is equipped with two nodes. We denote q_{kl} as the bidirectional link quality between nodes k and l. We define the car closeness metric as the combined link quality measurements between all the nodes on N_i and N_j, such that,

$$d_{ij} = \sum_{\substack{k=2i,2i+1 \\ l=2j,2j+1}} q_{kl} \tag{3}$$

where d_{ii} is scaled to four times the maximum link quality value. Under ideal conditions, for each pair of cars, this summation has four terms. When any of the pair wise measurements are missing, q_{kl} is set to zero. Because the link quality measure is asymmetric and the scale may vary across nodes, we apply two conditioning operations. First, we scale the rows of $\{d_{ij}\}$ so that the diagonal becomes all ones. Second, we make the matrix symmetric by computing the average of the original matrix and its transposition.

$$\phi(\{d_{ij}\}) = \{d'_{ij}\}, \tag{4}$$

$$d'_{ij} = \frac{\sum_{k=1}^{n} d_{ik} d_{jk}}{\left(\sum_{k=1}^{n} d_{ik} \sum_{k=1}^{n} d_{jk}\right)^{1/2}} \tag{5}$$

Step 2: We could construct the graph using the metric directly and perform the maximum Hamiltonian path search. However, this often leads to unstable results, as the metric is based on noisy measurements. We introduce a refinement operator on the metric to improve the search algorithm's stability. Consider the function $g_i(k) = d_{ik}$. It can be viewed as a distribution of the closeness with regard to N_i over the entire set $\mathbf{N}$. The correlation between g_i and g_j gives a second-order measurement of the closeness between N_i and N_j. It makes the graph searching algorithm more stable as it incorporates a number of d_{ij}. The operator is thus formally defined by (4) and (5).

Step 3: The general case of maximum Hamiltonian path problem is equivalent to the Traveling Salesmen Problem, which is NP-Hard. For simplicity, we use a greedy algorithm to construct a maximum Hamiltonian path. The algorithm starts from a known starting node (the gateway on the locomotive). Each successive node is found by following the edge with the maximum weight. This simple algorithm is $O(n^2)$.

4 Evaluation

We implemented SEAIT in TinyOS [18] and the gateway software in Java. We deployed 32 WSN platforms along the front metal railing of the roof at our facility.

Each platform contained a TmoteSky sensor node, a sensor board, two 5.4 Ah batteries, an embedded antenna, an input/output connection board, and a weatherproof enclosure. The sensor board included temperature, light, and accelerometer sensors. The WSN node code fits into 38 KB of ROM and 7 KB of RAM. On average, freight railroad cars are about 60 feet long, ranging from as little as 40 feet up to 90 feet. Deploying two sensor platforms per car, we emulated a train of fifteen cars plus a locomotive, with each car 60 feet long and a node spacing of 10 feet between adjacent cars. The entire deployment spans about 0.25 miles across the roof.

The consist identification application (Section 3.7) was run for almost two hours using the 16-car deployment on the rooftop. The application was invoked via an electronic clear-to-go message. The periodic synchronous reporting paradigm was setup with a period of 2 minutes and a slot time of 384ms. To evaluate the ordering algorithm, we used the following criteria: 1) the latency before reaching a stable consist and 2) the accuracy of the consist when compared with the expected consist. We define the error as the number of cars that must be moved to achieve the expected consist. We define the ordering algorithm output to be stable when the error over the last 5 cycles is 2 cars or less. We use the term flip to denote the case when the algorithm transposes the order of two cars that are physically adjacent; 1 flip equals 2 cars in error.

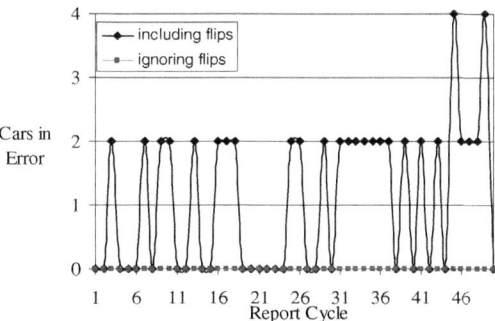

Fig. 4. Within 5 cycles, the algorithm is stable and 100% accurate, when ignoring flips

Fig. 4 shows details of a consist formed during 50 cycles of the application. When flips are included, the algorithm output is stable by the fifth cycle and the accuracy averaged over 50 cycles is 93%. A close look at the data revealed that all of the errors where the result of flips, which is encouraging because a small number of flips are tolerable; no single cycle had more than 2 flips (4 cars in error). If we ignore flips, the generated consist still stabilizes within five cycles, but the accuracy increase to 100%. During this experiment, the end-to-end reliability for each node ranged from 95.4% to 100%. Overall, these results satisfy the application requirements, namely accurate and timely detection is possible using only RF measurements. Because consist identification occurs on a stationary or slow moving train (5 mph or less), we expect similar results on a real train.

We conducted a series of experiments to study the most challenging requirements for the outlier detection class of applications. Namely, for asynchronous alert reports, can we achieve high reliability (95% or better) and low latency (10s of seconds to minutes) over the worst-case length train (150 cars). We configured trains of 5, 7, 10, 12, and 15 cars. For each configuration, one node was configured to simulate a critical event by sending a maximum length (128-byte) alert message every 30 seconds. To stress the system, we imposed a worst-case routing strategy (from a reliability perspective) by forcing all nodes to attempt a route through their closest neighbor first. If the initial route fails, then up to three alternative routes are tried in succession, each alternative being one car closer to the gateway. This routing strategy assumes consist ordering is known. The alerting node was configured at 10, 15, 20, 25, and 30 hops. We conducted 120 trials at each configuration using a wakeup burst of 2 seconds. The preamble sampling duty cycle was 1.2% (wakeup 12 ms every second). For robustness over long distances in a sparse deployment, we found the duty cycle should be large enough to capture 6 wakeup packets per wakeup burst.

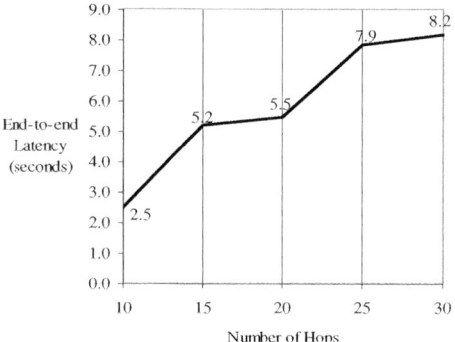

Fig. 5. Every 5 cars, an additional wakeup is required to continue forwarding the alert

Fig. 5 shows the relationship between the number of hops and the latency to reach the gateway, averaged over 120 trials. As expected, more hops to the gateway increased the delivery latency, in a monotonically increasing fashion. Roughly every 10 hops or 5 cars (350 feet) one additional wakeup is required to continue forwarding the alert. Extrapolating these numbers out to a 150-car train, we would expect the alert latency for a node on the 150th car to be about 75 sec, roughly 2.5 seconds per 5 cars. These extrapolated results satisfy requirements and are encouraging because they suggest an upper bound on latency, performance optimizations are certainly plausible. In particular, at the expense of more energy, shorter wakeups (or pulsed wakeups) can be made more robust by doubling the duty cycle to 2.4%. Additionally, in practice, we would use more efficient routing, such as choosing the furthest neighbor first. We also measured the reliability or packet success rate (PSR). The end-to-end PSR was virtually 100% over all experiments; only one packet was lost over all trials. This result is very encouraging because we do not expect to exceed 30 hops for a 150-car train.

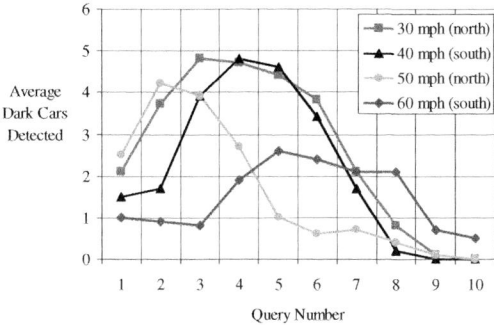

Fig. 6. The response distribution is sensitive to speed and direction. Shapes are similar for the same direction, with lower speeds having greater peaks and tails.

To measure the performance of the dark car detection application we performed experiments using a mobile gateway and five emulated dark cars, each car equipped with two nodes. The spacing between nodes was approximately 49 feet, and each node was elevated about 5 feet from the ground. The dark cars were placed on the wayside of a busy four-lane street. The perpendicular distance from the wayside to the path of the mobile gateway was 48 or 56 feet, depending on which direction the mobile gateway was traveling (northbound or southbound). In practice, waysides are typically no more than 30 to 40 feet from the track. The mobile gateway was a laptop secured in a car with an antenna mounted on the roof. Each WSN node employed a preamble sampling duty cycle of 2.4% (wakeup 24 ms every second). The dark car application was implemented using the following query sequence: 1) send 1-second wakeup, 2) send 3 on-demand RepReq messages for the dark car block, one every 200ms, and 3) wait 400ms for RepResp messages. This sequence was repeated every two seconds as the mobile gateway moved past the dark cars. We conducted 10 trials each for this experiment at 30, 40, 50, and 60 mph. Note, dark cars use a different duty cycle compared to cars associated to a train (i.e., alert latency experiments) because we wanted to use a 1-second wakeup to enable faster detection. Wakeup type WAKE_DARKCAR was used to indicate only dark cars should wakeup. A dark car information block contains nodeId, carId, the last known TWSN ID, and a timestamp indicating when the car went dark. Additionally, to stress the system, the block was zero padded to the maximum packet length.

The application detected the majority of cars at each speed, with the average ranging from 4.4 to 5.0 detected dark cars per trial. A detailed look at the distribution of responses to each query is presented in Fig. 6. The most effective queries occur when the mobile gateway is close to the center of the deployment. The figure also reveals sensitivity to speed and direction. For the 30 and 50 mph trials, the more effective queries are slightly biased towards the beginning of the run because the gateway was moving northbound, where the center of the deployment was reached during the initial part of the run. In contrast, data for the other speeds was taken from the southbound direction, where the center of the deployment was reached midway through the run. We also note that the slowest speeds have a higher peak and longer tail, which suggests a slower query rate could have been used. In contrast, the highest speeds require the fast query rate because more queries are needed to successfully detect most of the dark cars. These results satisfy the applications requirements because it shows that a high detection rate is achievable, over a range of plausible train speeds and over a typical track to wayside range.

5 Conclusion

This paper presented SEAIT, a WSN-based architecture and system built as a part of feasibility study to determine if WSN technology is viable for use in advanced freight railroad applications. While this study is ongoing, the set of experiments reported in this paper suggest the affirmative. The results showed that application requirements can be met for several exemplar scenarios, citing four key results: 1) reliable message delivery of 95% or better is achievable over a long train, 2) low-latency delivery of tens of seconds over 150-car train is plausible, 3) accurate and timely (within minutes) identification of a train consist is achievable when only using RF measurements, and 4) accurate dark car detection is achievable at a range of typical train speeds and for the characteristic distance form track to wayside.

While these key results are the culmination of substantial investigation, there are still areas for continued exploration. Since this work was part of a feasibility study, where satisfying application requirements are foremost, some aspects of the implementation can benefit from optimization. The specific routing protocols used in the experiments were deliberately made sub-optimal to help project an upper bound on latency and a lower bound on reliability. Similarly, we hope to improve the complexity of the ordering algorithm to $O(n)$. Automation of the scenarios, while maintaining stability, is another area for further study. For example, the dark car scenario should be automatically invoked when a train approaches a waypoint and automatically stopped as a train moves away from a waypoint. Using GPS and predetermined waypoints, a gateway can identify these proximity events and invoke the scenario at the appropriate time.

Acknowledgements. We would like to thank Lynden Tennison and Dan Rubin of Union Pacific Railroad for providing invaluable industry insights and support, Keith Dierkx for making this effort possible, and Maria Ebling and Paul Chou for their guidance and support throughout the project.

References

[1] Arora, A., et al.: ExScal: Elements of an Extreme Scale Wireless Sensor Network. In: Proc. of the 11th IEEE Intl. Conf. on Real Time Computing Systems and Applications, Hong Kong, August 17-19, pp. 102–108 (2005)
[2] Krishnamurthy, L., et al.: Design and Deployment of Industrial Sensor Networks: Experiences from a Semiconductor Plant and the North Sea. In: SenSys 2005: Proc. of 3rd Intl. Conf. on Embedded Networked Sensor Systems, November 2005, pp. 64–75 (2005)
[3] Kim, S., et al.: Wireless Sensor Networks for Structural Health Monitoring. In: SenSys 2006: Proc. of the 4th Intl. Conf. on Embedded Networked Sensor Systems, Boulder, Colorado, October 31-November 3, pp. 427–428 (2006)

[4] Werner-Allen, G., et al.: Deploying a Wireless Sensor Network on an Active Volcano. IEEE Internet Comp. 10(2), 18–25 (2006)
[5] Stoianov, I., Nachman, L., Madden, S., Tokmouline, T.: PIPENET: A Wireless Sensor Network for Pipeline Monitoring. In: IPSN 2007: Proc. of the 6th Intl. Conf. on Information Processing in Sensor Networks, Cambridge, MA, April 25-26, pp. 264–273 (2007)
[6] Priyantha, N.B., Kansal, A., Goraczko, M., Zhao, F.: Tiny Web Services: Design and Implementation of Interoperable and Evolvable Sensor Networks. In: SenSys 2008: Proc. of the 6th Intl. Conf. on Embedded Networked Sensor Systems, pp. 253–266 (2008)
[7] Mills, D.: Network Time Protocol Version 4 Protocol And Algorithms Specification (September 5, 2005), http://www.ietf.org/internet-drafts/draft-ietf-ntp-ntpv4-proto-11.txt
[8] Ping, S.: Delay measurement time synchronization for wireless sensor networks. Tech. Rep. IRB-TR-03-013, Intel Research, Berkeley, CA (June 2003)
[9] Chebrolu, K., Raman, B., Mishra, N., Valiveti, P.K., Kumar, R.: BriMon: A Sensor Network System for Railway Bridge Monitoring. In: MobiSys 2008: Proc. of the 6th Intl. Conf. on Mobile Systems, Applications, and Services, pp. 2–14 (2008)
[10] Maroti, M., Kusy, B., Simon, G., Ledeczi, A.: The Flooding Time Synchronization Protocol. In: SenSys 2004: Proc. of the 2nd Intl. Conf. on Embedded Network Sensor Systems, pp. 39–49 (2004)
[11] Ganeriwal, S., Kumar, R., Srivastava, M.B.: Timing-Sync Protocol for Sensor Networks. In: SenSys 2003: Proc. of the 1st Intl. Conf. on Embedded Network Sensor Systems, pp. 138–149 (2003)
[12] Intl. Telecommunication Union. Codes and formats for radio paging, ITU-R Rec. M.584-2 (11/97) (November 1997), http://www.itu.int
[13] El-Hoiydi, A.: Aloha with Preamble Sampling for Sporadic Traffic in Ad Hoc Wireless Sensor Networks. In: Proc. IEEE Intl. Conf. on Communications (2002)
[14] Polastre, J., Hill, J., Culler, D.: Versatile low power media access for wireless sensor networks. In: SenSys 2004: Proc of the 2nd Intl. Conf. on Embedded Networked Sensor Systems, pp. 95–107 (2004)
[15] El-Hoiydi, A., Decotignie, J.: Low power downlink MAC protocols for infrastructure wireless sensor networks. ACM Mobile Networks and Appls. 10(5), 675–690 (2005)
[16] Buettner, M., Yee, G.V., Anderson, E., Han, R.: X-MAC: a short preamble MAC protocol for duty-cycled wireless sensor networks. In: SenSys 2006: Proc. of the 4th Intl. Conf. on Embedded Networked Sensor Systems, pp. 307–320 (2006)
[17] Hui, J.W., Culler, D.E.: IP is Dead, Long Live IP for Wireless Sensor Networks. In: SenSys 2008: Proc. of the 6th Intl. Conf. on Embedded Network Sensor Systems, pp. 15–28 (2008)
[18] Levis, P., et al.: TinyOS: An operating system for sensor networks. In: Ambient Intelligence. Springer, Heidelberg (2004)

OneBusAway: A Transit Traveler Information System

Brian Ferris[1], Kari Watkins[2], and Alan Borning[1]

[1] Computer Science & Engineering, University of Washington, Seattle, WA
[2] Civil & Environmental Engineering, University of Washington, Seattle, WA

Abstract. Public transit is an important tool for those looking to ease their commutes, reduce their car dependence, or perhaps minimize their environmental impact. Unfortunately, the usability of transit systems often leaves much to be desired, to the point of deterring new riders. Tools on the web and mobile devices are increasingly being used to help tame confusing transit systems. OneBusAway is one such set of tools, providing access to real-time transit information for Seattle bus riders through a variety of interfaces, including web (http://onebusaway.org), phone, SMS, and mobile devices. We describe the current system, and then discuss current and planned research that builds on it to use increasingly-powerful smart mobile devices to provide location and context-aware tools for navigating transit systems.

1 Introduction

Public transit systems play an increasingly important role in the way people move around their communities. By helping travelers move from single-occupancy vehicles to transit systems, communities can reduce traffic congestion and the environmental impact of transportation. While there are significant benefits to using transit, many choice riders (that is, riders for whom transit is not the sole option) are ultimately reluctant to make the switch. Riders are often confused or intimidated by the complexity of large transit systems. Transit agencies often do themselves no favors by failing to provide information about the systems they maintain in simple, understandable ways.

Increasingly, smart mobile devices are being used to help manage the complexity of using transit. Whether it be a simple phone or SMS interface, or a more complex native mobile application, these systems can provide schedules, routes, real-time arrival information, and service alerts to users where they need it most: out and about, using their transit systems. We refer to these tools that help riders understand their transit systems as "transit traveler information systems." We already have some experience working on these systems, as we run a system called OneBusAway (http://onebusaway.org). OneBusAway provides real-time transit information and commuter tools for Seattle-area bus riders through a variety of interfaces, including web, phone, SMS, and mobile devices (see Figure 1).

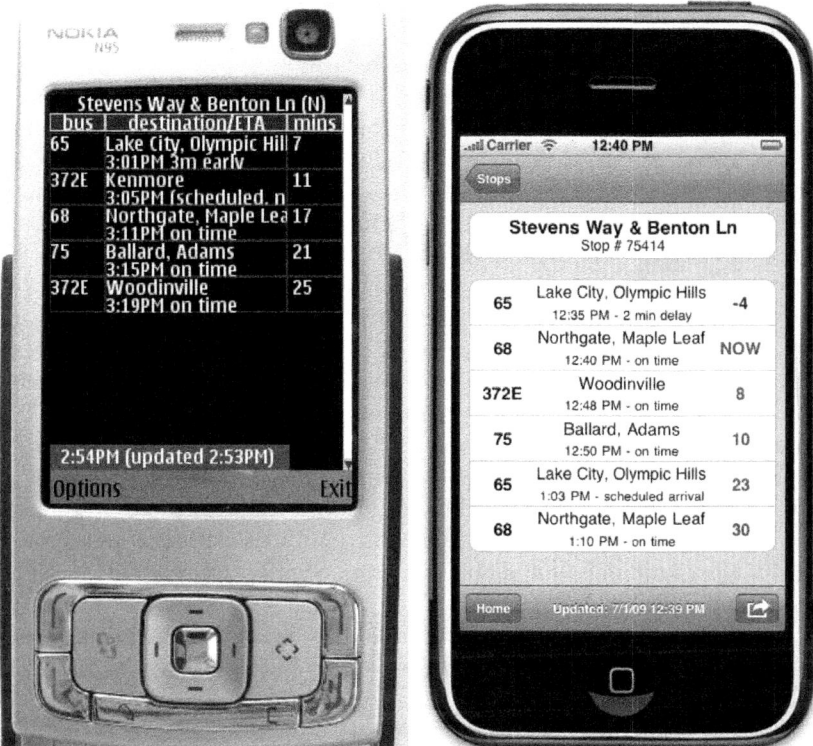

Fig. 1. The current OneBusAway mobile interfaces show real-time arrival information for both Nokia (left) and iPhone (right) platforms, among others

In the remainder of this paper, we first describe related work in the area of mobile transit tools. We then describe the OneBusAway system as it currently exists, and conclude with directions for current and future research that build on it.

2 Related Work

Displays that provide real-time arrival information for buses, subways, light rail, and other transit vehicles are now available in a significant number of cities worldwide at places such as rail stations, transit centers, and major bus stops. However, it is likely to be prohibitively expensive to provide and maintain such displays at (for example) every bus stop in a region.

With the increased availability of powerful mobile devices and the public availability of transit schedule data in machine readable formats, there have been a significant number of tools developed to improve the usability of public transit, especially mobile tools. One motivation is that, as noted above, it is unlikely that real-time transit information will be available on a public display at every stop. Another is that personal mobile devices can also support additional, personalized functionality, such as customized alerts.

One of the first online bus tracking systems, BusView, was developed by Daniel Dailey and others at the University of Washington [8]. More recently, Google Transit, which was started as a Google Labs project in December 2005 [7], is now directly integrated into the Google Maps product and provides transit trip planning for more than 405 cities around the world [5]. In addition to providing trip planning through a web-based interface, interfaces to Google Transit exist on a variety of mobile devices as well, making use of location sensors such as GPS and WiFi localization on the device to improve the usability of the transit app.

While Google Transit has been useful to transit riders around the world, it is also significant for establishing a *de facto* standard for exchanging transit schedule data: the Google Transit Feed Specification [6], or GTFS for short. The upshot is that many of the transit agencies participating in the Google Transit program have also released their transit scheduling data in the GTFS format for third-party developers to work with. Development ecosystems have grown out of the public availability of this data. The Portland TriMet third-party applications page [10] lists over 20 applications using Portland's transit data, many targeted at providing transit data on mobile devices. Similar ecosystems exist in San Francisco and the Bay Area, Chicago, and other major cities.

An example of a mobile application that makes use of GTFS transit data is the Travel Assistance Device (TAD) developed at the University of South Florida [1]. The TAD uses the GPS on a mobile device to detect the current location of a bus rider and to prompt the rider when his or her stop is near. Routes and desired stops are manually entered into the system for later detection. The application is specifically targeted at riders with cognitive impairments to increase the usability of public transit for these users.

Another example of a mobile application to improve the usability of public transit can be found in previous work at the University of Washington. The Opportunity Knocks system [9] also provides a mobile application to provide cognitive assistance to transit riders. Like the TAD system, the Opportunity Knocks system uses GPS data to model a user's current location. Unlike the TAD system, the Opportunity Knocks system automatically detects the user's current mode of transportation from GPS traces and learns the important places a user travels to, such as home and workplace, without manual labeling. Based on these learned models, the application can automatically predict with high confidence where a user is headed given only a small amount of tracking data, correcting the predicted destination as more data becomes available. Additionally, the system can detect when the user does something unexpected, such as forgetting to get off the bus at the regular stop, and then automatically notify the user. The main drawback to the Opportunity Knocks system is that it does not currently run on-line. That is, a week or two of GPS data must be collected, at which point a model of the user's important places and travel patterns are learned and used by the system. The system does not update this model when new places are added or travel routines change.

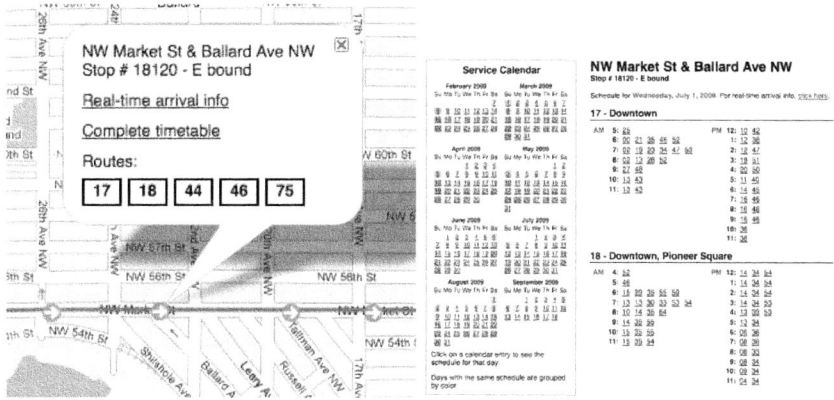

Fig. 2. Web-based route map and timetable interface, with routes and stops displayed on a map with direction of travel indicated. The full schedule for the stops is listed as well.

3 The OneBusAway System

In this section we describe the components that make up the OneBusAway transit traveler information system, along with its history and current usage.

3.1 Route Maps and Timetables

At a bare minimum, a modern transit website must provide static route maps and timetables to users. Our system is no different, though we enhance the usability of this static data by presenting it in novel ways, including a variety of Web 2.0 enhancements to make searching for stops, routes, and trips easier. Figure 2 shows an example route map and timetable. Routes and stops are displayed in a visual maps interface, with stop travel-direction indicated on the map, and routes servicing a stop shown in a pop-up dialog. When the timetable for a stop is examined, we display the complete service calendar, highlighting different service due to weekends and holidays. We also display specific route timetables in stem-and-leaf format to highlight frequency of service over the course of the day.

3.2 Real-Time Tracker

As with trip planners, real-time tracking has become accepted as an integral part of transit user information. The ability to determine when the next vehicle is coming brings travelers' perception of wait time in line with the true time spent waiting [3]. Transit users value knowing how long their waits will be or if they have just missed the last bus.

As noted in the related work section, one of the first online bus trackers was BusView, developed by Daniel Dailey and others at the University of Washington. We have built upon this system to develop a more user-friendly version of

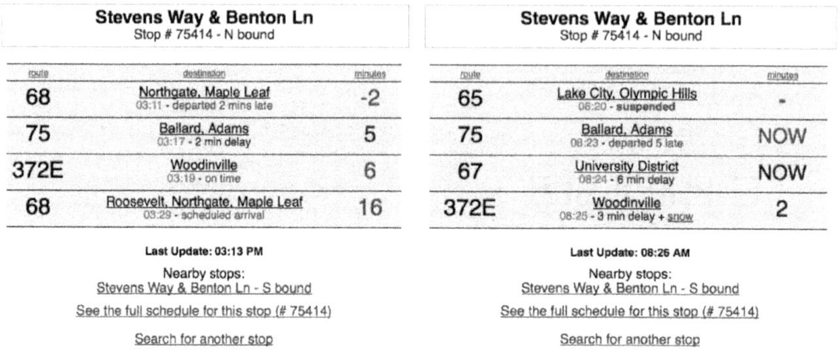

Fig. 3. Web interface to real-time arrival information. On the right, see example service alerts indicating cancellations and temporary reroutes due to snow-related adverse weather conditions.

the interface, available on the OneBusAway website. It includes various interfaces to real-time tracking data, including a telephone number users can call to have arrival information read to them, an SMS interface for receiving arrival information as text messages, a standard web interface (see Figure 3), and a website optimized for internet-enabled mobile devices.

3.3 Service Alerts

While fixed transit schedules change infrequently, the world in which those schedules must be kept is in constant flux. Temporary incidents such as construction, detours, accidents, severe weather, or special events often mean temporary service modifications in the forms of reroutes, skipped stops, or canceled service. Keeping users informed of these temporary service modifications is an essential task for any transit agency, but most traveler information systems do not use the full complement of communication modalities to notify their users. For this reason, service alert notification is a major component of our system, with an eye towards tight integration with the route timetables, trip planner and real-time tracker components so that service alerts are pushed to users across all communication channels.

Our current service alert infrastructure was put to the test during a major snowstorm in the Seattle area last winter. For more than a week, upwards of half the bus service in the Seattle area was cancelled due to icy road conditions. What service remained was often on detour to avoid iced-over hills or stalled vehicles. With conditions changing rapidly, it was difficult to keep riders up to date on cancellations and reroutes. We used OneBusAway to keep riders notified of service status when they accessed information for a particular stop on the web (see Figure 3) or through the phone system.

In the future, we wish to work on supporting better standardization of machine-readable service alert information from transit agencies, so that additional agencies can be more easily supported. For agencies that cannot provide automated machine-readable service alert information, we are also considering alternatives such as crowd-sourcing, which would allow riders to share information about service changes as they happen. As with any crowd-sourcing solution, issues of trust and verification would play an important role.

3.4 Trip Planner

Trip planners use an origin and destination address to search for one or more scheduled trips that travel between the two according to the desired time-frame of the traveler. Trip planners have become relatively common with larger transit agencies in the past several years. Although these trip planners are useful tools, they remain closed-source. This means that programs expanding and extending their content and functionality cannot be easily developed. As part of OneBusAway, we developed our own trip planner engine so that we might explore interesting planning applications.

One example of such an application is the Explore tool on the OneBusAway website (Figure 4). The Explore tool is a nearby attractions search tool, which

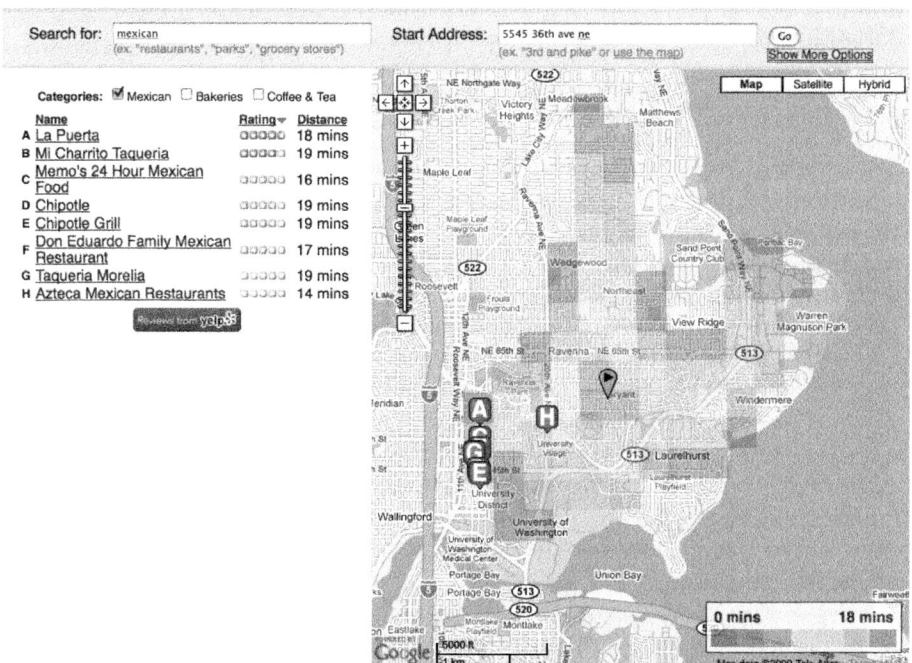

Fig. 4. The Explore tool. This search shows areas reachable in less than 20 minutes by transit from a given starting point, Mexican restaurants in that area, and the Yelp ratings for those restaurants.

attempts to answer another common transit rider question: "I'm looking for a nearby restaurant / park / library that's close by when taking public transit. What are my options?"

For first-time and infrequent riders who are not familiar with what is accessible using their local routes, this can be a difficult question to answer. The Explore tool aims to make answering that question easy by combining the functionality of a trip planner with online databases of local restaurants, shopping, and other amenities. In our implementation, users specify their starting points along with what they are interested in searching for. Optionally, they might specify additional features, such as maximum trip length, number of transfers, or walking distance. When the search is submitted, we compute the total area reachable by transit given the specified constraints and then begin searching for local businesses and attractions as specified by the user in the target area.

In the interaction shown in Figure 4, the user has searched for nearby Mexican restaurants within 20 minutes by transit from home. The display of results includes the name of the restaurant, the average rating for that restaurant, and the minimum travel time to the restaurant, along with a display of all the results on a map.

Once a user has settled on a particular restaurant, he or she can select it for more information, including location and upcoming transit trip plans to that destination. Our prototype includes data from the Yelp (http://yelp.com) online database of reviews.

3.5 Interface Modalities

OneBusAway is accessible through a variety of interfaces. As mentioned in previous sections, the various tools described above can alternatively be accessed through the website, an interactive-voice-response (IVR) phone system, SMS text messages, mobile-optimized web pages, and native apps for popular mobile devices, including Nokia and iPhone (Figure 1). The native mobile apps are of particular interest because of the potential to integrate location sensing technologies, such as GPS and WiFi-localization, with the real-time transit information system (Figure 5). In the other interface modalities, much of the interaction involves trying to determine where the users currently are and what routes and stops they are interested in. With the mobile apps, the location information can make narrowing the context of interest much easier, so that relevant information can be found more quickly. As we describe in the next sections, the new affordances made possible by new mobile phone technology drive many of our new research applications.

3.6 History and Current Use

OneBusAway was originally developed by the first author after too many late nights wondering if the route 44, a notoriously unreliable bus in Seattle, would ever come. Despite not yet being an official service of the local transit agency, the website receives about 3500 visits per day, and the telephone interface about

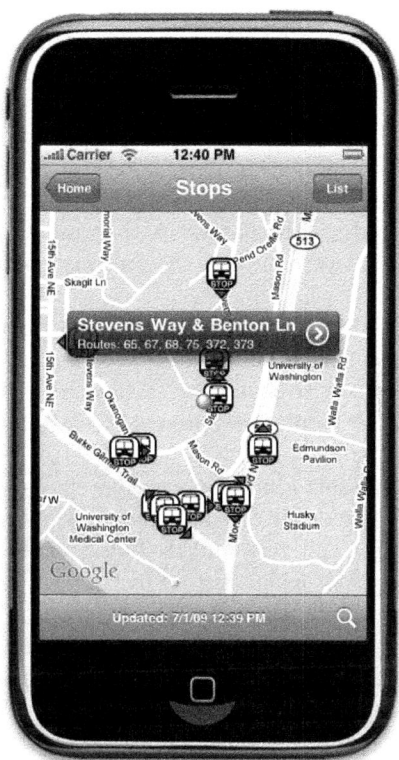

Fig. 5. Using a GPS sensor to automatically find nearby stops

1500 calls. The system has also received considerable publicity, including news stories and discussion on local transit blogs. We are currently discussing a more official status for the system with transit providers in the Puget Sound region, and also adding additional functionality.

Other transit agencies in Washington State and elsewhere in the U.S. have expressed considerable interest in the system as well. The system is open source, under the Apache 2.0 license. This supports one of our longer-term goals for deployment, namely, making the system easily available and customizable by different agencies. A closely related goal is to facilitate standardizing on an open standard for real-time and other transit information, to allow a rich set of transit applications to be prototyped and deployed.

3.7 Implementation

OneBusAway follows a standard multi-tiered architecture common to many web application projects. There are a few wrinkles, however, due to some of the large data structures involved in order to support fast trip planning, which we describe below.

One of the biggest implementation challenges in building OneBusAway was the trip planner engine. The first challenging aspect is the basic algorithmic details of building a "correct" trip planner. We represent transit data as a directed graph structure. Nodes are instances where a user could board or exit a transit vehicle at a particular time and stop, while edges are transitions between nodes as a transit vehicle moves along its route. There is a similar graph for street network data used to provide walking directions between stops. At a high-level, trip planning is just a directed-graph shortest-path problem over these graphs, but the implementation details of handling the various nuances of transit systems are quite complex.

The second challenging aspect is achieving high performance in the trip planner. Reasonable response times for modern websites are measured in the hundreds of milliseconds, so our trip planner engine must be able to very quickly compute trip itineraries. A common approach to speeding up computation is to keep the entire transit graph structure in memory. Keeping the transit graph in a standard database would introduce too much latency as nodes and edges in the graph are visited. For single-origin to single-destination planning tasks an A* directed shortest-path search can very quickly find a solution. The single-origin to any-destination planning task, as illustrated by the restaurant example described in 3.4, is more computationally complex. While essentially Dijkstra's algorithm is used to explore the graph, careful pruning and accounting is performed to limit the required search space.

The space complexity of the memory resident transit graph limits the size of trip planning tasks that can reasonably be attempted. A larger transit agency, such as King County Metro in Seattle, might have order-of-magnitude one million nodes in its transit graph and a memory foot print of 50-100 MB. While it is reasonable to build a graph for all the transit agencies in the larger Puget Sound region of Washington to support integrated transit planning, building and maintaining a combined transit graph for larger areas, such as the entire United States, is currently not attempted.

So that OneBusAway might grow to support data from other transit agencies, whether they be in the same county or around the world, we have adopted a federated data architecture that separates transit data into related regions that can be housed separately. The various transit agencies that make up the Puget Sound region of Washington might be combined into one data region, while the agencies of the Bay Area of California might make up another.

These data regions are represented by individual server processes exposed by a standard API using remote method invocation. Additionally, the various user interface components, such as the web interface, the interactive voice response (IVR) phone interface, and the external API interface are broken up into individual server instances as well, as shown in Figure 6. Each interface instance communicates with the appropriate transit data instance based on the particular region a user is interested in. The strong decoupling of the various instances allows for robust replication and failover across a cluster of machines.

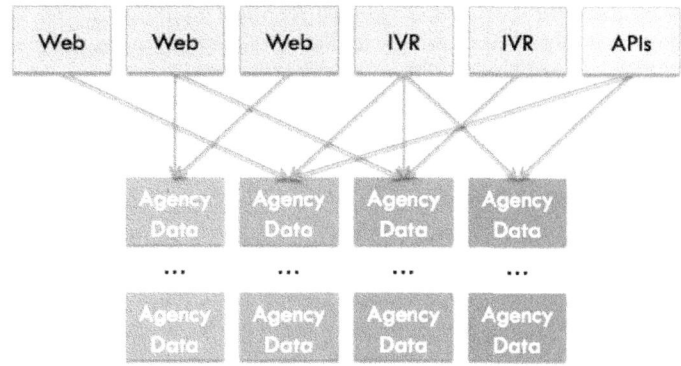

Fig. 6. OneBusAway architecture diagram. There is a strong decoupling between data sources and interface services, so that each service can be run as separate processes to support replication and failover across multiple machines.

OneBusAway is written in Java and uses a variety of standard open source development libraries and frameworks in its implementation. The system is composed of a number of service modules, each providing specific functionality, which are coupled together using the Spring inversion-of-control framework. Java object persistence to a relational database is handled by the Hibernate framework. The Tomcat servlet container combined with the Apache Struts MVC web framework does the bulk of the heavy lifting for web-based publishing. Client-side AJAX applications are written primarily using Google Web Toolkit, which compiles Java source code into optimized Javascript. For our telephony system, we use the Asterisk PBX server to pass incoming calls to handling code using the FastAGI interface. The only piece of non-open-source software in the entire system is our text-to-speech engine, which we license from Cepstral.

As mentioned, OneBusAway is open source software licensed under the Apache 2.0 license. The source code for the project, along with further implementation details, APIs, and documentation, can be found at our project site on the web: http://code.google.com/p/onebusaway/.

4 Current and Future Research

In addition to the goal of wider deployment of OneBusAway, we are actively working on a set of research questions opened up by the availability of this system as a base.

4.1 Automatic Learning of User Travel Patterns

We are working on two distinct types of travel pattern learning. The first type involves learning long-term models of riders' travel patterns as they interact with our tools, including which transit routes and stops they use on a regular basis, so that we might opportunistically notify them of service changes for future trips. The second type involves learning models for inferring the real-time transportation mode of a rider, so that we might advise them actively for a transit trip already in progress. This learning is similar to that of Opportunity Knocks [9], but we are extending the models to update continuously over time, so that changes in travel patterns can be integrated automatically.

To support both types of learning, we need data from riders about transit usage. To gather this information, we have instrumented the various tools in the OneBusAway system to support logging of key events. Collecting such data can of course reveal much personal information, so at this time we are only logging such event data for ourselves. In the future, we plan to also offer an instrumented version of OneBusAway to people who want to help us with user studies.[1]

As one example of data collection, every time a rider with the instrumented version of the system accesses real-time information at a particular bus stop, whether it be on the website, the IVR phone system, SMS, or a mobile app, we make a record. For our mobile app, we also log location and accelerometer data for supported devices. Riders also annotate their current activity, noting when they are walking, waiting for a bus, on a bus, biking, in a car, or stopped.

From this collected data, we can begin to build models of travel patterns. For long-term usage models, we collect simple frequency statistics about which stops and routes the rider typical uses. We also wish to build models of the places riders travel to and from, including where they live, where they work, and other places they frequently visit. These models are slightly more complex to build. While trip planner usage provides activity traces with clear start and end destinations, the high frequency trips that make up the bulk of travel for transit riders usually do not have trip planner activity traces because the rider is familiar enough with the trip to travel without explicit travel directions. Instead, we receive a number of real-time arrival data requests for the origin stop when a rider begins a trip, without any immediate indication of the final destination. Only by reasoning about sequences of these real-time arrival data lookups can we see the origin for a trip and then the destination when the rider takes the next trip.

Of course, there is no guarantee that the next stop accessed in a sequence is the actual destination, as a rider might have moved without leaving an activity trace in the meantime. Because of this ambiguity, we will make use of probabilistic inference to model the uncertainty in our modeling of travel goals. We will continue our work with directed and undirected graphical models such as dynamic Bayesian networks, conditional random fields, and Markov logic networks to tackle this modeling problem.

---

[1] We will follow standard Human Subjects protocols for such studies, including obtaining informed consent from participants and safeguarding the privacy of participant data. The version of OneBusAway offered for general download would of course not do any such logging, for privacy reasons.

In addition to long-term usage models, we also wish to build models for inferring the current transit activity of the rider. From the annotated activity traces of location and acceleration data, we build a simple activity model to recognize walking, waiting, riding a bus, riding a bike, riding in a car, and the default state of stopped. We are working to use a boosted classifier to take features computed from location and accelerometer data to build a local movement classifier. When we detect that a rider is in motion we will use high-level knowledge of transit routes, locations of streets, statistics about travel speed and path to infer the transportation type, jointly inferred and smoothed with a conditional random field.

All told, the focus of these models is to build and maintain an active model of how a rider typically uses transit and how he or she is actively using it right now. These models will be key to implementing our next application area.

4.2 Automatic Notification

When everything is going right, riders do not need a lot of help from their transit tools. Buses arrive on time and go where they are expected to. However, everything does not always go right. Buses and trains might be running late or not at all. Some routes might be redirected due to construction, weather, or accidents. A rider may not be familiar with a route, and may not know which stop to get off at, or what to do if he or she misses a bus. Ideally, when things go wrong, we can notify riders and help them pick appropriate actions.

Realizing this functionality requires two components: the models of travel, both historical and real-time, discussed in the previous section, and actual data from the transit agencies describing the status of their systems. The latter is actually trickier than it sounds. Only a handful of agencies in the United States have real-time tracking capabilities for their fleets. Even fewer push automated service alerts to their riders. We are actively working with agencies in our area, including King County Metro, Sound Transit, and Pierce Transit, to automate publication of services alerts in machine-readable standard-compliant formats so that they can be consumed by external systems such as OneBusAway.

Armed with accurate and timely status updates from the underlying transit systems, we can begin to integrate this information into trip-planner and real-time arrival information. However, additional services become possible when we consider pro-actively notifying riders about changes in the transit system. An example scenario would be automatically notifying a rider and suggesting an alternative when the bus he or she typically takes to work every morning is cancelled due to technical problems. This example combines our models of a rider's typical transit usage with real-time status updates in an innovative and useful way. We can also perform notifications once a trip has already begun, automatically detecting the rider's current mode of transit and providing assistance when we detect that the rider has missed the bus, did not get off at the proper stop, or got on the wrong bus.

4.3 Value Sensitive Design

At first glance, public transit might seem to be simply a tool for getting from point A to point B, and not have any particular relation to questions of human values. However, we believe this is not the case — rather, public transit touches on a number of important human values, including fairness, access, safety, health, community, social inclusion, and environmental sustainability. We will use Value Sensitive Design [4], a design methodology methodology for information technology systems that seeks to account for human values in a principled and comprehensive manner throughout the design process, as a way to help investigate this complex design space. The overall transit-using population (both current and potential) is a diverse group with diverse needs, many of which might be supported in different ways with technological solutions.

One specific topic that we are beginning to investigate is helping make transit information systems more accessible to blind and low-vision riders, and also deaf-blind riders. These riders are usually highly dependent on transit for travel within their communities, but transit traveler information systems can be difficult to use for the blind. We will start by conducting a set of focus groups and semi-structured interviews with blind and deaf-blind riders to help understand their values and needs, along with user studies of the OneBusAway interface, initially to simply improve the accessibility of OneBusAway for these riders, and subsequently to investigate tailoring information to their needs and interests. Combined with the increased capabilities of modern mobile phones, some truly innovative solutions may be possible.

Another topic is personal safety. This is an important value that influences the usage of transit: for example, some riders are uncomfortable waiting alone for a bus late at night. Our system might help support this value by providing real-time arrival information so that riders could minimize their wait time or by suggesting alternate stops that other riders report to be safer.

4.4 Transit Travel Behavior

Our primary goal in developing mobile transit tools is to help make transit more usable and enjoyable for a diverse range of riders. A secondary goal is to collect transit usage histories to help inform models of how people use both public transit and other transportation modes on a macro scale. One of the projects we are actively involved with is UrbanSim [2,11], a software-based simulation model for integrated planning and analysis of urban development, incorporating the interactions between land use, transportation, and public policy. UrbanSim includes a set of component models that simulate different actors and processes in the urban environment, such as residents looking for a place to live; real estate developers constructing new or renovated houses, offices, and other structures; and others. In its current implementation, UrbanSim is coupled with an external travel model, typically a commercial Four Step travel model.

One important area for improvement for UrbanSim is in the travel model. We want to replace the current external travel model with an activity-based travel model that we implement and maintain. An important aspect of calibrating and improving such models will be gathering richer information about people's travel behavior, including modeling under what circumstances a user will make a trip using transit, a personal vehicle, biking, or walking. This is one of the motivating reasons for being able to distinguish between travel using transit, bike, car, or foot in our travel classifier. The activity traces and transit models learned using our tool can help us better understand these transportation choices and build better models for our urban simulations.

4.5 Real-Time Arrival Prediction Accuracy

Although measures of travel time reliability on freeways and arterials are receiving increased attention, transit travel time reliability often continues to be viewed by transit agencies solely on the basis of overall on-time performance. We are currently working to increase our knowledge about the causes of travel time variability in transit. The objective of this research is to compare the on-time performance of routes based on specific characteristics of the service. Using historical Automatic Vehicle Location (AVL) data from King County Metro, we have begun an investigation of on-time performance and headway adherence on a segment by segment basis for routes throughout the transit system. We will develop a database of AVL data and link the data to characteristics of the segments, including:

1. Type of right-of-way exclusive right-of-way, exclusive lane, shared
2. Presence of transit signal priority
3. Underlying traffic volume
4. Route characteristics such as through-routing and stop spacing
5. Typical load factors on the route
6. Make-up of fare payment, such as percentage of monthly pass users

We will then analyze the effect of each of these factors on both the on-time performance and the deviation of transit travel times on that segment to determine which of these characteristics has the greatest impact on transit reliability. Ultimately, this information can then be used to improve the accuracy of the real-time arrival information from OneBusAway, as well as being useful to transit agencies in addressing bottlenecks in their systems.

5 Conclusion

Tools for enhancing the usability of public transit will continue to be an important application area for mobile application development. Our OneBusAway project is already providing a number of innovative transit tools, including providing access to real-time transit information for Seattle area bus riders through a variety of interfaces, including web, phone, SMS, and mobile devices. We have described the current suite of applications that constitute OneBusAway (including real-time arrival information, trip planning, and service alerts), and outlined a number of application research areas that we are actively pursuing to push the OneBusAway concept even further, using increasingly-powerful smart mobile devices to provide location and context-aware tools for navigating transit systems.

Acknowledgments. Many thanks to Nokia Research and to the National Science Foundation under Grant No. IIS-0705898 for their support of this research. Thanks also to Harlan Hile for his work on the Nokia native OneBusAway application, and especially to all of the users of OneBusAway for providing valuable feedback and suggestions.

References

1. Barbeau, S., Winters, P., Perez, R., Labrador, M., Georggi, N.: Travel Assistant Device, US Patent App. 11/464,079 (August 11, 2006)
2. Borning, A., Waddell, P., Förster, R.: UrbanSim: Using simulation to inform public deliberation and decision-making. In: Chen, H., et al. (eds.) Digital Government: E-Government Research, Case Studies, and Implementation, pp. 439–464. Springer, Heidelberg (2008)
3. Dziekan, K., Kottenhoff, K.: Dynamic at-stop real-time information displays for public transport: effects on customers. Transportation Research Part A 41(6), 489–501 (2007)
4. Friedman, B., Kahn Jr., P.H., Borning, A.: Value Sensitive Design and information systems: Three case studies. In: Human-Computer Interaction and Management Information Systems: Foundations. M.E. Sharpe, Armonk, NY (2006)
5. Google transit (June 2009),
   http://www.google.com/intl/en/landing/transit/#mdy
6. Google transit feed specification (June 2009),
   http://code.google.com/transit/spec/transit_feed_specification.html
7. Google transit partner program (June 2009),
   http://maps.google.com/help/maps/transit/partners/faq.html
8. Maclean, S., Dailey, D.: Wireless Internet access to real-time transit information. Transportation Research Record: Journal of the Transportation Research Board 1791(1), 92–98 (2002)
9. Patterson, D., Liao, L., Gajos, K., Collier, M., Livic, N., Olson, K., Wang, S., Fox, D., Kautz, H.: Opportunity Knocks: a system to provide cognitive assistance with transportation services. In: Davies, N., Mynatt, E.D., Siio, I. (eds.) UbiComp 2004. LNCS, vol. 3205, pp. 433–450. Springer, Heidelberg (2004)
10. Trimet 'unofficial' web and mobile applications (June 2009),
    http://trimet.org/apps/index.htm
11. Waddell, P., Wang, L., Liu, X.: UrbanSim: An evolving planning support system for evolving communities. In: Brail, R. (ed.) Planning Support Systems, Lincoln Institute for Land Policy (2008)

WebCall – A Rich Context Mobile Research Framework

Zhigang Liu, Hawk Yin Pang, Jun Yang, Guang Yang, and Péter Boda

Nokia Research Center
955 Page Mill Road, Palo Alto, CA 94304, USA
{zhigang.c.liu,hawk-yin.pang,jun.8.yang,guang.g.yang,
peter.boda}@nokia.com

Abstract. The ever-increasing capability of mobile devices enables many mobile services far beyond a traditional voice call. In this paper, we present WebCall, a research framework on how to share and utilize the rich contextual information about users, such as phonebook, indoor and outdoor location, and calendar. WebCall also demonstrates a few services (e.g. human powered questions and answers) that can be built on top of user context. Third-party services can be integrated with WebCall through a simple API and potentially benefit from context filtering. An invitation mechanism is introduced to enable bootstrapping user base. Privacy concern is addressed by giving full control to users on how to share their information.

Keywords: Mobile social networking, user context, location based services, phonebook.

1 Introduction

Mobile applications are becoming more and more widely available to users, often aiming to provide similar functionalities as online applications. The ubiquitous features mobility brings to these applications can provide a new level of user experience. Location and proximity information, sensor inputs, activity-related information, etc. mashed up with on-device data and shared with other users in a privacy preserving manner opens up entirely new dimensions to connecting people, sharing and consuming, and social networking in general.

The research framework we present in this paper takes the most central element of all on-device applications, namely the contact list (or phonebook), as the starting point. It is our initial assumption that most of the important persons are already present in the contact list and it may provide an additional value to the user if contextual information appended to contact list entries are available. Originally, we planned to provide a visual channel additionally to the voice channel when two persons are having a phone call, showing relevant contextual information through the visual channel, such as current location and weather information, the last picture taken, calendar occupancy. Later, we extended this "my mobile page" to off-the-call cases too, and to other visual representations around location: showing a group of people, e.g. colleagues' indoor locations within the work place, occupancy of meeting rooms, time zone and city level information, etc. Furthermore, we extended the framework with a simple plug-in mechanism where opt-in services can be provided to the user. Instead of focusing on known and popular service, e.g. micro-blogging and status reporting, we created a novel small-scale service that enables users to send anonymous questions to their 1st, 2nd, and further level of social groups. Answers provided anonymously by others are presented in a summarized way.

The developed research framework is used for experimenting various use cases and novel ideas. It has not been, neither planned to be introduced for wider use as a product or service. This research tool enables us to quickly test technologies, prove concepts, and learn user acceptance and experience feedbacks in small-scale trials. It was partially derived from our previous work on social proximity networks [1] but with more focus on system implementation.

The paper is structured as follows. Section 2 explains the overall WebCall system architecture. Section 3 describes features and their underlying technical details. Section 4 deals with lessons we learned through our project. Finally, we conclude the paper with future outlook and directions of research.

2 System Architecture

We chose the traditional client-server architecture to prototype our WebCall framework. Below are a few rationales behind the approach:

- The client-server architecture meets the basic requirements of authentication and authorization. A centralized repository also simplifies contextual data mashup, new service deployment and universal user access anytime from any device.
- Persistent data storage. This is needed to store users' historical data (e.g. past locations and communications) that can be useful for creating more personalized services. Although some of the data may be stored in mobile devices and processed in a distributed manner, a persistent storage on the server simplifies the system. In addition, it is particularly valuable as a user may lose or damage his/her mobile device.
- Easy "plug-ins" of third party services, e.g. advertisement or feeds to/from other social websites. The WebCall server is the center of intelligence and handles data filtering and aggregation.
- Ease of development and deployment. There are many off-the-shelf open source modules for us to prototype WebCall. We can avoid developing everything from scratch and focus more on the feature design and experiments.
- Ubiquitous HTTP connectivity. This avoids all the technical issues caused by Firewall/NAT traversal. In addition, it solves a practical problem that wireless operators may block non-HTTP traffic in their networks.

Fig. 1 shows the overall software system architecture with the main logical components.

In our prototype, we developed a WebCall client on Nokia N95 devices. The UI is implemented as a Web Runtime (WRT) widget [2] which produces good UI on mobile devices with existing web technology (e.g. basic JavaScipt or more advanced AJAX). However, since a widget cannot access local resources on device such as contacts, calendar and Bluetooth due to security constraints, we also implemented a local python module on top of Pys60 [3]. The python module acts as a local HTTP server. When the widget needs to access local resources, it sends an HTTP request to the python module which then processes the request and returns results in an HTTP response. This two-component client architecture is quite typical for developing web-based applications for mobile devices[1].

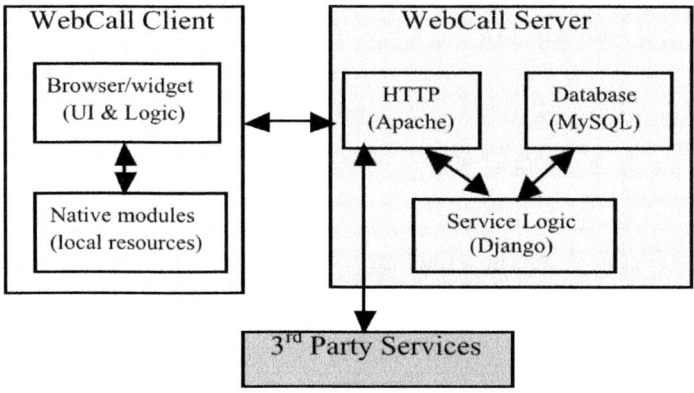

Fig. 1. WebCall system architecture

On the server side, we implemented the WebCall framework in a typical LAMP (Linux-Apache-MySql-Python) approach. Below is a quick overview on how the server processes each request from the client:

- The client sends an HTTP request to the server, triggered by either user interactions (e.g. menu selection) or background logic (e.g. location detection). Due to its simplicity and easy handling in Python, JSON [4] was chosen as the encoding format for the HTTP payload.
- The Apache HTTP server receives a request and dispatches it to the service logic unit if the URL prefix in the request matches that of WebCall.
- The service logic unit is the main part of the WebCall server. It is implemented in Python using the Django [5] framework. It parses each request received from the Apache HTTP server and dispatches it further to a particular Python function based on the request URL and request content (which is encoded in JSON). The python function processes the request and returns results in an HTTP response message.
- While processing a client request, the service logic unit will query and update MySql [6] databases as necessary. Fig. 2 shows a subset of tables in the database, in which virtual clipboard is a generic container for user related information (e.g. calendar, business card, messages, etc.)

---

[1] The latest version of WRT added a JavaScript device object through which a widget can access – and in some cases change – certain information on device. However, the capability is not without limit. A native proxy process is still needed for a widget if it needs to access certain resources such as writing to local file system.

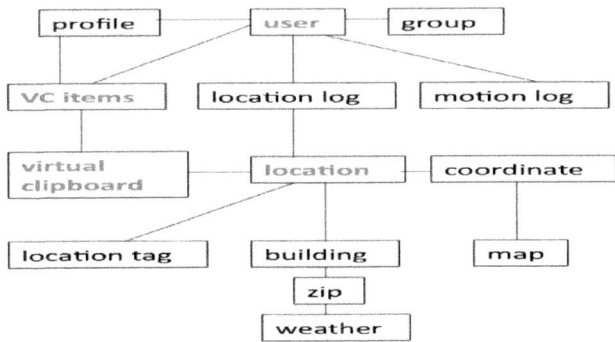

Fig. 2. Subset of database tables for WebCall

One important consideration in our system design is the capability to support third party plug-in services. As shown in Fig. 1, WebCall server can incorporate third party services through an HTTP interface. This is the way many features are implemented as described in the following section.

3 Features

The WebCall features are roughly categorized into channels: *People*, *Places*, and *Stories* (Fig. 3, left). After clicking on the people icon, a user will see a list of his/her contacts (Fig. 3, right), which the WebCall client reads from the local phonebook. A smiley face preceding a name indicates that the user is already registered with Web-Call. This allows a user to know whom he/she can invite into WebCall system (see next section). Users are identified in the system by phone numbers in the international standard format. Usernames are allowed but optional.

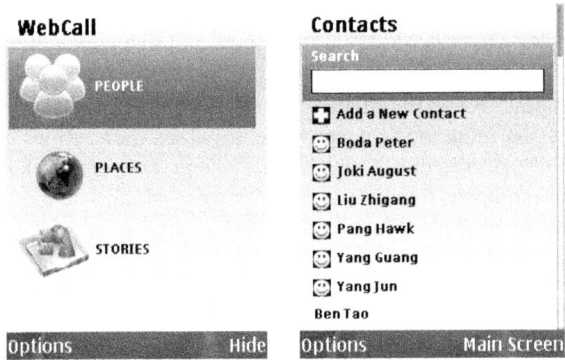

Fig. 3. WebCall top view and contact view

A user can initiate a phone call to a particular callee by clicking on his/her name. At the same time, information about the callee will be displayed on the screen as a dropdown list.

The "places" channel shows information about user locations, both indoor and outdoor. It can also show locations of a group of users.

The "stories" channel includes questions-and-answers and third party "plug-in" services.

The following sections will describe each of these services in detail.

3.1 Invitation and Registration

We provide an invitation mechanism in the WebCall framework to allow a registered user to invite his/her friends from the contact list who are not WebCall users yet. This is based on the concept of viral marketing where the number of users of a system could hit the tipping point and grow exponentially after initial success.

The invitation mechanism consists of two parts, one on the server side and the other on the client side. A WebCall user may select a contact from the phonebook and send an invitation to him/her if the intended recipient is not a WebCall user. The recipient will subsequently receive a Short Message Service (SMS) message with a short greeting text and an embedded link. Clicking on the link will lead the recipient to a registration/download page to be described shortly. Meanwhile the entire invitation process is recorded on the WebCall server such that various statistics can be retrieved later on.

3.1.1 Invitation

When the client sends out an invitation request, the server needs to handle this request accordingly. Under the Django framework, we first define a database schema for the invitation module that includes the InvitationID (a unique number to represent each new invitation), ApplicationID (optional), InviterName, InviterPhone, InviteeName, InviteePhone, Invitation_Status as well as different time stamps for status change. The invitation status can be "invitation started", "invitation cancelled", "invitation responded", "application installed" or "application removed", etc.

Then we divide invitation requests into five categories and handle them separately:

- Invitation_create: create a new invitation ID from inviter and the associated invitee information.
- Invitation_update: update an existing invitation ID with different status and corresponding time stamps.
- Invitation_retrieve: retrieve invitation status and corresponding time stamps information from an existing invitation ID.
- Invitation_track: for a given InviteePhone or InviterPhone, return a list of its inviters, invitees or a tree of invitees respectively.
- Invitation_stats: for a given InviterPhone, return some basic statistics like number of its direct invitees or indirect invitees.

On the client side, invitation functions are implemented as a library in JavaScript. The library exposes a single, uniform API that can be used not only in WebCall but also potentially in other third-party applications wishing for such functionality.

This single, uniform API is through an InvitationManager class, which internally consists of two building blocks: LocalComm and WebComm. LocalComm communicates via standard AJAX to the local HTTP server as described in Section 1. In the current version of InvitationManager we use LocalComm to mainly send invitations through SMS to the receivers.

The other module, WebComm, is architecturally similar but used to communicate to the WebCall server. It is also based on AJAX and adopts a similar syntax. The five basic functions it provides are invite, cancel, update, retrieve and stats. With the exception that cancel and update are tied to the same server-side request category of Invitation_update, the rest is one-to-one mapped to the server-side categories. Note that we have not implemented the counterpart of Invitation_track in the client library intentionally, because it is very difficult to efficiently and effectively present such complex data on the small screen. We believe it is more useful on a personal computer (PC).

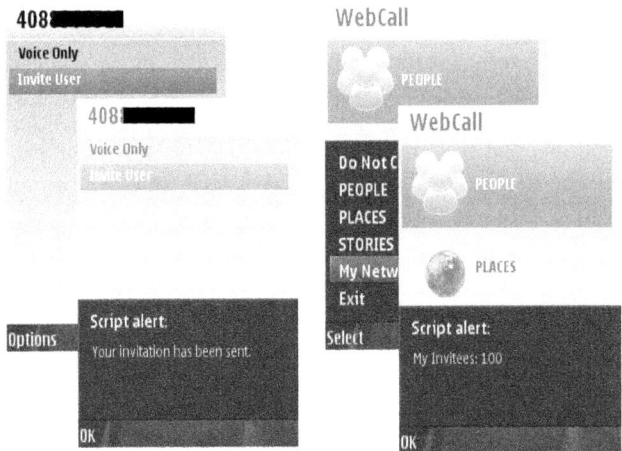

Fig. 4. Invitation module of WebCall

Fig. 4 shows several screenshots of the invitation module prototype in WebCall. On the left, a user is sending an invitation that involves notifying the WebCall server and sending an SMS message to the invitee. On the right, a user is checking statistics on his/her invitees, which involves checking the aggregated information with the WebCall server only.

3.1.2 Registration and Client Download/Installation

Before a new user can use WebCall, he/she needs to go through a simple three-step procedure: account creation, account activation, and client download/installation.

A new user can register with WebCall through a simple web interface from a browser either on a PC or on a mobile device. A registration can be triggered either by a user receiving an invitation in an SMS message, or by voluntarily visiting the registration web page. The only difference is that in the former case, the URL in the SMS message contains an InvitationID (as described in previous section) so that the system can track invitation statistics.

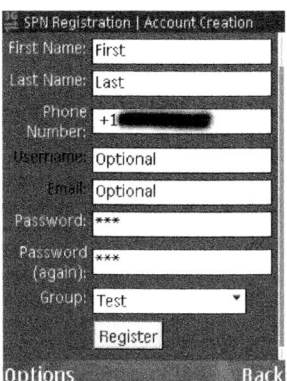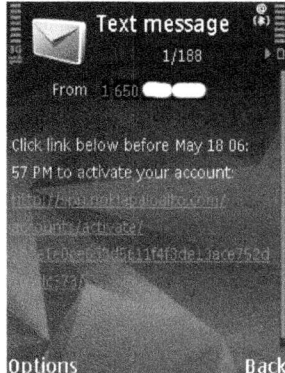

Fig. 5. User registration

The picture on the left-hand side of Fig. 5 shows a screenshot of the registration page as displayed in the Nokia N95 phone browser. To reduce the burden for a user to enter text from a mobile phone, all the fields can be dynamically pre-populated by the WebCall server in the case of an invitation-triggered registration. That is because the system can already retrieve information about the invitee from the inviter's contacts. Essentially the user only needs to fill in the password and group fields. Of course, the user may overwrite any field if he/she prefers to. A new user without invitation still needs to fill in the mandatory fields (in white). Note that the username field is optional. The system will generate a unique username from the phone number if the user does not provide one.

After the user fills in the registration form and clicks on the register button, the WebCall server will validate information received in the form and create a new user account. Then the server will send an account activation message (see Fig. 5, right) to the mobile device to verify both the phone number and the ownership. The final step of registration is for the user to click on the embedded URL in the activation SMS and launch the web browser. He/she can now proceed to download and install the Web-Call client.

Note that although the procedure is optimized for phone browsers, the first step (i.e. account creation form) can be also done through any browser on a PC.

3.2 Location Based Services

3.2.1 Indoor Location Based Services
Location based services are rapidly growing due to available radio infrastructure as well as increased capabilities of mobile devices. These technologies allow various applications and services to be built from guide and tracking systems to location specific advertising. However, while outdoor location based services have been growing because of availability of GPS in portable devices, indoor location based services have not been keeping the same pace with its outdoor counterpart. The reason for this is that no single good solution exists for indoor positioning in which it can achieve high position accuracy and scalability in the sense of minimal cost, time and effort for deployment.

Other than the obvious radio signal triangulation to determine the indoor location of people, which suffers from signal fluctuation due to moving objects, academia are exploring the possibility of tracking the user with low-cost, inertial-sensor-based devices attached to shoes, such as accelerometers, gyroscopes and magnetometers. These devices are small and can monitor the distances of foot travels from real time acceleration data as well as the angle of rotation and/or the change of direction through the gyroscope and magnetometer information. There has been much progress in this field and results seem quite promising [7]. But this too is not a perfect system and requires frequent recalibration of the user's location due to limitation of the hardware performance.

The approach taken for the indoor location system was to find the quickest deployable technology to establish an experimental testbed [8] in our work environment that allows the demonstration of indoor location based services. Our indoor location system consisted of Bluetooth tags situated at all points of interest in the building. Each office, meeting room, laboratory and common area was tagged with a total of 64 Bluetooth tags. The Bluetooth tags acts as positioning beacons where the radiated powers from the tags are adjustable through software API control. Having the ability to control the radiated power allows for adjusting the range/granularity for Bluetooth discovery between the Bluetooth tags and mobile device. In this particular setup the mobile device does not use the Bluetooth received signal strength indicator information (RSSI), but rather senses by scanning to find whether it is within the range of a Bluetooth tag.

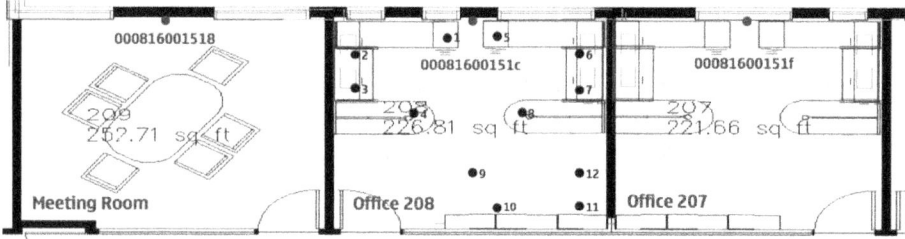

Fig. 6. Mobile device various measurement. Locations marked in blue for Bluetooth tag signal. Red dots represent locations of Bluetooth beacons in each office.

A number of measurements have been made to calibrate and verify the robustness of the indoor positioning system. Fig. 6 shows three Bluetooth tags [9] marked in red placed in each office and a mobile device positioned in 12 different locations within the center office marked in blue. Each of the Bluetooth tag's transmitted power adjustment was made in steps of 2dB to determine the level of Bluetooth signal penetration through the walls of the offices. In the case for Fig. 6, an optimal transmitted power from the Bluetooth tags was found to be -26dBm, equivalent to 2.5uW. Different room dimensions and wall material in buildings have a direct impact on the calibration of the transmitted power from the Bluetooth tags. Table 1 provides some approximations of the attenuation values [10] through common office construction at 2.4GHz (802.11b/g), the frequency range shared by Bluetooth and WiFi. A 3dB attenuation of radio signal is equivalent to a loss of half the power.

Table 1. Attenuation of radio signal at 2.4GHZ

Material	Attenuation
Plasterboard Wall	3 dB
Glass Wall with Metal Frame	6 dB
Cinder Block Wall	4 dB
Office Window	3 dB
Metal Door	6 dB

Shown in Fig. 7 is one set of measured results from the mobile device in location 5 of Office 208. Out of the 12 measured positions, the worst dataset has been selected as an illustration. The mobile device performed 5 Bluetooth scans searching for only the MAC addresses of the Bluetooth tags. These scans are then repeated 10 times, which are represented by legends as "try" in Fig. 7. For example for "try2", out of the 5 Bluetooth scans from the mobile device, there are 5 discovery of office 208, 2 discovery of the neighboring meeting room and no missed discovery. "No discovery" only occurs when the Bluetooth tag's radio signal are too weak for the mobile device to detect. As expected, variations are observed in each "try" due to movement of objects, including movement outside the room as well as possible interference from other radio signals in this frequency band. In addition, other scenarios have been taking into account such as the mobile device located in the user's pocket and when holding the mobile device with respect to the calibration of the Bluetooth tag's transmitted power.

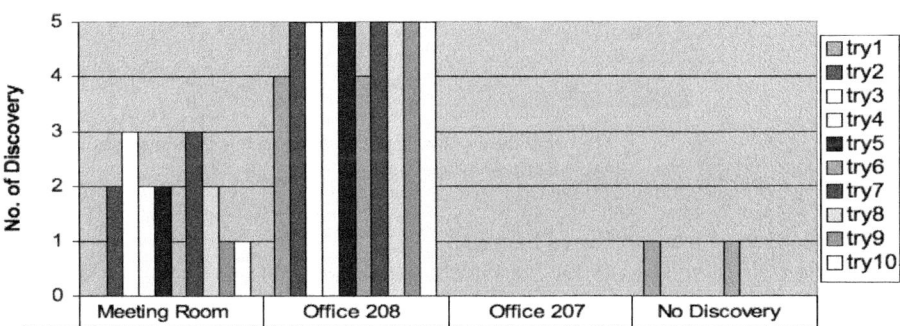

Fig. 7. One set of measurement results from mobile device at location 5 (Fig. 6, middle)

From these results, it does not seem reasonable to only make a single Bluetooth scan for the discovery of Bluetooth tags to determine the user's location, since there would be times when false or no location is detected. Instead, several Bluetooth scans are required to statistically compare the MAC addresses discovered and determine the correct location of the user.

To conserve the battery life on the mobile device, time interval for Bluetooth scanning and to report the user's location to the WebCall server can be adjusted by the user. In addition, the accelerometer sensor in the mobile device has been utilized to monitor whether the user is transitioning from one location to another. If this activity is detected, Bluetooth scanning from the mobile device is automatically initiated, ensuring the user's location information is recent rather than waiting for the set interval time to elapse.

One main disadvantage of Bluetooth tagging is that the battery life on the tags only last from days to weeks. Another disadvantage is the Bluetooth discovery time, which is specified at minimum time of 10.24 seconds. To achieve low maintenance of the indoor location system, we need significantly lower power consumption as well as improvement in discovery time. It is anticipated that in the near future ultra low power (ULP) Bluetooth would be available. ULP Bluetooth in many cases makes it possible to operate low cost sensor-type devices, namely radio tagging for more than a year without recharging and also improves on discovery time. Hence, future stand-alone ULP Bluetooth tags will be able to operate without intervention for long periods of time.

Powered by the room-level accuracy in our indoor positioning mechanism, we developed three simple features in WebCall: map showing a user's or his/her team's location (Fig. 8), messages attached to locations, and meeting room occupancy.

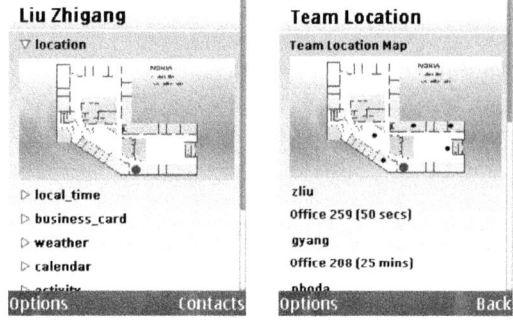

Fig. 8. Indoor location map. Red dot shows the target user's latest location while black dots (picture on the right) shows latest locations of his/her team.

3.2.2 Outdoor Location Based Services

We use a Cell-ID method as the basic technique to provide outdoor location services and applications. The method relies on the fact that mobile networks can identify the approximate position of a mobile device by knowing in which cell the device is operating at a given time. The main benefit of the technology is that it is already in use today and supported by most (if not all) mobile devices. The power consumption is much lower than GPS. However, the accuracy of the method is generally low (in the range of 200 meters), depending on the cell size. Generally speaking, the accuracy is higher in densely covered areas like urban places and lower in rural environments.

A Cell-ID usually includes four parameters: cell tower id (cellid), mobile country code (mcc), mobile network code (mnc), local area code (lac). When the WebCall server receives an outdoor location request from a WebCall client, it extracts these four parameters from the client's request. Then the WebCall server uses three APIs plus a local Cell-ID lookup table to obtain city-level location information and return it to the client. The ZoneTag API [11] allows registered developers to access the ZoneTag location services and get the best-known location directly (such as country, state, city, zip code). The OpenCellID API [12] can return the geo-coordinates information (latitude and longitude) of a specific cell. Another reverse geo-coding API [13] can parse the latitue/longitude and return a list of postal codes and places.

In addition to querying above servers directly, the WebCall server also creates a local lookup table to cache the mappings between Cell-IDs and location information. A search in the local lookup table is always performed first by the WebCall server when processing a request. This can significantly improve the average response time. If a Cell ID is neither in the local lookup table nor in the three APIs' databases, a failure response is sent back to the WebCall client. The whole procedure is described as follows.

Procedure of query logic in outdoor location
```
Get a request of outdoor location with cellid parameters;
Outdoor_loc = cellid_local_lookup(cellid, mnc, mcc, lac);
If ourdoor_loc == None:
    OpenCellID_xml = OpenCellID_API(cellid, mnc, mcc, lac);
    Lat = OpenCellID_xml.lat;
    Long = OpenCellID_xml.long;
    If lat == '0.0' or long == '0.0':
        ZoneTag_xml = ZoneTag_API(cellid, mnc, mcc, lac);
        If ZoneTag_xml == []:
            Return('Fail');
        Else:
            Outdoor_loc =xml.dom.parse( ZoneTag_xml);
            Update_cellid_lookup_table(outdoor_loc);
            Return(outdoor_loc);
    Else:
        Geonames_json = Geonames_API(lat, long);
        If Geonames_json == {}:
            Return('Fail');
        Else:
            Outdoor_loc = json.read(Geonames_json);
            Update_cellid_lookup_table(outdoor_loc);
            Return(outdoor_loc);
Else:
    Return(outdoor_loc);
```

3.2.3 Local Weather and Time Services

Once the WebCall server obtains outdoor location information of a WebCall client, it can deliver a couple of additional services such as weather and local time. The outdoor location information includes city, state and country as well as zip code. An XML search API from weather.com [14] can be used to get a unique weather location ID. If the country is US, the WebCall server passes city and state as parameters to call the API; otherwise it passes city and county instead. Based on this weather location ID, the Yahoo Weather API [15] enables the WebCall server to retrieve up-to-date weather information for the corresponding location, as shown in Fig. 9.

There is also local time information contained in the Yahoo API XML response, which includes date, time and timezone as well. This allows a caller to know a callee's local time, a small but quite useful feature in practice (for example to avoid making a phone call during midnight at the callee's local time).

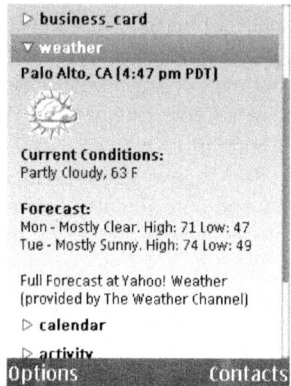

Fig. 9. Local weather information

3.3 Human Powered Questions and Answers

Another feature we added to WebCall is to enable users to ask questions from their mobile devices. Although people can get answers to many questions using services such as Google or Wikipedia, there still remains a large amount of information that is too transient or specific to be captured in the general web. The screenshot on the left-hand side of Fig. 10 shows a few examples of such questions. It is obvious that some of the questions are relevant only during certain time frames and to specific communities. The right-hand side screenshot in Fig. 10 shows a simple interface allowing a user to reply to a question and view the answers.

Besides the transience and specificity, a more important reason we offer the questions/answers (Q/A) feature in WebCall – in stead of using "traditional" web chat rooms or email – is to use contextual information collected by WebCall for optimal question routing. (Note: this is also how Q/A in WebCall is different than simple blogging applications such as Twitter). Such contextual information includes users' current and past locations, call logs (from which the system can infer social distances), calendar events, and even the Q/A exchange log itself. For example, a question about availability of a coffee machine in the kitchen can be routed only to people who are in the kitchen. Or, a simple question about whether you have seen John today can be routed only to people who is close to John, in terms of either social distance (from call logs) or physical distance (sitting in neighbor offices). Or, the same question can be routed to people who may have had meetings with John according to the calendar. In terms of the Q/A log, the WebCall server can analyze who replies to which questions most and use that information to rank users and construct target groups for question dispatching.

Another benefit of this Q/A mechanism is anonymity. Although the system tracks people who raise and answer questions, the identity is not shared among users unless a user opts to do so. This is to encourage answers from strangers.

We have implemented an experimental set of question routing rules including time, location, and group. Question routing is a critical factor for a successful Q/A service on mobile devices, given the constrained UI and limited user attention span on-the-go. It deserves more study in the future.

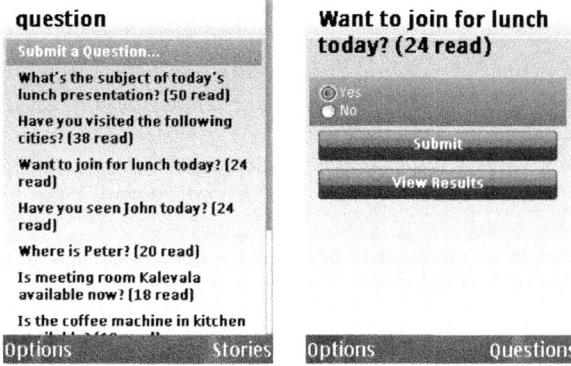

Fig. 10. Human powered question and answers

3.4 Other Features

WebCall also supports other features such as a user picture, business cards, and calendar event (Fig. 11, left). Another feature we want to emphasize is the capability to support third party "plug-in" services, such as news alert (Fig. 11, right) and advertisement. As many have discovered, content relevance is key to mobile services. One way to achieve that is to exploit user contextual information. This is the rationale why we combine third-party service with other WebCall features. We have implemented a simple framework in which a third party can submit content in the form of HTML and associated metadata (e.g. tags, target user criteria) to the WebCall server. The content will then be pushed to a user's mobile device after server validation and filtering. On the client side, hooks are already in place such that a new content item can be added to the menu and presented to users upon user selection.

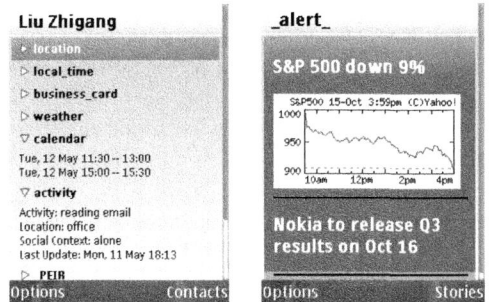

Fig. 11. Screenshots for calendar, activity, and news alert

WebCall can also act as the aggregator of user contextual information from sources other than mobile device. For example, Fig. 11 (left) shows a user's activities on a computer, as detected by context-sensing software named Pennyworth [16].

3.5 Privacy Control

Many of the contextual information described in previous sections are very sensitive in terms of privacy. The WebCall framework provides a few layers of privacy protection. First of all, only a user authenticated by the system can access other users' information. Second, we provide a simple privacy control interface (Fig. 12) that allows a user to specify what information can be shared with whom. User input will be translated into an entry to an access control table and enforced by the WebCall server when processing each request from a client. Lastly, every transaction is logged by the WebCall server so that any accident – if happened – can be analyzed afterwards.

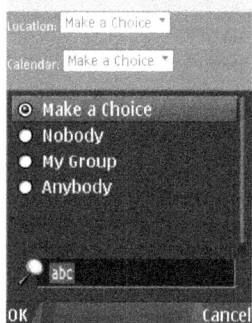

Fig. 12. A simple interface for privacy control

4 Lessons Learned

While trivial as it seems to use Bluetooth tags as location beacons, there are issues where Bluetooth signal is not completely attenuated by walls depending on the walls material. Hence, neighboring Bluetooth tag signal can easily penetrate the wall and give a false reading of a user's location. In addition, orientation and location of the mobile device can adversely affect the discovery of a Bluetooth tag that already operates at low output power.

As in any prototype of research concepts, we found that it is beneficial to have a clean and extensible system architecture. This is particularly crucial in our case as new ideas and features kept emerging even as we were prototyping the system. For example, one key component in WebCall is the database. The schema we chose at the beginning allows us to add new tables without loss of database integrity or having to modify existing tables. We found that the Django framework is quite capable and flexible, although documentation could be better and occasionally version upgrades break backward compatibility.

One thing we should have done but could not due to resource limitation is to involve real users for feature design. Although we selected WebCall features through brainstorming and common sense, it would be better to collect input from real users even before we implement a particular feature. Different perspectives, particularly from non-technical users, can provide balance to our probably biased view.

In terms of implementation, we found that the development environment on mobile devices still falls behind that on PC. While providing a good UI toolkit built on top of Javascript, WRT exhibits performance limitations as the WebCall client grows with more complexity. In addition, current version of WRT does not support dynamic loading of embedded Javascript. We had to work around the problem by creating a custom Javascript loader function that creates a DOM node on the fly. Hopefully the next release of WRT will address those issues and a few bugs we found during our implementation.

It requires extra effort to debug mobile applications. Even though an emulator on PC is very helpful, it does not cover all the test cases such as accessing local resources on device or setup network connections. The WebCall server logs every client request and server response in a table. It not only provides history of a user's transactions, but also serves as a good debugging utility for tracing down bugs. We also found that carrying query parameters in HTTP GET message header, rather than in the body of a POST message, is much more convenient for unit testing. One can simply tests the server with a web browser. Of course, it is a balanced judgment between API consistency and convenience.

5 Conclusions and Future Work

We developed WebCall as a generic research framework that extends traditional voice call with rich contextual information about users. The people component centers on the contact list on mobile devices, arguably the largest social networks in the world. The places channel provides location and location related information. Besides outdoor locations, we developed reliable indoor positioning system based on variable-powered Bluetooth beacons that can achieve accuracy to room level. The stories channel demonstrates a few promising features utilizing contextual information about people and places.

WebCall includes simple and intuitive invitation and registration mechanism to help grow the user base. We pay special attention to privacy and give users the full control on how to share their contextual information. Last but not the least, WebCall provides a simple API that allows third parties to integrate their services.

Still, there remain many interesting topics we did not have time to explore in depth and worth further study:

- First of all, as a research framework, we did not test the WebCall system with real users other than team members. It would be good to conduct user studies to validate the concepts and have a better understanding on whether and how user contextual information can benefit regular users without pushing them out of their comfort zone in terms of privacy.
- Similarly, we have not tried to build a real service out of this research framework. Although we believe the framework is quite robust and flexible, it remains to be tested on how it can be adopted by the industry and academic communities. In particular, we did not study in-depth the scalability of the system.
- Some components of the WebCall system are still "shallow". For instance, we only scratched the surface on the issue of Q/A routing based on users' historic

data. The topic is actually quite big and requires much more efforts. In addition, more study is needed on the third-party plug-in API.
- There are a few topics we left out from our system due to resource constraints and lack of real user data. One of them is user behavior predication based historical data. It would be a natural extension to the WebCall research framework by applying the latest machine learning techniques to the contextual information that could be collected by the system.

Acknowledgments

We would like to thank August Joki, who implemented most part of the widget UI on device and contributed to many of the ideas described in this paper. We also wish to thank Chris Karr for his assistance setting up Pennyworth for our system. The anonymous reviewers of this paper provided very good and constructive feedback, which we appreciate very much.

References

1. Yang, G., Liu, Z., Seada, K., Pang, H.-Y., Joki, A., Yang, J., Rosner, D., Anand, M., Boda, P.: Social Proximity Networks on Cruise Ships. In: MIRW 2008, Mobile HCI Workshop, Amsterdam, The Netherland (2008)
2. Web Runtime (WRT), http://www.forum.nokia.com/ResourcesInformation/Explore/WebTechnologies/WebRuntime
3. Python for S60 (PyS60), http://wiki.opensource.nokia.com/projects/PyS60
4. JavaScript Object Notation (JSON), http://www.json.org
5. Django, http://www.djangoproject.com
6. MySQL, http://www.mysql.com
7. Jeda, L., Borenstein, J.: Non-GPS Navigation with the Personal Dead-Reckoning System. In: SPIE Defense and Security Conference, Unmanned System Technology IX, Orlando, Florida (2007)
8. Cheung, K.C., Intille, S.S., Larson, K.: An Inexpensive Bluetooth-Based Indoor Positioning Hack. Massachusetts Institute of Technology (2006)
9. BodyTag BT-002. Bluelon, http://www.bluelon.com
10. Beating Signal Loss in WLANs, http://www.wi-fiplanet.com/tutorials/article.php/1431101
11. ZoneTag, http://developer.yahoo.com/yrb/zonetag/locatecell.html
12. OpenCellID, http://www.opencellid.org
13. GeoNames WebServices, http://www.geonames.org/export/ws-overview.html
14. weather.com API, http://xoap.weather.com/search/search?where
15. Yahoo weather API, http://developer.yahoo.com/weather
16. Pennyworth, http://pennyworth.aetherial.net

Lively Mashups for Mobile Devices

Feetu Nyrhinen[1], Arto Salminen[1], Tommi Mikkonen[1], and Antero Taivalsaari[2]

[1] Tampere University of Technology, Korkeakoulunkatu 1, FI-33720 Tampere, Finland
{feetu.nyrhinen,arto.salminen,tommi.mikkonen}@tut.fi
[2] Sun Microsystems Laboratories, P.O. Box 553 (TUT), FI-33101 Tampere, Finland
antero.taivalsaari@sun.com

Abstract. The software industry is currently experiencing a paradigm shift towards web-based software and web-enabled mobile devices. With the Web as the ultimate information distribution platform, mashups that combine data, code and other content from numerous web sites are becoming popular. Unfortunately, there are various limitations when building mashups that run in a web browser. The problems are even more challenging when using those mashups on mobile devices. In this paper, we present our experiences in building mashups using *Qt*, a Nokia-owned cross-platform application framework that provides built-in support for web browsing and scripting. These experiences are part of a larger activity called *Lively for Qt*, an effort that has created a highly interactive, mobile web application and mashup development environment on top of the Qt framework.

Keywords: mobile web applications, mashup development, Qt, Lively for Qt.

1 Introduction

In the past few years, the Web has become a popular deployment environment for new software systems and applications such as word processors, spreadsheets, calendars and games. In the new era of web-based software, applications live on the Web as services. They consist of data, code and other resources that can be located anywhere in the world. Furthermore, they require no installation or manual upgrades. Ideally, applications should also support user collaboration, i.e. allow multiple users to interact and share the same applications and data over the Internet.

An important realization about web applications is that they do not have to live by the same constraints that characterized the evolution of conventional desktop software. The ability to instantly publish software worldwide, and the ability to dynamically combine data, code and other content from numerous web sites all over the world will open up entirely new possibilities for software development.

In web terminology, a web site that combines ("mashes up") content from more than one source is commonly referred to as a *mashup*. Mashups are content aggregates that leverage the power of the Web to support instant, worldwide sharing of content. Typical examples of mashups are web sites that combine photographs or maps taken from one site with other data (e.g., news, blog entries, weather or traffic information, or price comparison data) that are overlaid on top of the map or photo.

Mashups usually run inside a web browser. However, because the web browser was originally designed to be a document viewing tool – not an environment for highly interactive applications – there are challenges when running web applications and mashups that behave in a highly interactive fashion. Support for user interface widgets can also be limited. Furthermore, poor performance of the web browser can be a major issue especially when running mashups in mobile devices. On mobile devices, usability issues cannot be ignored either [9].

In this paper, we present our experiences in developing practical, compelling web mashups, with a special emphasis on making those mashups work well on mobile devices. The work reported here is part of a larger activity called *Lively for Qt* (http://lively.cs.tut.fi/qt) – a project that has created a highly interactive, mobile web application and mashup development environment for the Qt cross-platform application framework (http://www.qtsoftware.com/). The Qt framework was recently acquired by Nokia, and versions of the framework have already been announced for Nokia's device platforms.

The rest of the paper is structured as follows. In Section 2 we summarize the existing environments and tools available for mashup development. In Section 3 we provide an overview of the Qt platform from the viewpoint of web application and mashup development. Section 4 contains a description of the most interesting mashups and other applications that we have written, including some source code of one of the applications. In Section 5, we discuss our experiences and lessons learned during the development of those mashups. Finally, Section 6 concludes the paper and outlines some future directions.

2 Existing Mashup Development Environments and Tools

The landscape of mashup development technologies is still rather diverse, reflecting the rapidly evolving state of the art in web development. Since mashups are usually built on top of existing content available on the Web, mashups can be composed manually using the classic DHTML technologies available in every commercial web browser: HTML, Cascading Style Sheets (CSS), JavaScript and the Document Object Model (DOM) [6]. However, since the actual representation of data, behavior and content can vary dramatically between different web sites, manual mashup construction can be extremely tedious, fragile and error-prone. For instance, since web sites do not generally present any well-defined interfaces that would clearly separate the public parts of the sites from their implementation details, there are usually few guarantees that the behavior and the data representations used by those web sites would remain the same over time.

To facilitate mashup development, a number of tools are available. In principle, mashups can be developed using general-purpose web application development platforms such as Adobe AIR [15], Google Web Toolkit [8], Microsoft Silverlight [12] and Sun Microsystems' JavaFX [1]. However, in practice the capabilities of these general-purpose web programming environments are still somewhat limited when it comes to the flexible extraction and combination of data from different web sites. The same comment applies also to general-purpose web content development tools including for example Adobe Creative Suite (http://www.adobe.com/products/creativesuite/) and Microsoft Expression (http://www.microsoft.com/expression/).

There are a number of existing tools that have been designed specifically for mashup development. Such tools include (in alphabetical order):
- Google Mashup Editor (http://code.google.com/gme/),
- IBM Mashup Center (http://www.ibm.com/software/info/mashup-center/),
- Intel Mash Maker (http://mashmaker.intel.com/),
- Microsoft Popfly (http://www.popfly.com/),
- Open Mashups Studio (http://www.open-mashups.org/),
- Yahoo Pipes (http://pipes.yahoo.com/).

We have reported our experiences in using these systems in an earlier paper [13]. In analyzing the systems, some common themes and trends have started to emerge. Such trends include:

- *Using the web browser not only to run applications/mashups but also to develop them.* For instance, Google Mashup Editor, Microsoft Popfly and Yahoo Pipes use the web browser to host the development environment and to provide seamless transition between the development and use of the mashups.
- *Using visual programming techniques to facilitate end-user development.* Visual "tile scripting" and "program by wire" environments are provided, e.g., by Microsoft Popfly and Yahoo Pipes.
- *Using the web server to host and share the created mashups.* Most of the mashup development tools mentioned above store the created mashups and applications on a web server that is hosted by the service provider.
- *Direct hook-ups to various existing web services.* Since the Web itself does not provide enough semantic information or well-defined interfaces to access information in web sites in a generalized fashion, most of the mashup development tools include custom-built hook-ups to existing web services such as Digg, Facebook, Flickr, Google Maps, Picasa, Twitter, Yahoo Traffic and various RSS newsfeeds.

So far, very little attention has been put on optimizing mashup development for mobile devices. It should also be mentioned that most of the above listed mashup development tools are still under development, e.g., in beta or some other pre-release stage, reflecting the rapidly evolving state of the art in mashup development. Nevertheless, many of the systems are already quite advanced and capable, and – perhaps most importantly – a lot of fun even for children to use.

3 Qt as a Mashup Platform

As part of the broader *Lively for Qt* (http://lively.cs.tut.fi/qt) activity mentioned earlier, we have created a dynamic, cross-platform mashup environment based on the Qt application framework. The broader *Lively for Qt* activity is reported in a separate paper [10]. In this section we provide an introduction to Qt, with a particular emphasis on its suitability for mashup development.

3.1 Introduction to Qt

Qt (http://www.qtsoftware.com/) is a mature, well-documented cross-platform application framework that has been under development since the early 1990s. Qt supports a rich set of APIs, widgets and tools that run on most commercial software platforms, including Mac OS X, Linux and Windows. In addition, Qt is available for mobile devices based on Nokia's Maemo Linux platform (http://maemo.org/) and Series 60 Symbian platform (http://www.s60.com/). Qt has been used in various commercial applications before. Examples of desktop applications built with Qt include Adobe Photoshop Elements, Google Earth, Skype, and the KDE desktop environment for the Linux operating system. In addition, Qt has been used in various embedded devices and applications, including mobile phones, PDAs, GPS receivers and handheld media players.

Trolltech, the company developing Qt, was acquired by Nokia in 2008. Nokia is currently in the process of making Qt libraries available on their phone platforms. Nokia's market share will make Qt an extremely interesting target platform for mobile applications as well.

From the technical viewpoint, Qt is primarily a GUI framework that includes a rich set of widgets, graphics rendering APIs, layout and stylesheet mechanisms and associated tools that can be used for creating compelling user interfaces that run in a wide array of target platforms. Qt widgets range from simple objects such as push buttons and labels to advanced widgets such as full-fledged text editors, calendars, and objects that host a complete web browser. Dozens and dozens of widget types are supported.

The GUI features of Qt adapt to the native look-and-feel of the target platform. For instance, on Mac OS X, all the widgets look like native Macintosh widgets, while on Windows applications utilizing the same widgets will look like native Windows applications. An essential part in enabling cross-platform GUI behavior is flexible support for widget positioning using *layouts*. Qt's layout components can adapt to different sizes, styles and fonts used by the host operating system. In general, automated layouts give significant advantage when a program is translated to other platforms and languages. The program adapts automatically to changed text sizes and resizes widgets in an aesthetically pleasant way. Additionally, since Qt supports full *internationalization*, all the locale-specific components (such as a calendar widget) automatically adapt to the current regional settings of the target platform.

In addition to its GUI capabilities, Qt has classes for networking, file access, database access, text processing, XML parsing and many other useful tasks. A multimedia framework called *Phonon* is included to support audio and video playback. Qt networking libraries provide support for *asynchronous HTTP communication* familiar from Ajax [2]. Asynchronous networking support is critical in building web applications that do not block their user interface while networking requests are in progress.

3.2 Qt and Web Development

What makes Qt relevant from the viewpoint of web development is that Qt libraries include a complete web browser based on the *WebKit* (http://webkit.org/) browser engine. The necessary DOM and XML APIs are also included to parse, manipulate and generate new web content easily. In addition, Qt includes a fully functional ECMAScript [4] (JavaScript) engine called *QtScript*. The presence of a JavaScript engine is important, since JavaScript – along with XML – is the *lingua franca* of the Web that is used by popular web service APIs such as the Google Maps API [5].

The web browser integration in Qt works in a number of different ways. For instance, it is possible to instantiate any number of web browsers inside a Qt application using the *QWebView* API. The *QWebView* class provides a widget that can be used to view and edit web documents inside applications. The data in web documents can be manipulated using the built-in DOM and XML APIs.

To support Qt applications in any web browser, a plugin called *QtBrowserPlugin* exists for embedding the Qt environment into any commercial web browser such as Mozilla Firefox or Apple Safari. The plugin makes it possible to run Qt applications inside a web browser, either as standalone Rich Internet Applications or alongside (or embedded in) conventional DHTML and Ajax web content.

JavaScript support in Qt is available both inside and outside the web browser. By default, the QtScript engine can only access those APIs that are part of the ECMAScript Specification [4]. However, by using a tool called *QtScriptGenerator* bindings to all the Qt APIs can be made visible to the JavaScript engine. This makes it possible to create JavaScript applications that combine classic DHTML behavior with widgets and other APIs offered by Qt.

3.3 Using Qt for Mashup Development

With Qt and its built-in web browser and JavaScript support, we have created a dynamic mashup environment that makes it possible to create mashups that can run inside the web browser as well as native desktop or mobile "phonetop" applications. The mashups are written in JavaScript, and they communicate with existing web services using asynchronous networking.

The mashups can leverage the rich Qt APIs for information visualization and processing. This is important since mashup development commonly relies on a plethora of data formats used by different web sites and services. In addition to binary image and video formats such as GIF, JPEG, PNG and MPEG-4, textual representations such as XML, CSV (Comma-Separated Value format), JSON (JavaScript Object Notation) and plain JavaScript source code play a central role in enabling the reuse of web content and scripts in new contexts. Qt provides excellent capabilities for processing such information, especially when combined with a dynamic language such as JavaScript that allows new object types to be constructed on the fly to accommodate the different data formats.

4 Sample Mashups

In this section we summarize the mashups that we have developed for our Lively for Qt system. First, we provide an example that includes source code as well. Then, we present two mashups that have been built on top of the Google Maps API [5]. Finally, we introduce some other types of mashups. All our mashups run on desktop computers (inside and outside the web browser), as well as in the Nokia N810 mobile device – a handheld WiFi webpad built around Nokia's Maemo Linux platform.

In the application descriptions below, some of the screen snapshots have been taken on the Nokia N810 device. For improved viewability, some of the snapshots have been taken on a PC. Further information on these applications is available on our website (http://lively.cs.tut.fi/qt).

4.1 QtFlickr: Animated Flickr Photo Viewer

In order to demonstrate mashup development with Qt, this section provides a simple example that includes source code. The application used here is called *QtFlickr* – a photo viewer application that fetches images from *Flickr* (http://www.flickr.com/) photo service based on keywords (photo tags) that are obtained automatically from the *Twitter* (http://www.twitter.com/) microblogging service, based on current Twitter trends (http://twitter.com/trends). Images are displayed using timer-based animation (rotation).

The general idea of this application is to automatically display images that reflect the most actively microblogged topics in the world. For instance, when the screenshot of the application shown in Figure 1 was taken, the most actively discussed topic in the world was the swine flu (H1N1).

Fig. 1. *QtFlickr* application running on a PC

When the *QtFlickr* application is started, it first obtains a list of the current microblogging trends from Twitter. Then, the application fetches images from Flickr using the trend names as photo tags. The actual loading of trends and images is performed asynchronously using *QNetworkRequest* and *QNetworkAccessManager* classes so that the user does not have to wait while data is being loaded. The image feed from Flickr is parsed using the *QXmlStreamReader* class. The image URLs contained within the feed are stored and the images to be shown are chosen randomly. Initially, each image is scaled according to the size of the application window. To simplify the implementation and to shorten the source code, only one image is displayed at a time.

Source code. The source code of the application's main class definition is shown in Listing 1. Note that this source code is ECMAScript (ECMA standard 262 [4]) code without any additional syntactic sugar. The *Lively for Qt* includes the option to also use the more class-oriented syntax defined by the *Prototype* JavaScript library (http://www.prototypejs.org/).

The main function of the application, *FlickrWidget*, defines the photo viewer class and its constructor. The class is defined as a subclass of Qt's class *QWidget*, allowing the application to flexibly behave both as a standalone main application (main window) as well as a widget that can be embedded in other Qt components.

The *FlickrWidget* constructor sets up the UI components and connects the components to the required actions. Two layout components are created to arrange widgets within the application window. A *QHBoxLayout* instance is used for horizontally lining up the *QLabel* widgets shown at the top of the application window. A *QVBoxLayout* object then vertically arranges the *QHBoxLayout* object and the *QLabel* object holding the image (*QPixmap*) to be displayed. Two separate *QPixmap* objects are used for images: the first one holds the current image and the second one the image to be displayed next.

```
function FlickrWidget(parent) {
    // FlickrWidget is a subclass of QWidget
    QWidget.call(this, parent);

    // The image references
    this.flickrUrl = 'http://api.flickr.com/'
    +'services/feeds/'
      +'photos_public.gne?format=rss2';
    this.imageUrls = new Array();

    // The visible UI components
    this.currentTagLabel = new QLabel("", this);
    this.imageLabel = new QLabel(this);
    this.imageLabel.setSizePolicy(
    QSizePolicy.Ignored, QSizePolicy.Ignored);
    this.imageLabel.alignment = Qt.AlignCenter;

    this.imagePixmap = new QPixmap();
    this.nextPixmap = new QPixmap();

    // The timers for downloading and rotation
    this.changeTagsTimer = new QTimer(this);
    this.changeTagsTimer["timeout"].connect(this,
    this.changeTagsTimerTimeout);
    this.changeTagsTimer.start(30000); // 30 seconds

    this.fetchImageTimer = new QTimer(this);
    this.fetchImageTimer["timeout"].connect(this,
    this.fetchImageTimerTimeout);

    this.rotTimer = new QTimer(this);
    this.rotTimer["timeout"].connect(this,
    this.rotTimerTimeout);

    this.angle = 90;

    // The layout components
    var hBoxLayout = new QHBoxLayout();
    hBoxLayout.addWidget(new QLabel("Tags:"),0,0);
    hBoxLayout.addWidget(this.currentTagLabel,1,0);
    this.layout = new QVBoxLayout();
    this.layout.addLayout(hBoxLayout);
    this.layout.addWidget(this.imageLabel,1,0);

    this.resize(300,300);
    this.getTwitterTrends();
}
```

Listing 1. The main function (JavaScript class) *FlickrWidget*

Three *QTimer* timer objects are utilized to execute functions in regular intervals. The first timer called *changeTagsTimer* handles the downloading of image tags from Twitter. The second *QTimer* called *fetchImageTimer* is used for downloading the next image from Flickr on the background after the current image has been displayed for five seconds. The third timer called *rotTimer* is used for rotating the current image. Qt's *connect* function is used for creating the connections between the timers and the callback functions that are invoked when the timers are triggered.

When the timer named *changeTagsTimer* timeouts, the function *getTwitterTrends*, shown in Listing 2, is called to downloads the current Twitter trends. At first a URL pointing to trend file (a JSON file available from Twitter's web site) is defined. The actual asynchronous HTTP GET request is sent using the class *QNetworkAccessManager*.

```
FlickrWidget.prototype.getTwitterTrends =
function() {
  var url = 'http://search.twitter.com/'
    + 'trends.json';
  var accessMgr = new QNetworkAccessManager(this);
  accessMgr["finished(QNetworkReply*)"].connect(
    this, twitterReplyFinished);
  accessMgr.get(new QNetworkRequest(
    new QUrl(url)));
}
```

Listing 2. Function *getTwitterTrends*

Parameter *twitterReplyFinished* defines the callback function that will be called when the asynchronous network request has been completed. The function *twitterReplyFinished*, shown in Listing 3, processes the JSON file that contains a list of Twitter trends. The JSON string is parsed and the tags in it are stored in an array. When the tags have been obtained, the function *loadFeed* is invoked to load images from Flickr.

```
twitterReplyFinished = function(reply) {
  var trendJSONString =
    reply.readAll().toString();
  var trendJSONObject =
    eval('(' + trendJSONString + ')');
  var tags = new Array();
  for(i=0;i<trendJSONObject.trends.length;i=i+1) {
    tags.push(trendJSONObject.trends[i].name);
    }
  this.loadFeed(tags);
}
```

Listing 3. Function *twitterReplyFinished*

The function *loadFeed*, presented in Listing 4, handles the loading of the Flickr XML feed containing image URLs. At first a URL for the HTTP GET request is constructed by adding the user's search terms (image tags) to it. This is the URL that is sent to the Flickr web service to obtain images. The network request and the callback functionality are created and handled in a manner that is analogous to the functions that were used for downloading Twitter data.

The function *flickrReplyFinished*, presented in Listing 5, reads the contents of the HTTP reply utilizing the *QXmlStreamReader* class. The image URLs found in the HTTP reply are parsed and stored in the *imageUrls* array. The actual images are then downloaded using a function called *showRandomImage*. Its behavior is analogous to the *loadFeed* function, so the code is not presented here.

```
FlickrWidget.prototype.loadFeed = function(tags) {
    this.imageUrls = [];
    var currentTag = tags[Math.floor(
      Math.random()*tags.length)];
    var url = this.flickrUrl +"&tags=" +currentTag;
    this.currentTagLabel.text = currentTag;
    var accessMgr = new
    QNetworkAccessManager(this);

    accessMgr["finished(QNetworkReply*)"].connect(
      this, flickrReplyFinished);
    accessMgr.get(
      new QNetworkRequest(new QUrl(url)));
 }
```

Listing 4. Function *loadFeed*

To support image animation (rotation), the *QTimer* object stored in the *rotTimer* variable invokes a function called *rotTimerTimeout* (shown in Listing 6) every 50 milliseconds. The function utilizes a *QTransform* object to rotate the currently displayed image. Qt's fast transformation mode is used instead of smooth transformation to improve animation performance on mobile devices at the cost of the quality of the displayed images. If the angle of rotation is 90 or 270 degrees, the image is projected sideways and is invisible to the user. At that point the image can be switched to the next one.

```
    flickrReplyFinished = function(reply) {
      var xml = new QXmlStreamReader();
      xml.addData(reply.readAll());

      while (!xml.atEnd()) {
        xml.readNext();
        if (xml.isStartElement()) {
          if (xml.name() == "enclosure") {
            this.imageUrls.push(
              xml.attributes().value("url").toString());
          }
        }
      }
      /* fetch next image after 3 seconds */
      this.fetchImageTimer.start(3000);
    }
```

Listing 5. Function *flickrReplyFinished*

Since image rotation is rather computation-intensive, it is not well suited to low-end mobile devices. We have used it in this application, because it gives a rather realistic view of the limited processing power and the graphics capabilities of the mobile device and its software stack.

```
FlickrWidget.prototype.rotTimerTimeout=function(){

  // When current image is drawn sideways,
  // switch to the next image
  if (this.angle % 90  == 0 ||
      this.angle % 270 == 0) {

    // Switch to the next image
    this.imagePixmap =
      new QPixmap(this.nextPixmap);
  }

  // Perform image rotation
  var trans = new QTransform();
  trans.rotate(this.angle, Qt.YAxis);
  trans.rotate(this.angle, Qt.ZAxis);

  // Display the image
  this.imageLabel.setPixmap(
    this.imagePixmap.transformed(
      trans, Qt.FastTransformation));
}
```

Listing 6. Function *rotTimerTimeout*

4.2 QtWeatherCameras: Live Road Weather

QtWeatherCameras is a mashup that utilizes the Google Maps JavaScript API and the road weather camera information available from Finnish Road Administration (http://www.tiehallinto.fi) – the government branch in Finland that is responsible for the highway network and road maintenance. The application utilizes the Google Maps API to calculate an optimal route between two chosen points on the map of Finland. The application then obtains information about the nearest road weather cameras along the route, and displays those cameras as markers on the map (see Figure 2). When the user clicks on any of the markers on the map, a live image and current weather conditions from the selected camera are fetched.

The displayed weather conditions include air temperature, road surface temperature and rain measurements. The weather camera image and weather conditions are displayed using a collapsible, semi-transparent widget placed over the map. The map underneath can be panned and zoomed freely.

At the implementation level, the *QtWeatherCameras* mashup uses Qt's *QWebView* web browser widget that has been placed in a *QVBoxLayout* layout component to allow smooth resizing of the application. The *QWebView* widget displays the map images from the Google Maps API.

After the user has selected a weather camera from the map, the application opens a semi-transparent widget called *ImageViewer* on top of the main widget. The widget consist of two *QLabel* components, a *QTextBrowser* and a *QPushButton* widget. The first *QLabel* is used for displaying the name of the weather camera. The image of the camera is loaded into the second *QLabel* widget. The *QTextBrowser* widget is used for displaying weather conditions from the nearest weather station. Although it looks like a simple text box, the widget is a full-fledged rich text browser that accepts any HTML-formatted string as a parameter, and also supports hypertext navigation. The *QPushButton* widget is used for minimizing and expanding the *ImageViewer* widget. In addition, the mashup utilizes a widget *styleSheet* property that defines the customizations to the widgets' style, including their transparency.

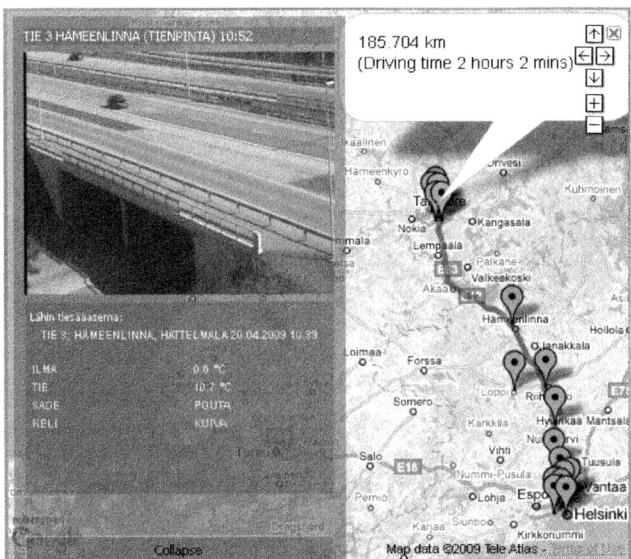

Fig. 2. *QtWeatherCameras* application running on a PC

Images and weather information are downloaded from the web server of Finnish Road Administration. Because the Finnish Road Administration does not provide any well-defined API for accessing the weather camera information, rather heavy parsing is needed in the *QtWeatherCameras* application to digest the information. Locations of road weather cameras are loaded into an array upon application startup. When a new route is created by the user, the application performs the selection of the nearest weather cameras and weather stations locally using the preloaded information. When the user clicks a marker representing a camera, web page containing camera data is loaded and parsed. Weather camera image will be passed on to the *ImageViewer* widget. The data from the nearest weather station is also parsed and passed to the *ImageViewer*.

4.3 QtMapNews: Geotagged RSS Feed Viewer

QtMapNews is a mashup that displays geotagged news items and other geotagged information utilizing the Google Maps API (see Figure 3). The application includes a *QTreeWidget* (tree view) component that lists a selection of predefined geotagged RSS feeds:

- *Earthquakes*: All the Magnitude 5 or greater earthquakes in the world in the past seven days.
- *Emergencies*: The last one hundred incidents/emergencies in Finland based on information available from *Finnish Rescue Service* (http://www.pelastustoimi.fi).
- *News*: Geotagged news from CNN, Yahoo and Yle (Finnish Broadcasting Service).

Fig. 3. *QtMapNews* application running on Nokia N810

The user can add more RSS feeds by pressing a QPushButton labeled "Add...", which will open a simple dialog to enter a new RSS feed. If the new feed is not a geocoded GeoRSS feed, the QtMapNews application uses a publicly available RSS to GeoRSS converter service (http://www.geonames.org/rss-to-georss-converter.html) to geocode news items contained within the RSS feed. After the geocoding process, the items in the feed are displayed on the map as markers. When the user clicks on a marker, an overview of the news item is displayed on the map.

An interesting additional feature of the *QtMapNews* application is that it includes an embedded web browser to display more detailed information. Whenever the user clicks on a map item that contains an URL, a web browser view is opened inside the *QtMapNews* application (on top of the map) to display the contents of that web page. The web browser is implemented using a *QWebView* widget that is displayed on top of the map view when necessary.

4.4 QtScrapBook: Web Camera Scrapbook

QtScrapBook mashup (Figure 4) is a visual scrapbook that can be used for collecting and displaying static or dynamic images from the Web. The application is intended primarily for keeping track of the user's favorite web cameras. The application uses tabs (a *QTabWidget* object) to display multiple images from the web or local file system. To conserve screen space and to allow the application to be used on a mobile device, only a single image (a single tab) is displayed at a time. Images are updated on regular intervals based on a user-defined interval value that can be adjusted using a *QSlider* widget. A *QTimer* object is used for keeping track of time between updates. Images are fetched asynchronously utilizing the *QNetworkRequest* and *QNetworkAccessManager* classes.

Fig. 4. *QtScrapBook* application running on a PC

The user can add new images and webcams to the application by dragging them from a web browser or from the file explorer of the host operating system. At the implementation level, this is accomplished using Qt's built-in drag-and-drop mechanism that enables the sending of drop events to the application. Every drop event holds MIME data that can be used for determining if the application should handle the event. In this case, the application accepts only those drop events that contain a web address (URL) or a path to a local file.

Image rendering within the *QtScrapBook* application is performed using the *QPainter* class and its *drawImage* method. This makes it possible to resize and scale images flexibly. When the user changes the size of the application, a paint event is sent to the application implicitly. The *drawImage* method is then invoked to (re)render the current image. Rendering is performed in the *Qt.KeepAspectRatio* mode so that the aspect ratio of the original image is always preserved. Furthermore, we utilize bilinear filtering (*Qt.SmoothTransformation* mode) to ensure smooth resizing of graphics.

4.5 QtComics: Comic Strip Viewer

QtComics application collects and displays comic strips from all over the world based on RSS feeds published on the Web. When the application is started, the user can select from a set of feeds that contain multiple comic strips. The selection is performed using a popup list (a *QComboBox* object) that lists the available feeds. To conserve screen space, the application displays only one strip at a time, as shown in Figure 5. The comic shown in this figure is from XKCD (http://xkcd.com/); reprinted with permission.

Fig. 5. *QtComics* application running on Nokia N810

At the implementation level, the *QtComics* application uses the *QNetworkAccessManager* and *QtNetworkRequest* classes to download the RSS feed asynchronously. The feed is parsed with a *QXmlStreamReader* object. A typical comic strip RSS feed item contains an HTML formatted string. The HTML code found inside the RSS feed item elements is stored into an array. The first array element is then shown inside *QWebView* web browser component. The *QWebView* object downloads content defined in the HTML code asynchronously and displays it on the screen. Thanks to the maturity of the Qt APIs used for accomplishing all this, the source code of the *QtComics* application is very short, only about 180 lines of JavaScript code.

5 Experiences and Discussion

In addition to the mashups described in the previous section we have developed a number of other mashups and web applications. These applications range from various map-based mashups to sports news tracking, weather forecast applications, media players and games. For instance, one of the applications is a mobile audio player that automatically collects artist information and other related information from different web sites. In the Web era, most of such information is available on the Web, albeit not necessarily in an easily digestible form.

All the applications have been written in JavaScript, utilizing the web browser, the JavaScript engine and the rich APIs provided by the Qt platform. While writing those applications, we have gained a lot of experience that is summarized in this section. We start from general observations related to mashup development, and proceed to comments that are specific to mashups on mobile devices. Finally, we summarize our experiences in developing mashups programmatically using Qt.

5.1 General Experiences and Comments

As we have already discussed earlier [13, 14], the majority of problems in web application and mashup development can be traced back to the fact that the Web was not originally designed to be a platform for active content and applications. The transition from static web pages towards web applications is something that has occurred relatively recently, and the Web has not yet adapted fully to this transition. The problems in areas such as usability, compatibility and security are apparent when attempting to build web applications that run in a standard web browser.

From the viewpoint of mashup development, the two main problem areas are the lack of well-defined interfaces and insufficient security mechanisms. These two areas are discussed below.

Lack of well-defined interfaces. A key problem in mashup development today is the lack of well-defined interfaces that would describe the available web services in a standardized fashion. Although a number of web interface description languages exist, such as the Web Services Description Language (WSDL) [16] or the Web Application Description Language (WADL) [7], these languages are not yet in widespread use.

In general, only a fraction of the data, code and other content on the Web is available in a form that would make the content safely reusable in other contexts. Most web sites do not offer any public interface specification that would clearly state which parts of the site and its services are intended to be used externally by third parties, and which parts are implementation-specific and subject to change. In the absence of a clean separation between the specification and implementation of web sites, there are few guarantees that the reused services would remain consistent or even available in the future. This makes mashup development error-prone and the resulting mashups very brittle.

During the development of the mashups described in this paper, we found that only a small number of services, such as Google Maps and Flickr, offer a well-defined API through which these services can be used programmatically. In many cases, we had to parse HTML pages manually to scoop up the desired data from the web page. If there are subsequent changes in the format of the HTML page, the mashup that parses the page may suddenly stop working properly. This happened to us a few times, e.g., when developing the *QtComics* application.

Security-related issues. Another important problem in the creation of mashware is the absence of a fine-grained security model. The security model of the web browser is based on the *Same Origin Policy* introduced by Netscape back in 1996. The philosophy behind the same origin policy is simple: it is not safe to trust content loaded from arbitrary web sites. When a document containing a script is downloaded from a certain web site, the script is allowed to access resources only from the same web site ("origin") but not from other sites.

The same origin policy makes it difficult to build and deploy mashups or other web applications that combine content from multiple web sites. Since the web browser (the client) cannot easily access data from multiple origins, the mashing up of content must generally be performed on the server. Special proxy arrangements are usually needed on the server side to allow networking requests to be passed on to external sites.

The security problems of the Web present themselves in many other ways. Since there is no namespace isolation in the JavaScript engine, code and content downloaded from different web sites can interfere with each other. For instance, overlapping variable or function names in code downloaded from different sites will almost surely result in errors that are very difficult to detect. Vulnerabilities based on this characteristics – collectively known as *cross-site scripting* (XSS) issues – have been exploited to craft phishing attacks and other browser security exploits. The possibility of such vulnerabilities is the reason why the same origin policy restrictions were originally introduced.

In the mashup development work described in this paper, we managed to bypass the limitations of the same origin policy by using Qt's networking primitives which do not adhere to the same origin policy. However, the namespace problems could not be avoided, and in a few situations overlapping variable declarations causes us considerable debugging headache, in spite of the relatively advanced debugging capabilities offered by Qt.

The key observation arising from all these problems is that there is a need for a *more fine-grained security model* for web applications. Until a more fine-grained security model and proper namespace isolation are available, mashup development is unnecessarily tedious and unsafe.

5.2 Comments Related to Mobile Mashups

Mashup development for mobile devices is still a new area. In principle, there should be little difference between mashups developed for mobile devices and the general Web. Ideally, as described in the *Mobile Web Best Practices* document of the World Wide Web Consortium [11], there should be just "One Web", meaning that the same information and services should be available to users irrespective of the device they are using.

In practice, One Web is still a dream, although we believe that over time most of the issues will be resolved [9]. In this subsection we discuss the main issues today, focusing on usability, connectivity and performance issues.

Usability issues. The mashups that we developed were not written only for mobile use. Rather, we intended them to be practical on desktop computers as well. However, since our target mobile device (Nokia N810) is stylus-operated and has a significantly smaller, 800x480 pixel screen than a typical desktop computer, usability problems could not be avoided. For instance, many of the Qt widgets used in our mashups are so large that they used excessive amounts of precious screen space. Initially, some widgets ended up being outside the viewable area. Font size differences gave us some problems, too. Fonts that look nice on desktop computers are not necessarily readable on the small screens of mobile devices.

Since our target device had a stylus, applications that require the precise use of a pointing device (e.g., the *QtWeatherCameras* application in which the user has to choose precise points on a map) are quite easy to use even on a small screen. However, we suspect that on other types of mobile devices, such as on conventional "candybar" mobile phones with only a numeric keypad, the use of such applications could be challenging.

Connectivity issues. The availability of a reliable Internet connection is vital for mashups. In mobile devices the network connection can often be slow, unreliable or unavailable altogether. The application developer should take this into account in the design of the applications, and provide feedback to the user when problems do occur.

In our mashup development work with Qt, sporadic connection blackouts did not usually pose major problems. Since our mashups run mainly on the client, the applications remain active if the network connection goes down. When using the Qt networking classes, the network requests will remain active until they are successfully completed, or they will timeout eventually if something goes wrong.

Performance issues. One of the main factors separating mobile devices from desktop computers is performance. Not only are mobile devices considerably slower than their desktop counterparts, but they usually have significantly less memory and storage capacity as well. Although processor and memory limitations will decrease over time, performance issues still cannot be ignored in developing mobile applications today. In the development of the mashups described in this paper, performance differences played a significant factor. Mashups such as *QtFlickr* and *QtWeatherCameras* require a lot of computation, and they are very slow on our target device. The performance problems are caused partially by the slow JavaScript engine used by Qt. In the last year or so, several high-performance JavaScript engines such as Apple's *SquirrelFish Extreme* (http://webkit.org/blog/214/) and Google's *V8* (http://code.google.com/p/v8/) were released, with performance improvements of more than an order of magnitude over conventional JavaScript interpreters. Once such engines become widely available, application response and load times should improve considerably.

5.3 Comments Related to Qt

One of the characteristic features of our mashups and the Lively for Qt system is that the majority of software development is performed programmatically, using *imperative* development style familiar from desktop software development. This is in contrast with traditional web technologies, which rely heavily on *declarative* languages such as HTML and CSS. In this respect, our applications bear close resemblance to applications developed with Rich Internet Application (RIA) platforms such as Adobe AIR [15] or Microsoft Silverlight [12].

Given that Qt APIs have been in development and use since the early 1990s, the Qt APIs are on par with the best RIA systems today. For instance, the expressive power of Qt APIs such as the *QXmlStreamReader* class saved us a lot of work when parsing complex XML data. Furthermore, API documentation and the available development and debugging tools for Qt are in good shape. In general, it was very easy to get started with Qt.

More generally, the combination of an existing, mature application framework with a built-in web browser and popular, fully dynamic programming language (JavaScript) turned out to be a powerful combination. With a JavaScript engine and only a few thousand lines of JavaScript code, it is possible to turn an existing, mostly static, binary, desktop-era application framework into a highly interactive, dynamic web development environment supporting mobile mashup development. Applications require no compilation, binaries or explicit installation, and yet they can utilize the full power of the existing, mature application framework.

On the negative side, although Qt is intended to guarantee platform-independence, we did experience some portability issues. For instance, some widgets or fonts refused to render themselves correctly on some target platforms. Apart from rendering errors, event-handling differences and some performance-related issues on mobile devices, no other major issues were encountered, though.

The use of the JavaScript language for developing real applications is still a relatively new topic. When JavaScript is used as a programming language for developing full-fledged applications – as opposed to the conventional use of JavaScript as a *scripting* language – one has to be aware of its caveats and peculiarities. These topics have been summarized well by Crockford [3].

6 Conclusions and Future Work

In this paper we have presented an overview of the mobile web mashups that we have implemented in JavaScript on top of Qt – a cross-platform application framework recently acquired by Nokia. This work is part of a larger activity called *Lively for Qt*, an effort that has created a highly interactive, mobile web application and mashup development environment on top of the Qt framework. Here, we summarized our experiences in developing these applications, and included some source code to illustrate the development style.

Plenty of interesting avenues remain for future work. We are especially excited about the possibility of building *location-aware mashups* that have been customized to take into account the user's current location, utilizing the GPS satellite position information available in modern mobile devices. With increasingly reliable and inexpensive network connections, *collaborative mashups* that allow real-time collaboration between multiple mobile users are also becoming a reality. In general, the use of mashups and web applications in mobile devices offers entirely new possibilities that are beyond the reach of web services on desktop computers. Although various obstacles still remain, we are inspired by these possibilities and hope that this paper, for its part, encourages people to continue the work in this exciting new area.

Acknowledgments

This research has been supported by the Academy of Finland (grant 115485).

References

1. Clarke, J., Connors, J., Bruno, E.: JavaFX: Developing Rich Internet Applications. Java Series. Prentice Hall, Englewood Cliffs (2009)
2. Crane, D., Pascarello, E., James, D.: Ajax in Action. Manning Publications (2005)
3. Crockford, D.: JavaScript: The Good Parts. O'Reilly Media, Sebastopol (2008)
4. ECMA Standard 262: ECMAScript Language Specification, 3rd edn. (December 1999), http://www.ecma-international.org/publications/standards/Ecma-262.htm
5. Gibson, R., Erle, S.: Google Maps Hacks. O'Reilly Media, Sebastopol (2006)
6. Goodman, D.: Dynamic HTML: The Definitive Reference. O'Reilly Media, Sebastopol (2006)

7. Hadley, M.: Web Application Description Language Specification (November 9, 2006), https://wadl.dev.java.net/
8. Hanson, R., Tacy, A.: GWT in Action: Easy Ajax with Google Web Toolkit. Manning Publications (2007)
9. Mikkonen, T., Taivalsaari, A.: Creating a Mobile Web Application Platform: The Lively Kernel Experiences. In: Proceedings of the 24th ACM Symposium on Applied Computing, SAC 2009, Honolulu, Hawaii, March 8-12, pp. 177–184 (2009)
10. Mikkonen, T., Taivalsaari, A., Terho, M.: Lively for Qt: A Platform for Mobile Web Applications. In: The Proceedings of the Sixth ACM Mobility Conference, Mobility 2009, Nice, France, September 2-4 (2009) (to appear)
11. Mobile Web Best Practices 1.0. World Wide Web Consortium Recommendation Document (July 29, 2008), http://www.w3.org/TR/mobile-bp/
12. Moroney, L.: Introducing Microsoft Silverlight 2.0, 2nd edn. Microsoft Press (2008)
13. Taivalsaari, A.: Mashware: The Future of Web Applications. Sun Labs Technical Report TR-2009-181 (February 2009)
14. Taivalsaari, A., Mikkonen, T.: Mashups and Modularity: Towards Secure and Reusable Web Applications. In: Proceedings of First Workshop on Social Software Engineering and Applications, SoSEA 2008, L'Aquila, Italy, September 16 (2008)
15. Tucker, D., Casario, M., De Weggheleire, K., Tretola, K.: Adobe AIR 1.5 Cookbook. O'Reilly Media, Sebastopol (2008)
16. Web Services Description Language. World Wide Web Consortium (W3C) Specification (March 15, 2001), http://www.w3.org/TR/wsdl

Ads Go Mobile: Assessing the Opportunities and Challenges of Personalised Ads in a Mobile Search Service

Sigmund Akselsen, Bente Evjemo, and Calicrates Policroniades

Group Business Development and Research, Telenor
N-1331 Fornebu, Norway
{name.surname}@telenor.com

Abstract. Since highly relevant content is linked to click-though rates (CTR) and eventually to an increase in revenue potential, key players in the search industry are devoting resources to profile and exploit the context in which a search is performed. In this paper we present the pilot implementation and evaluation of a mobile operator centric search service that enables the provision of personalised advertising. We have deployed a true ecosystem solution comprised of a mobile federated search platform, a mobile advertiser platform, advertisers and advertising campaigns, and a number of content providers. The attitude towards personalised advertising has been evaluated with 175 ordinary mobile phone subscribers across Norway during the summer of 2008. Advertisements are accepted if relevance is high. Non-utilitarian factors as humour have considerable impact on end-users' acceptance of advertisements, and a personalised approach is welcomed due to its potential to filter undesirable content. Based on our experience in this study, we also discuss technological challenges and strategic opportunities for a mobile operator in this area.

Keywords: Mobile search, mobile advertising, personalisation.

1 Introduction

It is believed that the high mobile phone penetration and the increase in average time spent on mobile phones will subsequently shift parts of advertising from traditional channels to the mobile channels. The *personal and always-on* characteristics of mobile handsets represent a unique opportunity to personalise mobile services and advertisements. If done correctly, personalisation might give extra value to the customers and possibly strengthen the positive attitudes towards mobile ads, so far being moderated due to general reluctance to spam and information overload [1].

Mobile ads can be adapted for different mobile spaces. They can be deployed on the handset idle screen (e.g. as part of the screen background), portal pages (e.g. WAP banners), message services (e.g. SMS tail), or integrated in basic functionality of the phone (e.g. as ring tones). Increased flexibility, and probably increased impact, may be achieved if advertising is combined with particular mobile services. In this context, mobile search stands out as a good candidate for various reasons [2]. Search services fit the advertisement requirement for dynamic allocation of screen space. In addition, while searching the user is assumed to be in the mood to receive information (including advertisements). Intertwining mobile search services and advertisements also opens for new business models and from the user's point of view, opportunities for reduced service costs. By controlling advertising spaces tied to mobile Internet content, the monetisation potential for a mobile operator is considerable.

In this paper we present a mobile operator perspective of a mobile search service that enables the provision of personalised advertising in a mobile ecosystem. A mobile operator has several resources that can be useful when developing and offering personalised mobile services such as means for user identification, customer relationship and customer knowledge, access channels and ad spaces, service development platforms, and third party service providers. In order to obtain a practical experience that allows us to precisely evaluate the strategic and commercial significance of these mobile operator assets, we have deployed a true-ecosystem solution with a number of business partners[1] comprising a mobile federated search platform, a mobile advertiser platform, advertisers and advertising campaigns, and a number of content providers. We have obtained feedback on the service based on a group of 175 pilot users comprised of ordinary mobile phone subscribers across Norway during the summer of 2008.

The contribution of this study is the analysis and discussion, from a mobile operator point of view, of the important aspects in the deployment of personalised ads in a mobile search service. This document is organised in the following way. In Section 2 we discuss relevant concepts to this study. In Section 3, we describe the service and the method used to gather customer insight. In Section 4, we discuss the results of this work; we cover issues such as user acceptance, deployment challenges, and strategic feasibility. Finally, in Section 5 we conclude this paper.

2 Background

2.1 Mobile Advertising - Actors and Value Creation

Players in the mobile advertising value chain contribute in different ways to make the connection between advertisers and customers effective. Figure 1 shows the main players and helps us to distinguish their different motivation and focus. The advertiser is interested in scale, targeting and response, quality, creative potential, as well as information consumption and measurement. The advertiser may pay a fee to intermediaries for advertising inventory based on cost per impression or cost per clicked ad. Naturally, if targeted or personalised ads are offered more value is present for advertisers and consequently the price per impression can be higher.

---

[1] Business partners will remain anonymous.

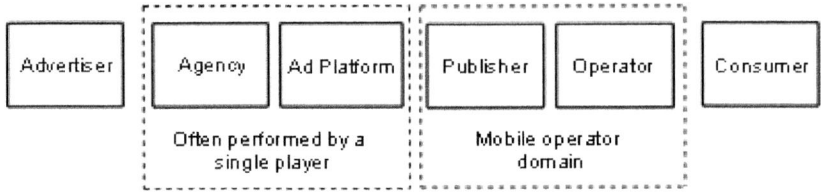

Fig. 1. Roles within the mobile advertising value chain

The actors filling the agency (ad sales) and ad platform roles are involved in creating the advertising campaign and placing them in the right media according to the goals of the campaign. Ad sales actors include agencies such as advertising agencies (creative), media agencies (placing campaigns in different media), direct marketing agencies and Web agencies (online advertising and marketing). Ad networks and platform providers include actors such as Google, Yahoo, MS Nokia Ad Business, MADS, Amdocs, etc. The publisher (including content owner) and mobile operators will offer different media according to the needs of the advertisers and the target for the campaign. The actors filling these roles may receive a fee from the ad sales and ad platform actors.

The target group for the advertisers is the users or customers of different media houses and media channels. Thus, when a mobile subscriber browses the operator portal, or off-portal content with approved content partners, the mobile operator obtains income from the generated data traffic and from the advertisements displayed through the operator channels. The consumer interests include content relevance, value return, control, quality, and privacy.

One of the key operators' assets in this ecosystem, and to a great extent still an under-exploited one, is the detailed knowledge of their consumers (e.g. user identification, customer relationship, customer knowledge, roaming patterns, etc). This data can be used to offer the consumer more relevant ads and consequently adding value to the ecosystem [3]. Mobile ads represent a potential revenue channel and its success depends on how well operators are able to leverage on their existing assets to add value to the mobile search business.

2.2 Mobile Advertising - Targeting and Personalisation

As in classic advertising, online advertising can be split into brand advertising whose goal is to create a distinct favourable image for the advertiser's product, and direct-marketing advertising that involves a direct response to buy, subscribe, vote, donate, etc, now or soon. In terms of delivery, there are two major types [6]:

- **Search advertising** refers to the ads displayed alongside the organic results on the pages of the Internet search engines. This type of advertising is mostly

direct marketing and supports a variety of retailers from large to small, including micro-retailers that cover specialized niche markets.
- **Content advertising** refers to ads displayed alongside some publisher produced content, akin to traditional ads displayed in newspapers. It includes both brand advertising and direct marketing.

Search platforms enjoy a huge advantage over other online advertising platforms, because they can use a customer's own search terms to match customers' interest with advertisers [8]. The advertisement is considered far less intrusive by the consumers than online banner advertisements or pop-ups [7]. Previous studies have shown that more relevant ads lead to improved user satisfaction and higher response rates [3,5]. Google, Yahoo, MSN, and other online advertising platform providers can charge higher prices for these ads than for conventional non-tailored ads. Since listings appear when a keyword is searched for, an advertiser can reach a more targeted audience on a much lower budget.

In content advertising, the first technologies for obtaining relevancy simply extracted one or more phrases from the given page content, and displayed ads corresponding to searches on these phrases, in a purely syntactic approach. However, due to the vagaries of phrase extraction, and the lack of context, this approach leads to many irrelevant ads. One way to deal with this problem is through contextual ad matching based on a combination of semantic and syntactic features [4]. The problem of selecting contextually relevant ads so that they are both relevant to queries and profitable to the search engine has been addressed by Radlinski et al [9]; surprisingly, they show that optimising ad relevance and revenue is not equivalent.

The so called intelligent portals are automatically personalised based on the user's usage pattern, e.g. ChangingWorlds' from ClixSmart technology[2]. Tests show that considerable improvements are made to reduce the average number of clicks to reach content. However, the personalisation effect on mobile portals usage is, in contrast to ordinary Web portals, moderate. This is explained due to the impatience of mobile users and their tendency to access content that is only within a short distance from the mobile portal home page. This tendency may be explained in part due to the limited user interface (including user input) of mobile devices, as well as the lack of facilities for keeping track of browsing context, such as tabbed-browsing and rich history logs, of mobile browsers. Thus, the impact of any automatic adaptation of mobile portals will be limited when compared to similar efforts on Web portals [10].

Our approach to personalisation differs from previous ones since it is based on user identification through the MSISDN (Mobile Subscriber Integrated Services Digital Network Number) number, which is used to obtain customer data kept within the mobile operator premises. This approach brings promises of improved targeting but also challenges tied to technological aspects, user experience, and privacy.

---
[2] http://www.changingworlds.com/solutions.htm

2.3 Mobile Search - Federated vs. Centralised Approach

The distinction between federated and centralised search solutions are relevant to this study since the service pilot employs a federated search platform. The federated search approach has been suggested as an alternative that may counteract some of the problems of traditional search engines such as content hidden behind operators' portal, difficulties of query formulation (small display and keyboard), and lack of appropriate content and services. This approach should ensure real-time results through a decentralised system (vs. centralised indexing system) where every entity in the federation is responsible for its own data and for responding with its best real-time results at the moment they are needed. This might eliminate the need for expensive centralised indexes and re-indexing procedures.

Furthermore, the federated search approach integrates the power of vertical search engines. It produces and executes a search execution plan, optimized for the query against an array of vertical back-ends, and aggregates the search results of the verticals and blends the results in a manner optimal to the query. This strategy is promising in order to minimise the number of clicks and the number of round trip requests to the servers over the wireless data network. New vertical search engines appear in key markets (image search, people search, health search, etc.), for particular content (e.g. Flickr, YouTube) and are gradually integrated in the big one-stop search engines. From a mobile operator's point of view, the federated search approach makes it easy to adjust, add or remove premium content. However, it also brings challenges with regard to timely gathering and presenting search results from different content sources.

3 The Pilot Service and Methodology

3.1 Technological Setup and Provisioning of the Solution

As shown in Figure 2, the mobile search and advertising pilot comprised a true ecosystem covering partners in all the steps of the value chain:

- **Federated search platform**. It provides a federated search service tailored for mobile devices. In practice this element act as the aggregator of the different content sources and relevant advertising. It produces the final presentation of the search results including personalised advertisements in designated ad spaces.
- **Telenor Playground**[3]. The Telenor Playground laboratory offers access to a number of mobile operator services such as messaging, location, payment, call control, and subscriber data.
- **Mobile advertising platform**. It supplies the necessary technology and infrastructure to manage advertising campaigns and to control different aspects such as publishing, ad placement, and dynamic allocation and customisation of

---

[3] http://playground.telenor.com

ads based on different criteria. Recruiting advertisers was conducted through existing relationships with the mobile advertising platform.

- **Content providers**. The service searches and aggregates the results of different content sources. New content providers can be added to the solution as needed.
- **Advertisers**. A number of advertisers participated in the implementation of the service.

The parties cooperate and exchange messages at different phases in order to solve a search request; these messages are illustrated in Figure 2 with arrows and numbers. When a user types a query in the search box of the mobile search service a request is generated (1, 2). An important part of the solution was tied to the question of how subscriber identity could be made available through the domains of the different partners while still preserving security and control over the identity of the subscriber. There are several ways in which identification can be performed. The solution adopted included the encrypted MSISDN as part of the HTTP header in the search request; the MSISDN resides on the SIM (Subscriber Identity Module) and is in turn facilitated in an encrypted form by

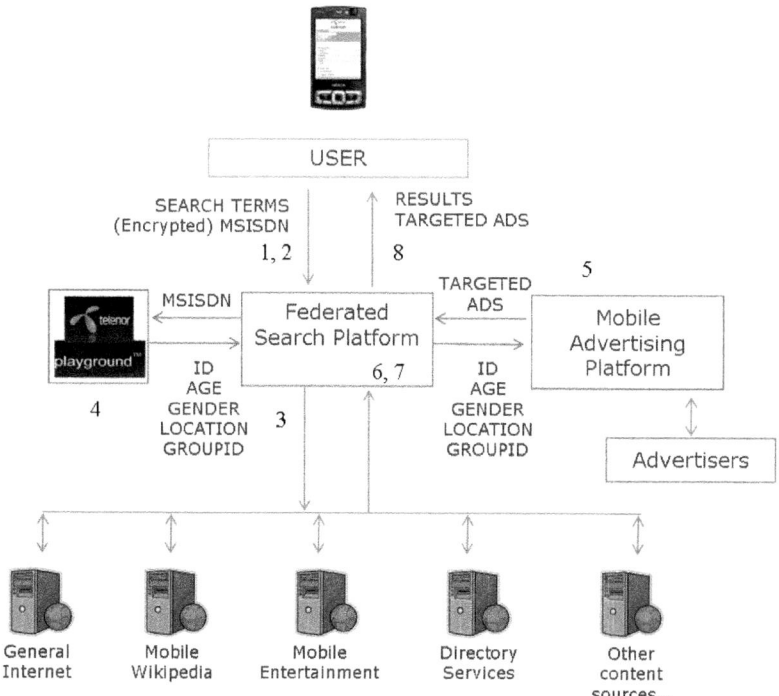

Fig. 2. High-level architecture of the mobile search and ads service

the Telenor mobile gateway. In Section 4.2 we discuss the relative advantages of this and other potential ways of user identification.

Once the search request is delivered to the federated search platform, the search term is distributed to the set of content providers participating in the pilot (3), who in turn generate a reply with the results of the query. The approach taken by the federated search solution, in which each of the content sources are handled independently, enables the mobile operator to strategically chose the best available content sources among different search verticals (e.g. music, games) or content sources that are more appropriate for a local market (e.g. directory services, location-based services), and in this way implementing a best-of-breed approach for each of the search verticals. In addition, the MSISDN is used to retrieve subscriber data from the Telenor Playground premises such as age, gender, and location (4). This data is in turn used to request the mobile advertising platform for the most relevant campaign and ads (5). Once the results from the content providers and the mobile advertising platform are received, the federated search platform builds the result page (6, 7) and delivers it to the user (8).

3.2 Service Description

From an end-user point of view, the interaction with the service takes place in the two screens shown in Figure 3. The service was made available as a search box placed on the Telenor WAP portal as exemplified in the screen on the left side of the figure. The service allowed users to type in one or several terms in the search box, and then to execute a search by pressing the accompanying search button (labelled 'Søk'). It enables the user to search on a specific content category or to start a search on all the categories available in the service. At the top, it also provides space for advertising in the form of banners. A more detailed description of the content sources and the handling of advertising spaces is presented in Section 3.2 and 3.2 respectively. Once the user submits a query, it obtains the reply shown in the right hand side of Figure 3; in this particular case, the page shows results for the query term 'Volkswagen Golf'.

Two things are important to notice in the result page. The first one is the editorial design applied to the page. Results are organised in categories and only the top result is presented in each category; this was done for clarity purposes and to avoid crowding the page with content. The user can also navigate to obtain more results on a specific category domain through the clickable link labelled 'Flere resultater'. Then the user would be shown a page displaying the next 5 results in that category. Finally, a search box with the option to search in all or a specific category is placed at the bottom of the result pages. The second important aspect to note is the space dedicated to advertising. Banner ads are placed at the top of the page and sponsored links at the bottom. According to talks with our mobile advertising partner, it was suggested that the treatment of sponsored links as an additional content category at the bottom of the page (instead of at the top) obeys to user experience issues and not solely to revenue potential aspects.

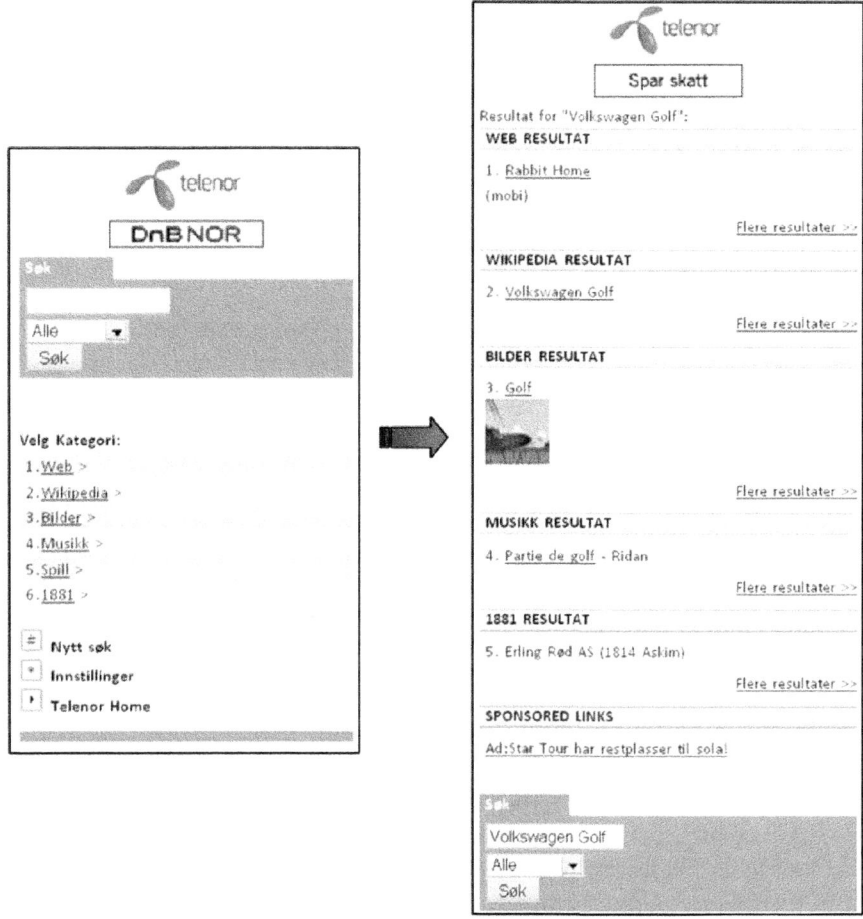

Fig. 3. WAP pages of the search service

Content Sources. The service searches and combines the results of the following content sources:

- **Web Index**. It comprises access to Internet content, and thus, constitutes off-portal Web pages. The content in this category was facilitated by one of the business partners in the pilot; it constitutes a transcoded version of the Web.
- **Wapedia**. This category includes results from searching an off-portal mobile index provided by Wapedia. Wapedia is a WAP site that optimises the contents of Wikipedia for mobile devices and provides auto detection of the best format for a given device.

- **Pictures**. This index was facilitated by one of the content providers and was restricted to image content that could be bought on the operator deck (i.e. on-portal images).
- **Music**. Similarly, this index was facilitated by one of the content providers and was restricted to music that could be bought on the operator deck (i.e. on-portal music).
- **Games**. Similar to the pictures and music categories, this index was restricted to game content that could be bought on the Telenor deck (i.e. on-portal games).
- **Directory services**. This category uses the results from searching an index provided by one of the partners covering directory services for the Scandinavian area.

Advertisements. Banner ads were displayed at the top of the search result pages. As a way of example Figure 3 shows an animated banner from a financial institution (DnB NOR) just below the Telenor logo (text on banner: 'DnB NOR' on the left side and 'Spar skatt' on the right side). An animated banner consists of three images as seen in Figure 4; each image is displayed for around a second in a result page. The format, aspect ratio, and banner dimension conform to the guidelines suggested by the Mobile Marketing Association [19].

Fig. 4. An example of two of the animated banner ads used in the service. Left: A mobile betting service. Right: DFDS Seaways cruise offers from the mobile.

The text ads, visible as sponsored links, were displayed at the bottom of the search result pages but always above the search box. The text ad 'Star Tour har restplasser til sola!', which is displayed in Figure 3 on the right side just below the *Sponsored links* label, is an example of a text advertisement. All banner and text ads were clickable and lead to a landing page for the particular advertising campaign.

In the practical implementation of the service a total of 20 campaigns from 6 different advertisers were defined. 9 of these campaigns were targeted by specifying associated search terms and demographics; details on the definition of campaigns are presented in the following paragraphs. As campaign and advertisement design often are associated with considerable costs, the ads and landing pages used in the pilot were adapted by the advertisers from other applications. The total set of campaigns covered a relatively limited range of products (travel, finance, betting, entertainment, catalogue information).

Ads Personalisation. The personalisation of advertising was done through advertising campaigns that took into account specific subscriber demographics such as age, gender, and place of residence, in addition to a set of key search terms. Table 1 shows the definition of DnB NOR BSU campaign as a way of example. For each campaign, explicit values are given for the three demographic variables and the set of associated search terms.

The target values were intended to match the life-phase groups of the Telenor Norway segmentation model, i.e. young (15-29 years), established (30-54 years), and senior (55+ years). However, the final segmentation was decided by the advertisers and did not exactly match the request for congruence to the Telenor segmentation model. The values used in the creation of campaigns were mainly based on the capacity of the mobile advertising platform to make practical use of the information exposed by the mobile operator.

Furthermore, the placement of ads was based on a set of business rules determining what campaign was selected if two or more campaigns matched the subscriber demographics and search terms. In practice, this implied a random choice of matching campaigns as long as the display settings (e.g. number of views and duration) in the advertisements management platform were set equal for all campaigns. In cases where no campaigns matched the customer demographics or search terms, a random campaign was displayed. During the execution of the pilot there were campaigns that covered all customer segments, thus always having appropriate ads ready for display.

Table 1. Definition of the DnB NOR BSU campaign

Campaign title	DnB NOR BSU
Age (years)	15-29
Gender (m/f)	All
Place of residence (postal code)	All
Search terms	dnb, dnbnor, mobilbank, bsu, lån

3.3 Research Design

The research design was guided by the following goals: i) explore how mobile operator assets can be utilised to personalise ads on mobile phones, ii) investigate user attitudes towards personalised ads, and iii) identify characteristics of an attractive mobile search service. To explore the first goal, a real-life ecosystem was set in place. To shed light into the second and third goals mobile phone subscribers were involved; their attitudes to the service were collected through Web questionnaires and focus groups.

Two Web based questionnaires became the main information sources of the project. The first one was filled out before the pilot service was launched and the second after the pilot was completed. The questionnaires, comprised mostly of multiple-choice questions using a Likert scale 1-5, could be answered within 10-15

> Find information about Burma
> What is the headline of today's newspapers?
> How many passengers can the coastal steamer Trollfjord carry?
> Which artists are performing on Roskilde festival this year?
> What is the current exchange rate for Euros?
> What is the duration of the music track "Don't Let Go" by Bryan Adams?
> When was Tivoli in Copenhagen opened?
> When was the betting game "Extra" launched in Norway?

Fig. 5. The test users received 8 tasks to stimulate the use of the pilot service

minutes. The questions addressed general attitudes towards personalised advertisements and the actual use of the pilot service. The first questionnaire also included questions about the informant's interests and their routines and familiarity with mobile phone and PC search services. In addition to the questionnaires, the focus groups were performed to better understand users' behaviour and attitudes. Traffic logs were analysed to verify users' reports and to get statistical material on page views and click-through rates; however, our study focused on gathering in-depth knowledge on attitudes rather than systematic usage of the solution. The analysis presented in Section 4 is based upon data collected from the respondents of both questionnaires and the insight obtained in the focus groups.

A total number of 175 ordinary mobile subscribers were recruited for the pilot running during the summer of 2008; they were randomly selected from a list of twenty-five thousand mobile subscribers whose logged activity showed recent use of Internet through their mobile phones. They were all located in Norway. Members of the focus groups were recruited among the pilot users. No installation or configuration was needed as the pilot service automatically appeared as a replica of the search service on the operator's WAP portal.

To meet the challenge of infrequent mobile search the test users were given a set of search tasks during a period of the trial; these tasks were sent one by one as an SMS (see Figure 5). The first message included a link to the pilot service and a request to create a bookmark. The tasks exemplified various types of queries and motivated usage beyond the first try. Some of them were chosen to make sure that ad campaigns relevant to search terms were triggered and presented together with relevant content. For instance, the task 'When was Tivoli in Copenhagen opened?' would presumable be solved by using terms like Tivoli and Copenhagen which purposely triggered the ad campaign for a ferry trip to Copenhagen to be displayed (see Figure 4).

4 Results and Discussion

4.1 Pilot Results: User Experience

Mobile search was characterised as *"reading the paper by looking through a key hole"*. Still, more than half of the informants found the service easy to use and easy to learn, and approximately the same share was satisfied or very satisfied with the service. Not surprisingly the younger informants displayed more positive attitudes to the service than the elder ones. Hands on experience impacted positively: *"Testing this service has opened up my mind"*. Similar results are reported from other studies [15].

Category Approach Counteract Problems with Small Displays. The category approach of federated search reduced the query formulation efforts and improved the overview of presented hits. The size of the display hampers hit list overviews and rich descriptions of the hits that otherwise supports the judgement of relevance. These problems were amplified among users 45+ and to those keeping old mobiles with very small displays. One informant stated: *"The service should use big fonts and bright colours. I should not be forced to put on extra glasses"*.

Accordingly, contextualised font sizes (e.g. taking into account display size and user age) and intuitive procedures for zooming could add significant value to the mobile search service. In addition to the small font issue, users reported problems related to finding the service on their phone. This is symptomatic to novice mobile users with a limited understanding of the mobile phone as an Internet access device.

Personalisation Triggers Privacy Concerns and Becomes a Welcomed Filter. Personalisation and privacy concerns are often mentioned together. Kobsa [11] phrases it this way: *"Online searchers who are pleased that a search engine disambiguates their queries and delivers search results geared towards their genuine interests may feel uneasy that this procedure entails recording all their past search terms"*.

In this study the surveillance and privacy issues were brought forward through the questionnaires, and not surprisingly we received comments in line with *"I would not like anyone to track my interests and behaviour"*. They stressed the importance of explicitly saying yes or no to ads, and to be informed of where, when and how personal information is utilised. This claim should be analysed in accordance to privacy laws that regulate routines concerning the anonymity of traffic and location data that claim user consent in the case of data transmission to third parties and assures the user's right to withdraw his previously-given consent.

In general the users preferred search services free from ads, but they seemed to recognise personalised ads as a filter to potential spam and unwanted information and, thus, considered personalisation as a way to reduce the negative effects of future increase of ad campaigns. When ads were considered relevant, e.g. they matched the topic searched for or the user's interests or immediate context, the ads were more likely to be appreciated and clicked on; a Finnish study confirms this tendency [17]. Further our study uncovered that general attitudes towards ads seem to be amplified when exposed on mobile phones. The mobile phone is a private and personal possession, and it is always there and it is always on. One of the users exclaimed: *"You cannot choose not to pay attention to ads on a mobile phone. It's not like reading a paper - just turning the page if it is embarrassing"*.

Kobsa categorises people into privacy fundamentalists, privacy unconcerned and privacy pragmatics, and claims that the size ratio between these clusters is roughly 1:1:2, with a slight decline of fundamentalists and the unconcerned over the past two decades and a corresponding increase in the number of pragmatists [11]. This development is promising as the attitudes uncovered in this study match the pragmatic person being described as privacy concerned, but still willing to disclose personal data if they understand the reasons for it, if there are any clear benefits and if there are privacy protections in place.

Non-utilitarian Factors are Important. Models constructed to explain adoption and acceptance of technology have been dominated by utilitarian factors [13]. In mobile settings the non-utilitarian factors as *pleasure of use* seem to be important [12,16,18,20]. In this study almost half of the informants reported that they are motivated to click on ads when they appeared humoristic. In particular the young respondents appreciated the scent of humour. Also the possibility to get rewards and discounts were welcomed. Other studies have shown increase of ads click-through-rates due to rewards and discounts [14]. Local affiliations, like ads presenting local services or products, seem to foster positive attitudes towards ads. An MMA study[4] claims that 69% of the informants preferred advertisements that were related to local products and services.

4.2 Technological Possibilities and Challenges

User Identification. There are several ways in which identification can be done, automatically, based on information stored in the mobile subscriber's handset, or manually, through explicit user log in. A method for identifying the user without the user's explicit interaction is preferable and can be enabled through information stored in the SIM card, one of the key assets of a network operator when it comes to mobile content and services.

The phone number, also known as the MSISDN number, is the most obvious way for user identification. The number resides on the SIM card and in the Home Location Registry (HLR) network element. This ID is typically used by call and message-based solutions, but can also be used for mobile Internet solutions, since some operators append the MSISDN to the HTTP headers in their WAP gateways. As the MSISDN resides on the SIM, it is associated with the subscriber and not the mobile phone itself. The operator can provide interfaces for third-party service providers which need access to the users' MSISDN or any of the alternative forms of subscriber identification. A relative disadvantage linked to the MSISDN is that access to this number needs to be requested with each operator. Common techniques of Web technology, like cookies, can also be applied to mobile services.

In our study, the MSISDN was chosen as the preferred solution for user identification since Telenor can easily include this number in HTTP headers. The lesson from the pilot on user identification was that identification by MSISDN

---

[4] See http://mmaglobal.com/?q=node/1518

worked well and strategically anchored the operator in the value chain. Our experience supports the belief that subscriber identity as an enabler of personalised services should be considered one of the important assets of a mobile operator.

Content Availability. A federated search service faces the challenge of having enough content sources to create perceived relevance for the customers within all categories. Also, copyright policies apply for digital content. Building an ecosystem for a mobile search service demands access to content via commercial agreements. The pilot project spent more time than expected to secure the necessary content for the pilot service. A better option would have been to use the already established content providers of the Telenor Norway's mobile entertainment offering (games, music and pictures), and thus avoiding the overhead of any additional commercial agreements or technological adaptations.

However, users expect results beyond premium on-deck content. For this reason, the service also included a general Internet index to enhance the search results. One of the challenges of generic mobile search indexes is the limited amount of mobile (WAP) pages, since the content itself is limited. All providers of WAP indexes have this as a general problem. A way to increase the amount of content relies on transcoding Web pages to mobile format. By transcoding Web pages globally, a Web index with a broader content depth than ordinary WAP indexes can be included; we did this by including transcoded versions of a Web index and Wikipedia.

In our particular case, accessing the search results of the Web index provider had to be done via the provider's Web page before entering the requested search page result, making the solution suffer from a user experience point of view. Furthermore, the intention to provide indications of number of hits in each category turned out to be hard to fulfil. This is due to the fact that the result page is dynamically populated before the search of the content sources in the different categories is completed. Also the intention to provide snippets, i.e. short summaries or key words, for each hit was not met. Finally, the federated search concept makes it easy to adjust, add or remove content and content categories. However, the study uncovered challenges with regard to gathering and presenting search results from different content sources timely.

Subscriber Data. As noted before, operator's subscriber data was used in order to facilitate targeted advertising and personalisation. Subscriber data was dynamically queried by the other elements of the search solution. The information for each of the pilot users was the following: MSISDN, first name, last name, gender, date of birth, postcode, and postcode name. This data set is relatively simple and the data process extraction was relatively straightforward.

In our experience two points are worth noticing. First, there are occasions in which the information stored in the databases does not correspond to real usage, and this may compromise the interpretation of the experimental results. For example, it is not unusual that a subscription is shared among the members of a family. Thus, the conclusions drawn from the information in the databases in this scenario may lead to inappropriate personalisation of advertisements. In the pilot, these cases were detected and corrected manually but a more practical way of solving these issues is required.

Second, due to time constraints, it was not possible to have a totally automated and standardised solution to share information between the different parties involved in the process of managing and exposing customer data. It is still necessary to study in detail the mechanisms and best-practices to expose this information to third parties. This represents a practical challenge due to the cross boundary relationships and shared responsibilities that have to be built outside the operator. The implications and challenges of facilitating customer data and generic identity management solutions is receiving attention from the mobile industry. Currently on development, the GSMA OneAPI[5] aims to provide a commonly supported API so that applications and content, including among others user profile and user identification, are portable across operators.

Location Information. The various types of location information considered for targeting ads in the pilot included the user's current position (made available through the operator's positioning service), most used base station (made available through the operator's databases) and, place of residence (made available through the operator's databases). In some cases a combination of these could be interesting for advertisers. In practise, it turned out that the response time for obtaining the user's current position from the positioning service was on average around 5 seconds; the main reason for this latency was due to the experimental implementation of the location service in the Telenor Playground premises. This was too slow to be used in the piloted search service. The operator's positioning techniques have higher accuracy than pure cell identification, but the response times that we obtained made it inconvenient for practical use. Neither available time nor the low interest from the advertisers allowed pursuing this issue any further; this might change quickly as scenarios for using location information become widely available. Of the other techniques for providing location information, i.e. address of the most used base station or place of residence, we made a pragmatic decision of using the latter. On-device GPS (Global Positioning System) positioning was discarded due to the rather limited footprint of devices supporting this feature not only among the pilot users but also in the Nordic market in general.

Using Telenor's Playground Lab for acquiring location information proved to be difficult due to response times. However, this situation improves in the operator's professional service development platforms. Thus, our experience supports the belief that mobile operators' ability to make available various location information constitute an enabler of personalised services and should be seen as a key important assets of an operator.

State of Tools for Targeted Search and Ads. Providers of mobile search solutions offer ranking capabilities that could be of interest to mobile operators and mobile portal owners. These capabilities include placement of categories on the result page as well as internal ranking within each category. An operator

---
[5] http://oneapi.aepona.com/

focusing on monetising search by driving usage of premium content and services would be interested in this option. For us, this aspect was not fundamental. Instead, the ambition of creating the best possible user experience determined the order of the categories in the result page. No special in-category ranking was applied, i.e. the content sources' original ranking of search results was employed. The use of search history was not applied to rank results, neither were tools for doing this explored.

Mobile marketing companies typically offer ad servers for mobile campaign management and targeting such as ad management (inventory and rendering), campaign management (targeting, channel selection and ad spend), device recognition and content adaptation, user identification and profiling, personalisation and, reporting. The targeting of ads could take into account a number of factors such as: bid price, time, distribution and inventory status, frequency of clicks, countries, devices, operator, and user demographics.

The particular platform used for mobile marketing has been built around the key concepts of campaigns and ad zones and it supports advertisers' needs well. However, our approach from the start was developed around the personalisation perspective. We found that the platform had some limitations and was not very flexible when it came to defining groups of users receiving different advertising treatments and the logic to be applied when computing the best ad for each individual visitor. Also the addressing of different ad spaces (or ad zones) had some limitations. We recognise that the mobile advertising industry is evolving and the infrastructure for targeted ads based on search terms and demographics are still immature; an observation even more valid during the set up phase of the pilot.

Campaign management turned out to be challenging and consistent handling of campaigns through the search process was not happening in all cases during the trial. The reason for this was the limitations in ad space addressing, i.e. campaigns are separate entities in the advertisement management system, and no functionality to link campaigns across ad-spaces was present. Finally, the pilot showed us that with the current state of tools the development of new campaigns is costly and adaptations of existing ads could be done only in some cases.

5 Conclusions: The Way Ahead

The operator perspective on mobile search is that it can be a driver of revenues from: i) advertisements, ii) usage of premium content, iii) usage of operator services. In this paper we have focused on the possible revenues from advertisements and explored value creation and revenue sharing among participants in a mobile search and advertising ecosystem. The pilot study has revealed a strong interest among advertisers, ad platform and search solution providers to work together with operators on mechanisms for targeting advertisements and personalisation.

As relatively small amounts of money are currently spent on mobile advertising, the operators face a challenge in growing the mobile advertising business. In this respect measurement of advertising effects may provide the most convincing argument, and the operators are well positioned in the mobile advertising ecosystem for producing such measurements and to facilitate ad personalisation. Mechanisms for management of user privacy are assets worth to explore for operators in the area of mobile advertising. In general, users preferred search services free from ads, but recognised personalised ads as an effective way to filter spam and unwanted information. Customer identification through MSISDN is a true enabler of personalised services.

The study uncovered immaturity in both internal and external systems and processes for offering a personalised search service including ads. There are several challenges in order to exploit the information stored in the operators' subscriber databases. Location is a vital mobile tool that mobile handset manufacturers, operators, and content providers can exploit commercially. Therefore, the provisioning of location based services should be improved to make local ad campaigns attractive. As the market for location based services grows over the following years, mobile operators should take a strong position to secure a strategic place in this market.

Another limitation relies on the systems used for defining, managing, and monitoring campaigns when personalisation and not just segmentation is a key objective. The mobile operator position as ad space and access channel provider brings a potential for supporting multi channel campaigns across platforms that should be further explored. Regarding opportunities for improving the quality of ad personalisation, operators could exploit social network analysis of their subscribers to provide added value and ultimately facilitate a better search experience for their customers.

References

1. Eurescom. Mobile Advertisement - Threats and opportunities. Sollund, A.M. (ed). In: EURESCOM P1654 D1 (May 2007)
2. Bughin, J., Henryson, D., Parvizi, P., Rudolph, S., Taqqu, Y., Wilkins, J., Wilshire, M.: Mobile search: The On-Ramp to mobile Advertising, pp. 49–56 (2006)
3. Wang, C., Zhang, P., Choi, R., Eredita, M.D.: Understanding consumers attitude towards advertising. In: Americas Conference on Information Systems, pp. 1143–1148 (2002)
4. Broder, A., Fontoura, M., Josifovski, V., Riedel, L.: A semantic approach to contextual advertising. In: SIGIR 2007, pp. 559–566 (2007)
5. Chatterjee, P., Hoffman, D.L., Novak, T.P.: Modeling the clickstream: Implications for Web-based advertising efforts. Marketing Science 22(4), 520–541 (2003)
6. Dominowska, E., Josifovski, V.: Internet monetization - sponsored search. In: WWW 2008: First Workshop on Targeting and Ranking for Online Advertising, Beijing, China, April 21-25, pp. 219–225 (2008)
7. Ghose, A., Yang, S.: Analysing search engine advertising: firm behaviour and cross-selling in electronic markets. In: WWW 2008: Workshop summary, Beijing, China, April 21-25, pp. 1269–1270 (2008)
8. Goldfarb, A., Tucker, C.: Search engine advertising. Communications of the ACM 51(11), 22–24 (2008)
9. Radlinski, F., Broder, A., Ciccolo, P., Gabrilovich, E., Josifovski, V., Riedel, L.: Optimizing relevance and revenue in ad search: A query substitution approach. In: SIGIR 2008, Singapore, July 20-24, pp. 403–410 (2008)

10. Smyth, B., McCarthy, K., Reilly, J.: Relevance at a distance - An investigation of Distance-Biased Personalization on the Mobile Internet. In: Proceedings from 14th Irish conference on Artificial Intelligence and Cognitive Science, AICS (2003)
11. Kobsa, A.: Privacy-Enhanced-Personalization. Communications of the ACM 50(8), 24–33 (2007)
12. Bergvik, S., Svendsen, G.B., Evjemo, B.: Beyond Utility. Modelling User Acceptance of Mobile Devices. In: Workshop User Experience, NordiCHI 2006, Oslo, October 14-18 (2006)
13. Davis, F.D.: Perceived usefulness, perceived ease of use, and user acceptance of information technology. MIS Quarterly 13(3), 319–340 (1989)
14. Drossos, D., Giaglis, G.M., Lekakos, G., Kokkinaki, F., Stavraki, M.G.: Determinants of effective SMS advertising: An experimental Study. Journal of Interactive Advertising 7(2) (2007)
15. Heggtveit, P.O., Yttri, B., Stenvold, L.A., Akselsen, S., Spilling, A.: Aida - A pilot study on active desktops for mobile phones. Fornebu, Telenor Research and Innovation, R&I N9/2005 (2005)
16. Hibberd, M.: Selling space. In: Mobile Communications International, pp. 42-46 (September 2008)
17. Kesti, M., Ristola, A., Karjaluoto, H., Koivumaki, T.: Tracking consumer intentions to use mobile services: empirical evidence from a field trial in Finland. E-Business Review IV, 76–80 (2004)
18. Leung, L., Wei, R.: More than just talk on the move: Uses and gratifications of the cellular phone. Journalism and Mass Communication Quarterly 77, 308–320 (2000)
19. MMA. Mobile Advertising Guidelines. Mobile Marketing Association. Denver, CO, USA (2008), http://www.mmaglobal.com/mobileadvertising.pdf
20. Tsang, M.M., Ho, S.-C., Liang, T.-P.: Consumer Attitudes towards Mobile advertising: An empirical study. International Journal of Electronic Commerce 8(2), 65–78 (2004)

Mobile Visual Analytics for Datacenter Power and Cooling Management

Ratnesh Sharma, Ming Hao, Ravigopal Vennelakanti, Manish Gupta,
Umeshwar Dayal, Cullen Bash, Chandrakant Patel, Deepa Naik, A. Jayakumar,
Sairabanu Z. Ganihar, Ramesh Munusamy, and Vani Mohan

1501 Page Mill Road
Hewlett-Packard Co. Laboratories, Palo Alto, CA
{ratnesh.sharma,ming.hao,ravigopal.vennelakanti,manish.gupta,
umeshwar.dayal,cullen.bash,chandrakant.patel,deepa.naik,
jayakumara,sairabanu-z.ganihar,ramesh.munuswamy,vanim}@hp.com

Abstract. The demand for data center solutions with lower total cost of ownership and lower complexity of management is driving the creation of next generation datacenters. The information technology industry is in the midst of a transformation to lower the cost of operation through consolidation and better utilization of critical data center resources. Successful consolidation necessitates increasing utilization of capital intensive "always-on" data center infrastructure, reduction in the recurring cost of power and management of physical resources. In this paper, we describe a tool that allows the data center facility managers and administrators to view and analyze the Key Performance Indicators (KPIs) associated with their data centers using pixel cell-based [10,11] visual analytics. The basic idea of our technique is to use the smallest element in the display to present the detailed information of the poser and thermal data records. Administrators can quickly recognize the patterns, trends, and anomalies. Furthermore, we discuss case studies of mobile visual analytics for energy and thermal state monitoring utilizing data from a rich sensor network.

Keywords: Visual Analytics, Sensors, Data center, Thermal Management, Data mining.

1 Introduction

The design and operation of the data center infrastructure is one of the primary challenges facing IT organizations. Unprecedented growth in demand for IT services has led to development of large, complex, resource-intensive IT infrastructure to support pervasive computing. Emerging high-density computer systems and consolidation of IT resources into fewer data centers are stretching the limits of data center capacity [1] in terms of power and resource utilization. The large number of components in a data center including cooling systems, power systems, and computer systems and the diversity of these components make data center design and operation complex.

To improve customers' RoIT (Return on Information Technology) [2], it is critical to maximize the resource utilization efficiency of the data center and simplify its management. To accomplish this goal, data center administrator need an agile sensor network coupled with powerful analytics to discover events and plan remedial action. Currently no tools are available to administrators to reveal interactions between data center elements like servers, network switches, and air conditioning units. The problem is compounded with the diversity of IT equipment and the scale of deployment. Since most thermal problems are sensitive to physical location, an onsite deployable tool is of great advantage to administrators. Onsite problem identification and escalation prevention is much easier with a mobile platform.

2 Background

2.1 Overview

Visual analytics combines various traditional data mining techniques with applications that help to manage datacenter power and cooling performance [3]. The combination of automatic analysis with human perception and understanding is key to this process. Large quantities of raw or processed data may be displayed in various dimensions using color shades or shapes. The change in color or shape enables a human to identify anomalies or events for possible corrective actions. The technique covers wide variety of data including time series. In this paper, we use combination of a pixel cell-based spatio-temporal overall view with high-resolution time series.

We employ visual correlation queries to be performed in an interesting area to meet the challenge of root cause and correlation analysis. The method of our visual correlation queries is to interactively select focus areas in the visualization. Then, according to the characteristics of the selected areas, such as temperature hot spots, we employ data mining techniques to compute relationships to other attributes of the data set. Then, the visual correlation query generates visual representations for users to view and refine the results [4].

For correlation analysis we include the Pearson Correlation r into our framework [5]:

$$r = \frac{\sum_{i=1}^{n}(A_{1i} - \overline{A}_1)(A_{2i} - \overline{A}_2)}{\sqrt{\sum_{i=1}^{n}(A_{1i} - \overline{A}_1)^2 \sum_{i=1}^{n}(A_{2i} - \overline{A}_2)^2}} \qquad (1)$$

where n represents number of attributes in a sensor data set.

This function computes the pair-wise correlation between bi-variant data with attributes A_{1i} and A_{2i}. If two attributes are perfectly correlated, the correlation coefficient is 1; for an inverse correlation, -1.

2.2 Data Center Infrastructure

Figure 1 shows the basic data center building blocks from utility grid to the cooling tower [2] [5]. Switch gear comprising of transformers, static switches with associated panels distributes power to the cooling infrastructure and the IT infrastructure. Cooling infrastructure comprises of chillers, cooling towers, computer room air conditioning (CRAC) units and primary/secondary pumps. IT infrastructure includes servers, network devices and storage devices housed in standard racks. UPS maintains power quality during normal operation and provides energy storage to operate the IT infrastructure during brown outs or short power outages. Chillers provide chilled water to the data center room that houses the server racks and other IT equipment.

Figure 2 shows the detail of the datacenter room including CRAC units, server racks and air flow paths [5]. Data centers are typically air-cooled with a raised floor plenum to distribute cool air, power and networking. Figure 2 depicts a typical state-of-the-art data center air-conditioning environment with under-floor cool air distribution.

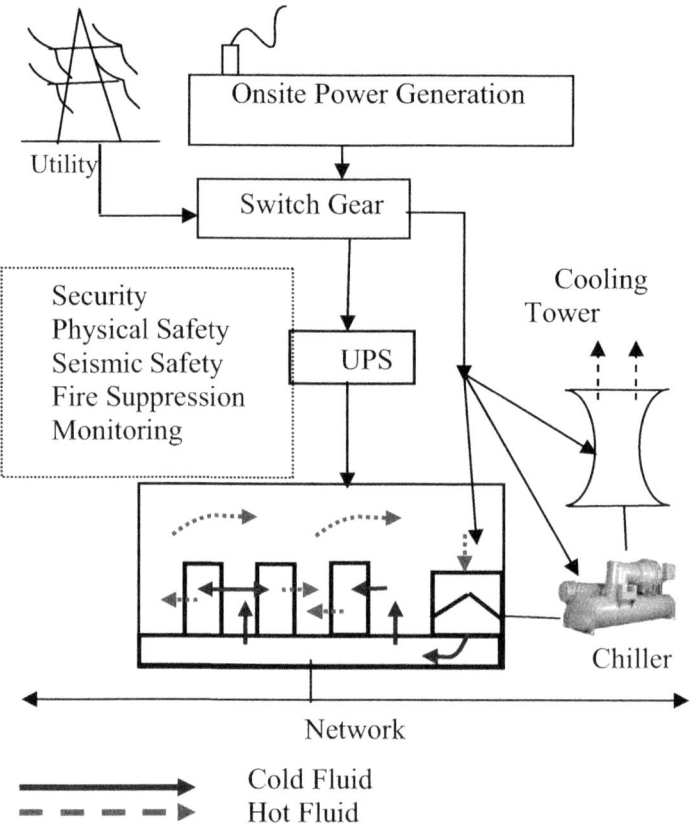

Fig. 1. Data Center Building Blocks

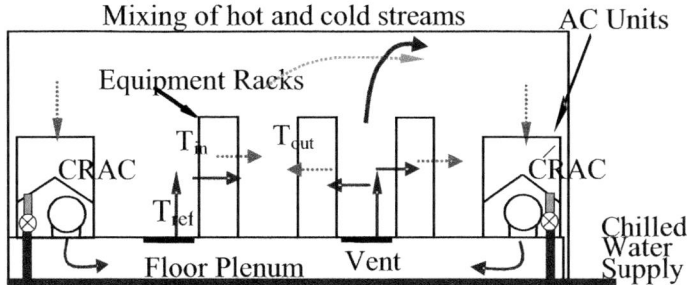

Fig. 2. Cross section of the datacenter

Computer room air conditioning (CRAC) units cool the exhaust hot air from the computer racks. Energy consumption in data center cooling comprises work done to distribute the cool air to the racks and to extract heat from the hot exhaust air. A refrigerated or chilled water cooling coil in the CRAC unit extracts the heat from the air and cools it within a range of 10 °C to 18 °C. The air movers in the CRAC units pressurize the plenum with cool air which enters the data center through vented tiles located on the raised floor close to the inlet of the racks. Typically the racks are laid out in rows separated by hot and cold aisles as shown in Figure 2. This separation is done for thermal efficiency considerations. Air inlets for all racks face cold aisles while hot air is expelled to hot aisles. A number of other equipment layout configurations and non-raised floor infrastructures also exist.

3 Architecture

The solution taps into a sensor grid distributed among racks and CRAC units [7] in the datacenter and implements data processing and analytics algorithms to extract information from it. It exposes a web-service interface to which mobile devices can connect to. The client application is implemented on HP iPAQ 210 mobile device, which runs Windows Mobile 6.0 and connects to the web-service over a Wi-Fi network (see fig.3). The mobile client application also provides an offline mode operation enabled with a rich set of visualization and analytics capabilities by enabling data mining on over 700,000 rows of data for every 24 hrs sensor data sampled on the mobile device. The choice of the mobile device was based on evaluating a number of physical and logistics constraints a datacenter operator would have to deal with while conducting either a routine or spontaneous operations workflow in a large data center environment including lack of desk space to operate more versatile large form-factor devices. This allows the data center administrators to use the solution in-the-field, i.e. inside the data center environment. The solution provides the administrator with up to the minute sensor information on any Data Center facilities infrastructure including information on a specific rack or set of racks that is being monitored by the sensors. The solution also enables a number of report generation capabilities that are available on demand backed by automated analytical engines. The data center administrator also has the ability to set specific filters based on thresholds for monitoring notification alerts. The client application automatically polls for the data and high-lights interest areas and creates canned reports for the operators use.

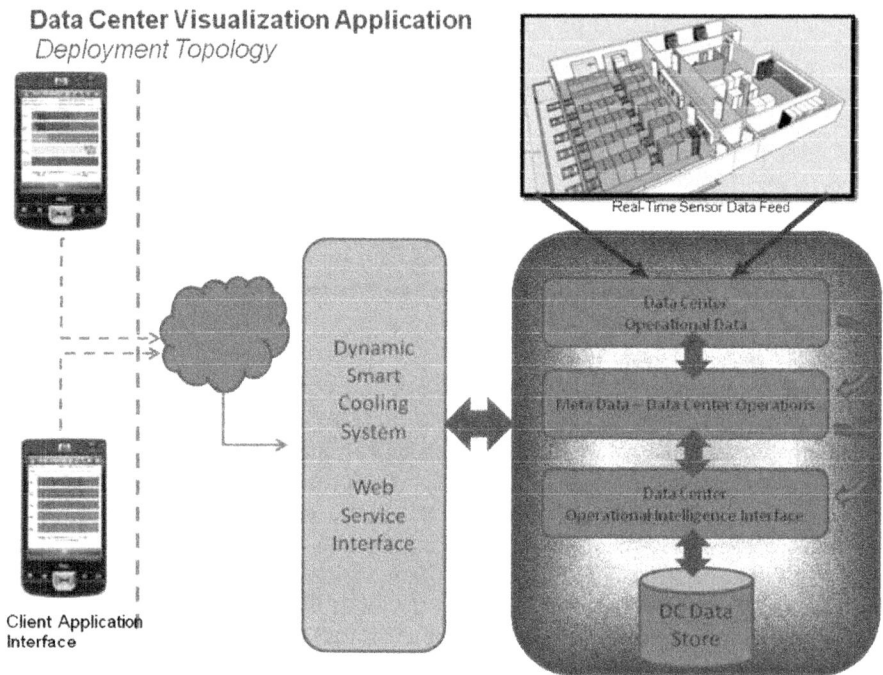

Fig. 3. Datacenter mobile studio – application architecture

Mobile devices have limited computing power, memory, storage space and connectivity options. Consequently, the design of the solution efficiently distributes the computation and data storage between the server and the mobile client device. Only the tasks that are dependent on the specific set of data under consideration and its spatio-temporal properties are executed on the device. Client component on the mobile device primarily consists of a correlation calculation component and visualization component. All the pre-processing requirements that are impervious to the dataset selected are performed on the server. Cleansing and transformations on the raw data being generated by sensor grid, which are compute intensive operations, are entirely performed on the server side web-service. The web-service also extends an interface to a repository of historical data staged and managed for its life cycle.

In the online mode of operation, which works over Wi-Fi, the client requests the web-service for specific data-sets based on user inputs, which get delivered in a ready-to-be-displayed form. Online mode allows for updated up-to-the-minute data to be available to the user. For the offline mode, the user can download a pre-packaged dataset containing complete but condensed 24-hour data for the entire data center on to the device. The pre-packaged datasets are periodically cached on the server side for client use. Implementation of visualization and correlation logic on the client, instead of the server allows the application to seamlessly and consistently work in connected and disconnected modes.

A reference implementation has been done and deployed in a 14,000 sq.ft. datacenter. This datacenter comprised sensor data monitored for over 500 racks, 11 CRAC units consuming 600KW of power for the IT load. The data center was monitored by over 1350 sensors sampled at 1 minute interval.[8] The thermal and operational events, key performance indicators (KPIs) and key risk indicators (KRIs) are now available at the click of a button on the mobile device for further analysis and report generation.

The visualization in the iPAQ uses a spatio-temporal layout. Figure 4 illustrates a sample mapping of 4 sensors of multi-dimensional thermal data to a 2-D space. The layout provides both a "top" view for the data center thermal performance and interactive "data exploration" views for troubleshooting. Each rectangle contains a sensor temperature time series in a datacenter. Hot spots are marked for thermal correlation analysis and root-cause queries.

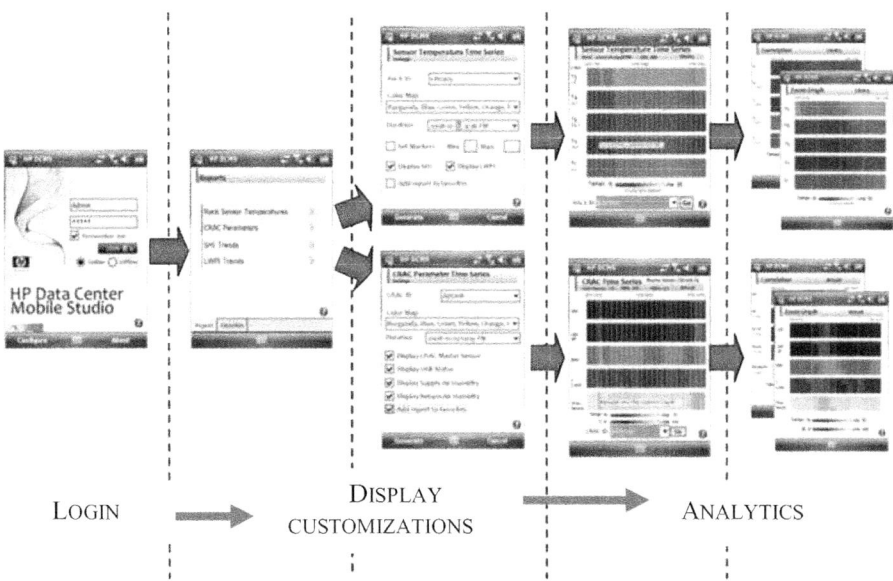

Fig. 4. Datacenter mobile studio – application pipeline

Figure 5 illustrate pixel cells of a time series which are arranged from bottom to top and left to right according to time. The color of a pixel is the temperature value of a sensor for each time interval. Some colors (i.e., red, orange, etc.) represent more severe conditions and alert the administrator to take immediate action. For example, T1 has a high temperature (over 35°C) at hour 11.

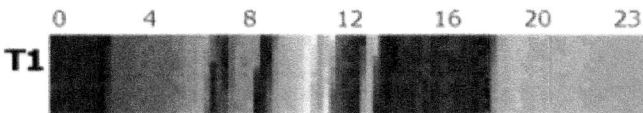

Fig. 5. A sensor temperature time series. The sensors are arranged in descending order with sensor T1 closest to the vent tile within a rack.

As described in the previous section internal datacenter infrastructure consists of racks and CRAC units. To visually analyze data center operations within the room, we use three different types of visual analytics queries: Finding correlation between inlet and outlet, at a change point, and among sensor attributes.

4 Case Studies

4.1 Thermal State Detection

Thermal state detection for distributed management and control of cooling resources is of great importance to datacenter operators, both on the facility and IT-side. High temperatures can adversely affect uptime and equipment performance and lead to premature end of life for equipment. Most data centers have four critical problem thermal states: (1) high sensor temperatures (e.g., temperatures exceeding a threshold), (2) high recirculation identified by temperature-based metrics such as Supply Heat Index (SHI) [2], (3) Large variation in temperature distribution identified by out-of-band sensor temperatures within a rack (e.g., T1>T2), and (4) Oscillations between the previously described states and the normal conditions [5]. Detection of hot spots is necessary to prevent temperature related failures while detection of abnormal air flows and metrics helps to identify energy inefficiencies.

Figure 6 shows the screenshot of inlet air sensor temperature time series obtained from groups of 5 sensors from two alternate racks over a period of one day. Each screen displays a total of 7200 records. The color scale denotes temperature range

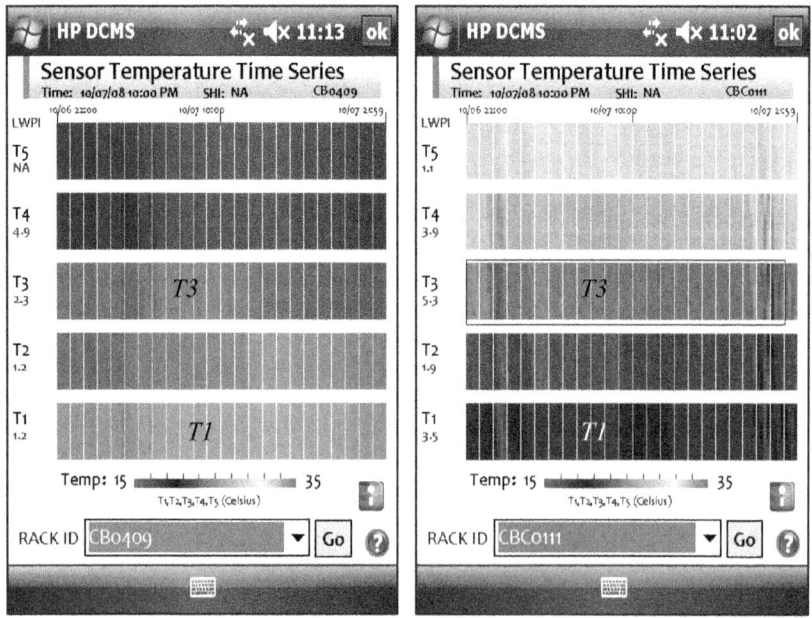

Fig. 6. Inlet air temperature display of alternate racks, the visualization on the left represents out of sequence temperatures (e.g.; $T5<T4<T3<T2<T1$). The visualization on the right has no out of sequence temperatures (e.g., T5>T4>T3>T2>T1) is a desirable time series.

between 15C and 35C. Observe the variation during the day and the reversal of temperature distribution between the two racks during the same time period. Such variations are prevalent in datacenters with heterogeneous equipment and knowledge of such states is important for onsite troubleshooting of thermal issues. Such a display can assist in normal operation by identifying mal configured racks or air distribution. A user can verify the rack assembly for potential air leakages as well. In many cases the server air flow path is not aligned with the air flow in the datacenter. Such mismatches can lead to temperature reversals as well.

4.2 Infrastructure Utilization

Utilization of infrastructure is often used as a yardstick to estimate return on investment [6]. The mobile datacenter studio provides methods to display utilization of power and cooling infrastructure, both historically and in real time. Users can apply analytics to obtain more information around inflexion points in the utilization display. Fan power utilization and chilled water utilization of CRAC units are a few of the several utilization metrics that can be displayed and analyzed. Among other things, infrastructure utilization can also assist in trouble shooting thermal state management issues. It can aid planning on addition and replacement of existing infrastructure and resource needs.

Figure 7 shows the CRAC unit time series display of data collected over a period of one day, from two units in the datacenter. On each screenshot, from top to bottom, air discharge temperature and set points are displayed followed by return air temperature, cooling load and blower fan power. While one of the units has 0% cooling load the

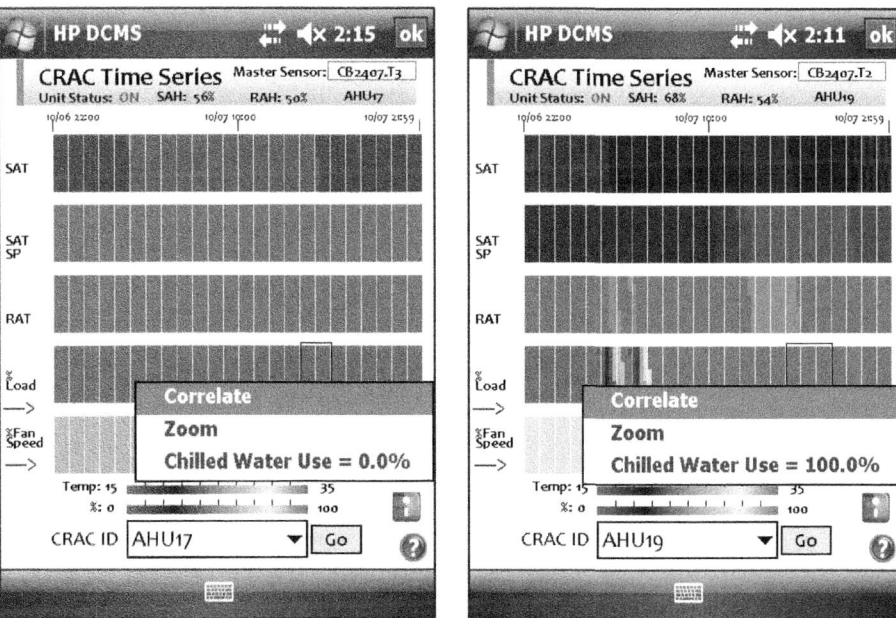

Fig. 7. CRAC unit displays - The pulldown window indicates the correlations between the supply air temperature and chilled water temperature. Figure on the left shows no chiller water utilization (i.e. zero). Figure on the right shows full utilization (100%).

other unit is at its maximum rated cooling load (100%) for a significant period of time. The demand variation between the units indicates the heterogeneity in heat load distribution in the datacenter. Users can investigate workload deployment in CRAC unit zones and balance out loads wherever possible. Such displays provide decision support related to cooling infrastructure maintenance/downtime schedules. Further analytics can help in linking CRAC unit events to IT power and cooling events.

Power and cooling accounts for significant portion of operating cost of datacenters. In addition to thermal management, energy consumption can be used to define operation state of the datacenter. Providing energy consumption enables a mobile user to factor it into management decisions. A high power usage level indicates that there is little margin for IT capacity expansion before power reduction can be achieved. Air flow studies can also be carried by data center operators to see real time impact of operation at different speeds.

4.3 Energy Consumption

Figure 8 shows the energy display on the CRAC unit time series screen. Tracking energy over time provides data administrator insights on time-of-use demand responses. Impact of rising power costs and routine downtimes can be reduced by appropriate demand response strategies. Variation in energy consumption can be correlated to thermal management states in the datacenter and power curtailment plans can be developed to manage demand during peak hours. Power and cooling delivery bottlenecks can also be identified using such displays.

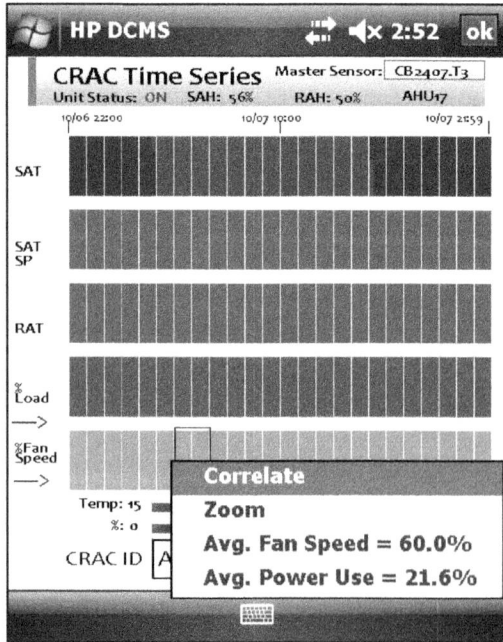

Fig. 8. CRAC unit energy display – Average Fan Speed and power consumption for the rubber-banded region are shown in the pull-down menu

5 Discussions and Future work

5.1 Root Cause Analysis

Apart from access to time series and analytics thereof, datacenter mobile studio provides metrics that assist in assessment of datacenter operation. While thermal management is an ongoing challenge in datacenters, it can be a manifestation of several anomalies in the datacenter like mal configuration of racks, mal-operation of CRAC units, obstruction of vent tiles (see fig 2) or high workload.

Mixing of hot and cold air streams caused by mal-configuration of racks is captured by Supply Heat Index (SHI) [2]. Plots of historical and real-time values can be obtained by rubber banding the hot spot region in the sensor display screen for further diagnostics.

$$SHI = \frac{T_i - T_{ref}}{T_o - T_{ref}} \qquad (2)$$

where T_i, T_o and T_{ref} are the rack inlet, rack exhaust and CRAC unit supply temperatures (see fig. 2), respectively.

Another important parameter in this process is the Local Workload Placement Index (LWPI) [8]. It is defined as ratio of thermal management and air-conditioning margin at that location over the degree of recirculation at that location.

$$\text{LWPI} = \text{AC Margin} + \text{Thermal Margin} - \text{Hot Air Recirculation} \qquad (3)$$

Recirculation limits the ability to deploy workload while air conditioning margin improves the ability of the server (at the location of interest) to accept workload. It

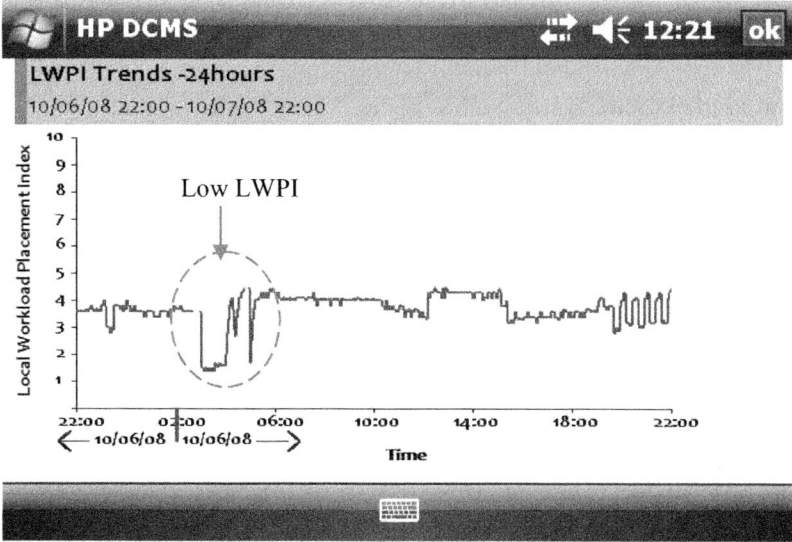

Fig. 9. LWPI plot for a sensor location over a 24 hour period – Lower LWPI indicates lower cooling effectiveness and less favorable area for workload deployment

can be used to gauge the effectiveness of cooling for workload placement. A detailed definition of eqn.(3) can be found in [8]. Administrators are able to rubber-band an interesting area and query for the correlations or detailed information. Varying LWPI can initiate migration of workloads.

Figure 9 shows the plot of LWPI for a single sensor over a period of 24 hours. The region of low LWPI indicated in the graph shows a period of reduced effectiveness of cooling in the region around the sensor. Such downtrends can be an early trigger for workload migration process to begin. Workload can be relocated to regions of high LWPI during the course of such downshifts.

5.2 Data Collection

Data collection for onsite changes and measurement of select operational parameters are important for routine management of datacenters. Operators can download relevant data for analysis based on visual clues. Other data collection applications include logging of infrastructure changes, asset management, system operational states, generation of alarms and initiation of remediation procedures.

6 Conclusions

Advanced data analytics and knowledge models are needed to identify inefficiencies in the datacenter infrastructure. In addition to troubleshooting, information about infrastructure states can enable involvement of knowledge experts for assessments and analysis at the right time. Mobile studio combines this power of visual analytics with onsite datacenter management. It assists administrators in identifying key power and cooling events followed by root cause analysis during runtime. Future work will focus on combining historical data analysis with trends to predict anomalies and initiate preemptive measures.

Acknowledgement

Authors wish to thank Vaibhav Bhatia, Rajkumar Velumani and Mohandas Mekanapurath for providing access to datacenter infrastructure for testing and evaluation.

References

[1] The Uptime Institute, Heat density trends in Data Processing, Computer systems and Telecommunications Equipment, White Paper issued by The Uptime Institute (2000)
[2] Sharma, R.K., Shih, R., Bash, C.E., Patel, C.D., Varghese, P., Mekanapurath, M., Velayudhan, S., Kumar, M.V.: On building Next Generation Data Centers. In: Proceedings of Compute 2008, Bangalore, Karnataka, India, January 18-20 (2008)
[3] Hao, M., Dayal, U., Keim, D.A., Schreck, T.: Multi-Resolution Techniques for Visual Exploration of Large Time-Series Data. In: Proceedings of IEEE VGTC Symposium on Visualization, EuroVis 2007 (2007)

[4] Hao, M., Dayal, U., Keim, D.A., Morent, D., Schneidewind, J.: Intelligent Visual Analytics Queries. In: IEEE Symposium on Visual Analytics Science and Technology, CA (2007)
[5] Sharma, S., Patel, S.: Application of exploratory data analysis techniques to temperature data in a conventional datacenter. In: Proc. ASME IPACK 2007, IPACK 2007-33170, BC, Canada (July 2007)
[6] Patel, et al.: Energy flow in the information technology Stack: Introducing the coefficient of performance of the ensemble. In: Proc. ASME Intl. Mech. Eng. Cong. Exp. (IMECE), IMECE 2006-14830
[7] Bash, C.E., Patel, S.: Dynamic Thermal Management of Air-Cooled Datacenters. In: ASME/IEEE ITherm 2006, San Diego, CA (2006)
[8] Marwah, M., Sharma, R.K., Shih, R., Patel, C.D., Bhatia, V., Mekanapurath, M., Velumani, R., Velayudhan, S.: Data analysis, Visualization and Knowledge Discovery in Sustainable Data Centers. In: Proceedings of Compute 2009, Bangalore, Karnataka, India, January 9-10 (2009)
[9] Bash, C.E., Forman, G.: Cool Job Allocation: Measuring the Power Savings of Placing Jobs at Cooling-Efficient Locations in the Data Center, USENIX, San Jose, CA (June 2007)
[10] Hao, M., Sharmal, R.K., Dayal, U., Keim, D.A., Vennelakanti, R.: Application of Visual Analytics For Thermal State Management in Large Data Centers. CGF Journal (2009)
[11] Hao, M., Keim, D.A., Dayal, U., Oelke, D., Tremblay, C.: Density Displays for Data Stream Monitoring. In: IEEE EuroVis 2008 (2008)

RFID-based Distributed Memory for Mobile Applications

Michel Simatic

Institut Télécom, Télécom & Management SudParis, 9 rue Charles Fourier,
91011 Evry Cedex, France
Michel.Simatic@it-sudparis.eu
http://www-public.it-sudparis.eu/~simatic/

Abstract. The goal of our work is to give a user equipped with an RFID-enabled mobile handset (mobile phone, PDA, laptop...) the ability to know the contents of distant passive RFID tags, without physically moving to them and without using a Wireless Area Network. The existing architectural patterns involving passive tags do not meet simultaneously all of these requirements. Our RFID-based distributed memory does. By associating vector clocks to tags, we replicate a view of this memory on each tag and each handset, and disseminate updates between all of the replicas. Thus a user can locally query the replica hold by their mobile handset without physically moving to a tag. We have developed a pervasive game as an application example. Using data collected during real game sessions, we evaluate the performance of our distributed memory. Then we discuss staleness and scalability issues. We conclude and give perspectives of our work.

Keywords: Distributed memory, RFID, NFC, Vector clocks, Gossip protocols, Pervasive application, Game.

1 Introduction

RFID tags should be increasingly used in (pervasive) mobile applications. They are easy to deploy and robust. In addition, the number of users equipped with an NFC-enabled mobile phone is sensitively increasing. For instance, according to Juniper Research, 700 million users will have such mobile phone by 2013 [5]. Most of the time, RFID tags take place in an architecture involving a Wireless Area Network (WAN) like Wi-Fi, UMTS, HSDPA... [16]. But using a WAN increases installation and operational costs. As an example, according to [10], the cost of data traffic is substantial in Germany, whereas negligible in Sweden. Our goal is to meet simultaneously the three following requirements: 1) users equipped with an RFID-enabled mobile handset (mobile phone, PDA, laptop...) can know the contents of any distant passive tag; 2) they do not need to be physically near the tag; 3) they do not use any WAN. This paper presents our answer: the RFID-based distributed memory.

To achieve this goal, we replicate the contents of the tags on each tag and each handset of the system. Thus a handset can get the value of the data stored on a distant tag. It just has to query its own replica. Whenever a mobile handset meets a tag (or another handset), their respective replica are made consistent by comparing the vector clock values coupled with the replicas. Thus the users are the cornerstones of what can be considered as a gossip protocol. We can stimulate this communication network by having an application design which incites users to meet repeatedly during a session. This is especially true if the application is a multi-user pervasive mobile game. If its game design does not already integrate this feature, we can adapt it to promote players to collaborate actively.

The work presented in this paper does not tackle security issues. Indeed a comprehensive analysis of security with respect to RFID/NFC represents a dedicated research field [12].

The rest of this paper is structured as follows. Section 2 details why standard RFID architectures do not meet our requirements. Then, Section 3 presents how we use vector clocks to get an RFID-based distributed memory. Afterwards, Section 4 presents the pervasive game we implemented as an application example. It describes how the game design stimulates the communication network implicitly made by the players. Section 5 analyzes some of the figures observed during our experiments. Section 6 is a discussion about this architecture: it presents related work and analyzes staleness and scalability issues. We conclude and give perspectives of our work in Section 7.

2 Passive RFID Tags in Existing Architectures

Passive RFID tags are pieces of memory which can only be accessed when activated by a transponder. These tags always contain an identifier specific to the tag. Some of them hold also a chunk of memory which can be read or written by the transponder. Passive RFID tags are commonly used following three architectural patterns.

The first architecture is a centralized one (as defined by GS1 EPCglobal organization [1]): each product is equipped with an RFID tag. When this product comes close to an RFID reader (either because the product has been moved near a reader, or an operator equipped with a mobile reader has moved near the product), the RFID reader reads the RFID identifier stored on the tag. Then it gives this information to a dedicated application. By accessing a central database, the application is able to link this identifier to the identity of the product and the procedure to handle it. Notice that to have this architecture operational, a WAN is required all the time in order to link the RFID reader and the central database. But, using a WAN introduces installation and operational costs which may not be compatible with requirements of some applications.

Thus, a second architecture can be considered. It consists in making local copies of the central database on each mobile handset. Doing so, when a mobile handset makes an update related to a product, it updates its local copy of the database. Regularly, the device is connected to the database in order to upload its updates. Possibly specific procedures are applied in order to reconcile updates coming from the different mobile handsets. It is this architecture which has been put in place by the city of Paris for the trees growing on the border of its avenues [13]. Each tree of Paris avenues is equipped with an RFID tag. During the night, in their storage room, tablet PCs are in contact with a central database in order to make a local copy of it. In the morning, each gardener takes one of these tablet PCs. Each time they does something to a tree, they reads its RFID tag with their tablet PC. The PC updates data on its local database. At the end of the day, the gardener puts back the tablet PC in its storage room. Finally, the tablet PC synchronizes its local database with the central database. The problem with this architecture is the following: user B cannot see the modifications done by user A on one piece of data unless user A has synchronized their device with the central database (after having modified the considered piece of data) and user B has also synchronized their device with the central database.

To solve this problem, a third architecture can be considered. It is based on RFID tags containing memory which can be written. In this case, the application reads/writes the updates concerning the product (*e.g.*, its history) on its associated RFID tag. Actually, the system made of all of such RFID tags corresponds to a distributed memory, each RFID tag holding piece of data of this distributed memory: there is no more need for central database storage. But this architecture experiences one severe limitation: a mobile handset has no way to have an idea of the value of the different tags of the system unless it is physically in contact with each tag. Thus the handset can never make queries on the whole contents of this distributed memory without physically moving to all of the tags.

None of the presented architectures meets simultaneously our three requirements. We propose a solution by introducing vector clocks inside the RFID tags.

3 Applying Vector Clocks to Passive Elements

In this section, we present how vector clocks are used to give to each handset the best up-to-date view of distributed memory (DM). This gives users the ability to make queries on a local view of DM.

Here are the main ideas of our solution. Each tag and mobile handset of the system holds a local view of DM. We associate a vector clock per local view. The vector clocks we use are plain vector clocks [15,8]. Whenever a mobile handset comes into contact with a tag (or another handset), these two elements make their own view of DM consistent by comparing vector clock values. Thus, each of them takes advantage of the knowledge of the other element to get more recent information concerning DM evolutions. Doing so, they get a more up-to-date view of DM. The next paragraphs detail our solution.

The system we consider is made of two types of elements: RFID tags and mobile handsets (See Figure 1). Each element e holds a local view DM_e of the distributed memory. We note $DM_e[r]$ the view element e has of the contents of DM hold by RFID tag r. Each element e holds also a vector clock VC_e which is used to propagate operations done on DM. $VC_r[r]$ holds the timestamp of the last update done on $DM_r[r]$ (which is the part of the distributed memory

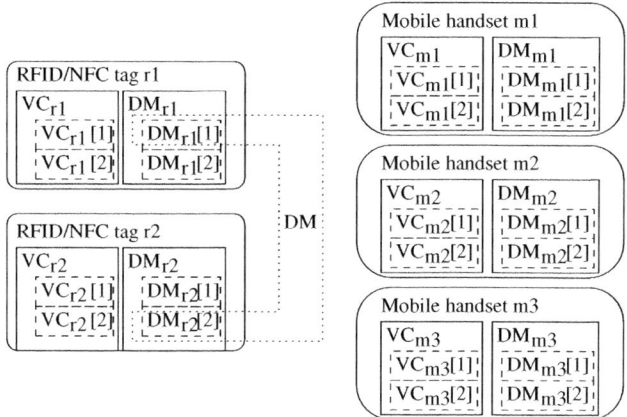

Fig. 1. Data present in a system made of 2 RFID tags and 3 mobile handsets

DM hold by RFID tag r). Meanwhile $VC_{e, e \neq r}[r]$ holds the timestamp of the last update of $DM_r[r]$ which element e is aware of.

At application initialization time, for all elements e of the system, DM_e is initialized with the initial value of the distributed memory. This initial value depends upon the application which will be using the distributed memory. On the contrary, VC_e is initialized to a value independent from the application: $(0, \ldots, 0)$.

Then, during the rest of application lifetime, whenever a mobile handset m changes the value stored in $DM_r[r]$ of a tag r, it applies Algorithm 1.

Algorithm 1. Update of $DM_r[r]$ on tag r by mobile handset m
1 $DM_r[r] \leftarrow update\ of\ DM_r[r]$
2 $VC_r[r] \leftarrow update\ of\ VC_r[r]$
3 $DM_m[r] \leftarrow DM_r[r]$
4 $VC_m[r] \leftarrow VC_r[r]$

As in [17], usually $VC_r[r]$ is a logical clock: line 2 of Algorithm 1 is $VC_r[r] \leftarrow VC_r[r] + 1$. But, as in [11], we may take advantage of two conditions to save space on each tag (and thus improve the scalability). First condition is "The real-time clock of the last update of the tag r is stored among the data of $DM_r[r]$". Second condition is "Two subsequent updates always have subsequent and different real-time clocks". If both conditions are satisfied, $VC_r[r]$ can hold directly the real-time clock of the update: line 2 of Algorithm 1 becomes $VC_r[r] \leftarrow$ *real-time clock of the update of $DM_r[r]$*. We save the space of one logical clock.

Another optimization can be coupled with the previous one. It can be considered if the real-time clock of the last query on $DM_r[r]$ is stored among the data of $DM_r[r]$. In the case there is no need to distinguish the real-time clock of the last update and the real-time clock of the last query, $VC_r[r]$ can hold the real-time clock of the last update or query: line 2 of Algorithm 1 becomes $VC_r[r] \leftarrow real-time\ clock\ of\ the\ update\ or\ query\ of\ DM_r[r]$. We save the space of one real-time clock.

In addition to Algorithm 1, by applying Algorithm 2, DM_e and VC_e may be updated whenever element e is able to exchange information with another element e'. In the context of an RFID/NFC-based application, this happens in two cases: a mobile handset is near an RFID tag (respectively another mobile handset) and is able to interact with it via RFID/NFC protocol (respectively NFC peer-to-peer protocol).

Algorithm 2. Making DM_e and $DM_{e'}$ consistent
1 **foreach** $i, 1 \leq i \leq$ number_of_tags_in_the_system
2   **if** $VC_e[i] < VC_{e'}[i]$ **then**
3   // Element e' holds a more up-to-date view of $DM_i[i]$
4     $DM_e[i] \leftarrow DM_{e'}[i]$
5     $VC_e[i] \leftarrow VC_{e'}[i]$
6   **elseif** $VC_e[i] > VC_{e'}[i]$ **then**
7   // Element e holds a more up-to-date view of $DM_i[i]$
8     $DM_{e'}[i] \leftarrow DM_e[i]$
9     $VC_{e'}[i] \leftarrow VC_e[i]$
10 **endif**
11**endforeach**

By applying Algorithms 1 and 2, each element e of the system has the best up-to-date view of the distributed memory it can have. But there is no guarantee that this *best* up-to-date view is the *most* up-to-date view of the distributed memory. As DM evolves during lifetime of the application, DM_e may contain stale data: there may be a tag $r, r \neq e$ for which $DM_e[r]$ does not correspond to $DM_r[r]$ currently hold by tag r. The only guarantee we have is that tag r did hold value $DM_e[r]$ at some point in time. Either it is its initial value. Or it is a subsequent value which induced the modification of $VC_r[r]$ (after application of Algorithm 1) and thus the propagation of this new value to $DM_e[r]$ (after application of Algorithm 2).

There are other staleness limitations. They are discussed in Section 6.

Despite them, an application running on a mobile handset m is able to make queries on DM_m: it has a view of the whole distributed memory without moving physically to the tags. Notice that the application can help reducing the staleness limitation. As in gossip protocols, if it stimulates information exchanges between elements of the system, propagation of updates will disseminate quicker in the system. Thus the gap between DM_e and DM will be thinner. Next section presents an application example and shows how its design promotes the dissemination.

4 Pervasive Game as an Application Example

The RFID-based distributed memory presented previously has been implemented in the context of pervasive game "Plug: Secrets of the museum" (PSM).

This section briefly presents PSM, makes the link between PSM and the data structures presented in Section 3, and describes how the game is designed to promote the dissemination of data.

PSM has been developed in the context of the PLUG project [3]. It is designed to let players discover, in a different way, meaningful objects of a museum [18]. In PSM, 8 teams of players are equipped with NFC-enabled mobile phones (Nokia 6131 NFC). Their main goal is to collect family of objects during a game session of at most 85 minutes. To do so, using a J2ME midlet we have developed, they exchange virtual cards (representing objects of the museum) either with one of the 16 RFID/NFC tags (ISO 14443, Mifare-NFC, 13.56 MHz, 1 Kbyte of RAM) located in different places of the museum, or with other teams (through NFC peer-to-peer communication). To reduce risks of deadlocks between players, every card has 3 instances in the game: there are 4 families and 4 cards by family times 3 instances, which makes 48 cards, in all. At the beginning of a game session, cards are shuffled and distributed between the 8 mobile phones (4 cards per handset) and the 16 RFID tags (1 card per tag).

PSM uses data structures presented in Section 3 as follows. $DM_e[r]$ (e being any element of the system, r being any tag) is a byte storing a value between 0 and 15 (each value corresponds to one of the 16 cards). $VC_e[r]$ is a short value (two bytes). It contains the real-time clock of the last update or query on $DM_r[r]$ as seen by e. Indeed we use the optimizations presented in Section 3. This is because our game has two interesting properties. The first one is that the clocks of the mobile phones are manually synchronized at the beginning of each day; we checked experimentally that the clocks do not drift away more than 1 second from each other during the day. The second property is that it takes at least 10 seconds for two players to exchange a virtual card with the same tag or to make a query on the same tag. As 10 seconds is much greater than 1 second: 1) we can apply the optimizations; 2) the real-time clock can be stored as the number of seconds since the beginning of the game (as a game session lasts at most 85 minutes —5100 seconds— there is no risk of overflow).

DM_e (respectively VC_e) is initialized to the 16 card values hold initially by the tags (respectively $(0, \ldots, 0)$).

At play time, whenever a team wants to exchange one of the cards hold by its mobile phone with the card hold by a tag, it must go physically near the tag. The application on the mobile handset applies Algorithm 1. Moreover, when a team wants to know what card is physically contained in a tag, it also has to go physically near the tag. We take advantage of this `read` operation to apply Algorithm 2 in order to make DM_{mobile} and DM_{tag} consistent. Thus, although the team believes there is only a `read` operation (to display the card stored in the tag), there is actually also a `write` operation which possibly modifies DM_{tag} and VC_{tag}. Algorithm 2 is also applied whenever two teams exchange cards between their mobile phones (via NFC peer-to-peer protocol): this makes DM_{mobile_1} and DM_{mobile_2} consistent.

To help players in their search, game design introduces a hint function: a team can ask its mobile phone for an indication of an RFID tag which contains a virtual card convenient for its collection. The hint function is implemented by analyzing DM_{mobile} and VC_{mobile}. It considers the virtual cards stored in DM_{mobile} which correspond to the family collected by the team. Among these, it selects a card which the team does not have already and for which the information in VC_{mobile} is the most recent (Intuitively, the more recent it is, the more probable it is that the virtual card is still in the RFID tag). Actually the hint function provides two kinds of result. The first kind is the indication of a tag r and the card $DM_{mobile}[r]$ it was containing $currentClock - VC_{mobile}[r]$ seconds ago. The second kind of result is "For the moment, no tag contains a card which is interesting for you". This message is displayed when none of the cards of DM_{mobile} satisfies team's need.

The following paragraphs describe the game design specificities which promote the dissemination of data done by algorithms of Section 3.

As seen previously, whenever a mobile phone is in contact with another element of the system (should it be a tag or another mobile phone), we apply Algorithm 2 in order to make DM_{mobile} and $DM_{tag\_or\_mobile}$ consistent. The only way a team can know the card stored in a tag is to physically go to that tag: team is promoted to physically move to tags. Thus we stimulate data propagation via tags. Moreover, a team gains points whenever it exchanges a card with another team: team is promoted to do such exchanges. Not only does it foster human interactions inside the museum [9], but it stimulates data dissemination via mobiles.

Concerning exchanges between teams, a rule prevents two teams from spending their whole PSM session doing exchanges of cards (and thus getting corresponding points). To do so, in PSM, a team can exchange at most two cards with another team in a sliding window of ten minutes. This rule limits the number of exchanges made by a team during a session. For instance, in the case of a session lasting 85 minutes, a team can make at most $(85/10 + 1) \times 2 = 18$ exchanges. Even though it is limiting number of exchanges, this rule has indeed a positive effect on dissemination of data. If it was not there, at least one pair of mobile phones would not contribute any more to the propagation of data updates to other mobile phones and to tags. Said in other words, at least 2 of the 8 mobile phones would not be used any more for this dissemination task: we would lose at least $2/8 = 25\%$ of our "network" capacity.

This section has presented the game we have developed as an application example. Next section presents the experimental results we observed with this game.

5 Experimental Results

In June 2009, five public sessions took place in the *Musée des arts et métiers* (Paris). Each of them was played with 6 teams. In this section, we present our methodology to obtain data out of these sessions. Next, we analyze some of the data.

During each game session, each mobile phone logs all of the events triggered by the team using it. We log systematically the date of the event. The other logged information depends upon the type of the event. Here are some details about the events concerning the rest of this article:

- for Application start event, we log the game session identifier, the family collected by the team, the virtual cards hold by the mobile phone, DM_{mobile} and VC_{mobile};
- for Consult a tag event, we log the tag identifier, the virtual card contained in the card, DM_{mobile} and VC_{mobile} after DM_{mobile} and DM_{tag} have been made consistent;
- for Exchange with a tag event, we log the tag identifier, the virtual card given, the virtual card received;
- for Exchange with another team event, we log the identifier of the other team, the virtual card given, the virtual card received, DM_{mobile} and VC_{mobile} after DM_{mobile} and $DM_{mobile\ other\ team}$ have been made consistent;
- for Hint request event, we log the family collected by the team, the kind of hint result, and the information given to the team (in the case of a result of the first kind).

At the end of each day of tests, all of the log files are transferred from the mobile phones to a laptop. On this laptop, we run a Java application. It orders all of the events according to their Date of event field. Then it counts the different events. Finally it builds a history of the state of the system. The state concerning cards contained by each mobile is initialized by analyzing Application start events. It evolves whenever there is an Exchange with a tag or Exchange with another team event. The state concerning tags is initialized when analyzing first Consult a tag event. It evolves whenever there is an Exchange with a tag event.

As we are in front of a fully distributed application, it is quite sure that the history we get does not correspond to what did happen in the reality. Section 4 has presented the properties concerning time in PSM. Thus, in the history of events we get, we have the guarantee that the chain of events concerning each tag and each mobile phone is correct. It is the interleaving of these events we obtain which may not have existed in reality [4]. Thus, the figures we present hereafter are only an approximation of what really happened during application execution.

During the 5 game sessions, 30 mobile phones played sessions of 70 to 85 minutes. 3889 events were logged. There were 590 visits to a tag, 279 exchanges between a mobile phone and a tag and 142 exchanges between one mobile phone and another one.

14 hint requests were logged[1]. As our event analyzer builds a history of the state of the whole system, whenever there is a Hint request event, we can evaluate the correctness of this hint. 8 (respectively 3) out of the 9 (respectively 5) hint results of the first (respectively second) kind were correct. This means $(8+3)/(9+5) = 79\%$

---

[1] This number of logged hint requests is low probably because of the game design: asking for a hint is costing points to a team. Moreover, the hint was presented to players as not being 100% sure. We suspect this resulted in players not asking often for a hint.

of the hints were correct. But this percentage is computed with only 14 data. To have a bigger sample, we have modified the event analyzer: whenever analyzing a Consult a tag event, the analyzer generates an artificial Hint request event. To do so, it applyies the algorithm which the mobile would have applied if the player had triggered the hint function at that moment. With the generated hints, we observe that 490 (respectively 30) out of the 540 (respectively 50) hint results of the first (respectively second) kind were correct. This means $(490 + 30)/(540 + 50) = 88\%$ of the hints are correct. Of course, 88% is much lower than the 100% accuracy we would have obtained with a centralized architecture. But, we do reach 88% without installation and operational costs of a WAN. The acceptability of this rate of correct hints is application dependant. In the case of our game, the hint function is presented to players as a kind of gambling: a rate of 88% is acceptable.

One may think intuitively that a recent hint has more chances of being correct than an old hint. The beginning of Table 1 (until line "[30, 34]") confirms this conjecture. But we have no satisfying explanation for the high percentages observed in the last lines.

Another factor of the correctness of hints is how many tags are notified of a given change in a tag (see Figure 2) and how long it takes to disseminate such information (see Figure 3). If it is rather frequent that at least 3 tags are notified of the change of a tag (it happens with a frequency of 39%, see Figure 2), the frequency of notification of at least 12 tags is less than 9%. This is because the notification of a tag change takes time to spread itself among other tags.

Figure 3 shows for instance that, to have 3 tags notified of a change, it takes at most 5 minutes in 50% of the cases. So while the information of a tag change

Table 1. Relation between rate of correct hint result of the first kind and hint age

Age of the hint (in minutes)	Number of hint requests	Number of correct hints	% correct hints
[0, 4]	176	178	99%
[5, 9]	59	65	91%
[10, 14]	50	61	82%
[15, 19]	55	64	86%
[20, 24]	32	35	91%
[25, 29]	7	8	88%
[30, 34]	14	18	78%
[35, 39]	15	15	100%
[40, 44]	12	12	100%
[45, 49]	7	8	88%
[50, 54]	7	8	88%
[55, 59]	10	10	100%
[60, 64]	6	7	86%
[65, 69]	10	17	59%
[70, 74]	11	13	85%
[75, 79]	10	12	83%
[80, 84]	5	5	100%
[85, 85]	4	4	100%

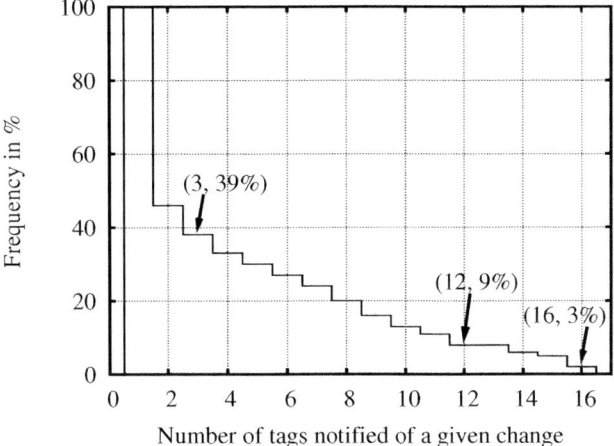

Fig. 2. Frequency of number of tags notified of a change (including the concerned tag)

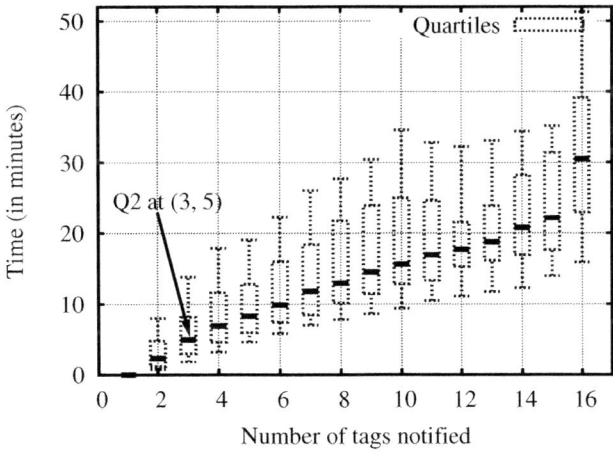

Fig. 3. Boxplot of times (in minutes) taken by a given number of tags to be notified of a change (decile D1, quartiles Q1, Q2, Q3, and decile D9)

is spreading, the probability that the concerned tag receives a new virtual card is increasing. This new information will also spread itself, replacing the previous one. This explains why it is very rare (frequency of 3%, see Figure 2), that all of the 16 tags get notified of 1 tag change.

The dissemination of the new value of the tag takes time. It is possible that, at some point in time, the new value of a tag is spreading among elements of the system, while its old value is still spreading among other elements. Thus, it may happen that a tag is notified of stale information. To evaluate this last phenomenon, each time an element e is notified of a value change in tag r, we

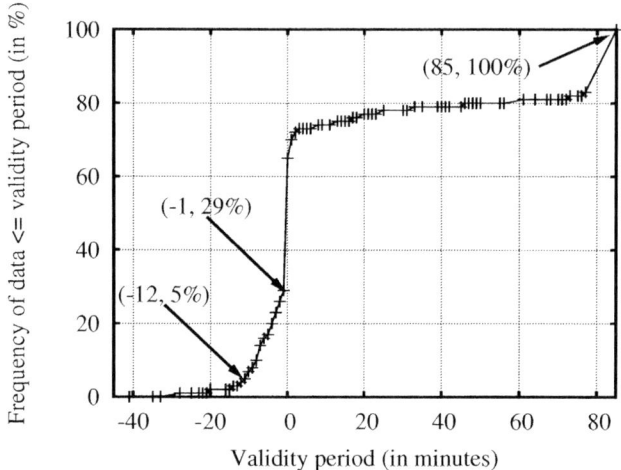

Fig. 4. Frequency of validity periods (T_{valid})

compare the date t_e of notification with the date t_r of next value change of tag r. Let's define the validity period as $T_{valid} = t_r - t_e$. If $T_{valid} \leq 0$, the value change of r which has been notified to e is stale. If $T_{valid} > 0$, the value change of r is an up-to-date information. The results are summarized in figure 4[2]. In $FrequencyOfValidityPeriods\{T_{valid} \leq -1\} = 29\%$ of the cases, we disseminate stale information.

Figure 5 evaluates the impact of stale information dissemination on the perception of the global state by the mobile. To get it, for each Consult a tag event, we count the number of differences (i.e. stale tags) between DM_{mobile} and the real state. Figure 5 shows that DM_{mobile} contains no stale tag in $FrequencyOfAtMostNStaleTags\{n = 0\} = 6\%$ of the cases. In other words, DM_{mobile} contains at least one stale tag in $100 - 6 = 94\%$ of the cases. On the other hand, DM_{mobile} contains at least 5 (respectively 7) correct tags in 100% (respectively 99.8%) of the cases. In the context of our game, these percentages help understanding the difference of rate of correct hint results between the two kinds of results. Indeed first kind (respectively second kind) has a rate of $490/540 = 91\%$ (respectively $30/50 = 60\%$). This is because, with first kind of results, we consider only the value of one card among the 16 cards. As DM_{mobile} contains at least 12 correct tags in 71% of the cases, the probability that the selected card is correct is high. Whereas, with second type of results, we take a decision based on the value of all of the cards in DM_{mobile}. The probability of a correct answer can only be lower.

This section has presented the experimental results from five public game sessions. In summary, at least 3 tags were notified of a given tag change in 39%

---

[2] Validity period +85 has a special meaning: the virtual card stored in tag r at moment t_r never changed again until the end of game session (which can last at most 85 minutes).

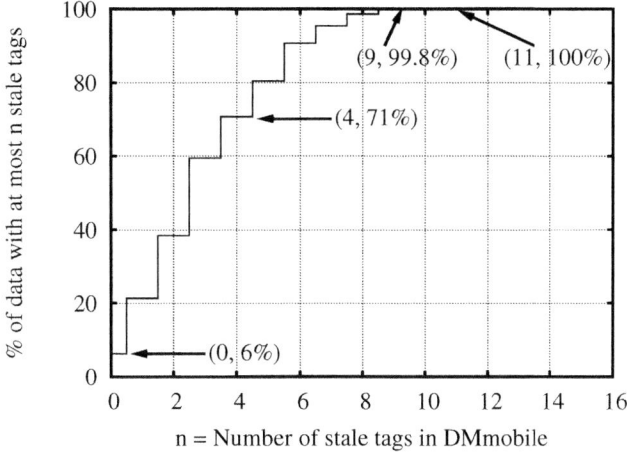

Fig. 5. Frequency (in %) of observation of at most n stale tags in DM_{mobile}

of the cases. The dissemination of this notification took at most 5 minutes in 50% of the cases. Moreover, in 29% of the cases, the notification received by a tag was stale. Despite this percentage, the view hold by each mobile had at most 4 stale tags in 71% of the cases. In addition to these "system" figures, we observed an "applicative" rate of hint success of 88%. This rate is acceptable by users in the context of our game (as the hint function is presented as a gambling function).

Next section analyzes our architecture with a more theoretical point of view.

6 Discussion

Section 2 presents how passive RFID tags are commonly used following 3 architectural patterns. In this section, we focus on scientific work related to our proposal. Then we analyze its limitations.

6.1 Related Work

As mentioned before, the work presented in this article relies on vector clocks theory [15,8]. Moreover, our usage of vector clocks can be considered as a gossip protocol. Indeed, our algorithms satisfy most of the conditions defined by [2] to consider a protocol as a gossip protocol. For instance, when a handset interacts with a tag or another handset, the state of both changes in a way which reflects the other. The style of gossip protocol implemented in this paper is an anti-entropy protocol for repairing replicated data [7]. The data is the contents of the distributed memory. It is replicated on the handsets. Tags correspond to the network: when a handset interacts with a tag r, it leaves a message on the tag. The message contains information about what the handset has seen on the other tags and what information it has received from the other handsets. This message will be received by the next handset to interact with this tag.

This work can also be considered as an implementation of opportunistic data flooding. Such flooding is implemented in several sensor networks. For instance, *ZebraNet* is a mobile, sensor network designed to collect data about zebras [14]. To do so, a sensor is attached to each animal. It collects data locally and transmits them whenever another sensor comes within range. Periodic zoologists drive-bys can then collect logged data from many animals despite encountering relatively few within range. Our contribution is to integrate passive tags into such peer-to-peer architecture. Of course, a passive tag cannot be used to collect data locally. But it can hold information without any energy saving issue. This offers two interesting opportunities. First, the tag can be used as a data cache. A moving sensor can store data to be transmitted to another moving sensor which will come within the tag in the future. The second opportunity is that the tag can give a networking capability to any moving entity. Suppose zebras are equiped with passive tags. When moving from one static sensor to another one, these zebras may transmit information via their passive tag. They become "network" elements.

[6] presents an RFID-based distributed memory which does not require any WAN. This distributed memory is illustrated through two applications: *Ubi-Check* and *Roboswarm*. In *Ubi-Check*, an RFID tag is attached to each of the traveler's items. Each tag is initialized with a value specific to the traveler. All of these RFID tags are read after special points (*e.g.*, after an airport security control). Their values are transmitted to an application which checks that they are consistent. If it is not the case, it means that, at some point, the traveler exchanged one of their items with the item of another traveler. An alarm is thus triggered to warn the traveler that one of their items is missing. In *Roboswarm*, RFID tags are placed throughout a physical space to give direction information to plain robots wandering in this space. These tags are initialized by dedicated robots before the standard robots are run. The major difference with our proposal is that, in *Ubi-Check* and *Roboswarm*, distributed memory data cannot be modified any more once the initialization process is over. In other words, [6] introduces an RFID-based distributed ROM which can be "flashed" (initialized), whereas we introduce an RFID-based distributed RAM.

6.2 Analysis of the Limitations

The architecture proposed in this paper has limitations concerning staleness and scalability. They are discussed in the following.

Staleness. For any element e of the system, DM_e may contain stale data of three types.

Firstly, there may be a tag $r, r \neq e$ for which $DM_e[r]$ does not correspond to $DM_r[r]$ currently hold by tag r. This happened in 94% of the cases in our PSM experiments (see Figure 5).

Secondly, our architecture experiences the problem inherent to vector clocks in distributed systems [15,8]: element e has no guarantee that DM_e corresponds to a value of DM which did exist at some point in time.

Thirdly, element e has no guarantee that it sees the whole history of changes of $DM_r[r]$ (for any tag r of the system). Suppose that, at time t_1, a mobile handset m_1 sets the value of $DM_r[r]$ of tag r to v_{t_1}. When applying Algorithm 1, $VC_{m_1} \leftarrow t_1$. Then, at time t_2, a mobile handset m_2 sets $DM_r[r]$ to v_{t_2}. Upon application of Algorithm 1, $VC_{m_2} \leftarrow t_2$. Afterwards, if m_2 comes to tag r' before m_1, when m_2 applies Algorithm 2, we have $DM_{r'}[r] \leftarrow v_{t_2}$ and $VC_{r'}[r] \leftarrow t_2$. When m_1 comes to tag r', because $VC_{m_1}[r] < VC_{r'}[r]$, $DM_{r'}[r]$ is never set to the value v_{t_1}.

In the following, we focus on the first type of stale data.

The rate of stale data and the period during which data remains stale is application dependant. How often do data evolve? How often are tags visited by application users? How long do users take to go from one tag to another?... In the context of PSM, the highest rate of stale data we observed is $11/16 = 69\%$ (see Figure 5). Concerning the period, 5% of the data remained stale more than 12 minutes (see Figure 4).

If the rate and/or period are too high for the application, three methods can be considered to reduce them. The two first ones are stimulating the "network". The third one introduces a WAN.

The first method consists in having a dedicated user who goes periodically through all of the tags just to read their contents and more important to update their DM_{tag} and VC_{tag}. By doing so, we reduce the risks of tags not being refreshed for a long time. In the case of PSM, this method would not be fruitful. Such a dedicated used would take 12 minutes to go through all of the tags. Thus this method would take off only $FrequencyOfValidityPeriods\{T_{valid} \leq -12\} = 5\%$ of negative dissemination (see Figure 4), reducing the total number of negative dissemination by only $5\%/29\% = 17\%$, which is not worthwhile the effort.

Another method consists in periodically asking all of the users to meet all together to synchronize their DM_{mobile}. This method introduces a constraint on users who may not accept it. In the context of PSM, it would not be applicable: it would reduce too much the fun during the game.

The final method consists in a hybrid approach. Each mobile applies Algorithms 1 and 2. But, sometimes, under some circumstances which need to be defined and which may be application dependant, it connects to a server through a WAN. By applying Algorithm 2, it synchronizes DM_{mobile} and DM_{server}. This approach introduces a WAN (and the cost of data plans). But it guarantees a moderate use of the WAN (these costs would be restrained). Because it requires a WAN, this method may not be always applicable.

In this section, we have studied the staleness limitations of our proposal. But how do architectures presented in Section 2 behave concerning this limitation? First architecture (Centralized) and third one (Data stored only on tags) do not experience any staleness issue. But the former one requires a WAN. And, with the latter one, it is impossible to have an idea of the contents of a distant tag. The second architecture (Local copy of a central database) has worse staleness limitations than our architecture. Each time a user modifies the contents associated with a tag, the other users have an additional stale data in their local database. In particular, a user reading a tag already modified by another user will not see these modifications (whereas user does see them in the case of our proposal). A synchronization with the central database is mandatory to cleanup any stale data (whereas Algorithm 2 cleans up some of the stale data in the case of our proposal).

Scalability. The number of data which can be managed by our architecture is limited by the hardware used.

First it takes time to read DM_r and VC_r from the tag to the handset, process Algorithm 2, and write back DM_r and VC_r to the tag. This time must be lower than an application dependant threshold TH. We suppose the processing time is negligible compared to the read and write time. Let TR be the transmission rate between the tag and the mobile, L the total length of one element of DM_r and one element of VC_r, and N the number of tag values which can be stored in DM_r. Then N is constrained by the following inequality: $N \leq \frac{TH \cdot TR}{2 \cdot L}$.

Moreover, the hardware has limited size. Let S be the size of the RAM in bytes. N is also constrained by: $N \leq \frac{S}{L}$.

We conclude:
$$N \leq \min(\frac{TH \cdot TR}{2 \cdot L}, \frac{S}{L}) . \qquad (1)$$

Inequality 1 quantifies the influence of the optimizations presented in Section 3 on the upper bound of N.

In the context of PSM, we have $TH = 0.5$ seconds[3], $TR = 106$ kbps (we use plain ISO 14443 tags), $L = 3$ bytes, and $S = 1$ Kbyte. By applying Inequality 1, we get $N \leq 341$.

It is not possible to apply our architecture if there are more than a hundreds of tags. This is all right for some applications such as our game. But we believe it does not fit most applications, and in particular the ones classically addressed by the centralized architecture (which is able to handle thousands or millions of tags).

Addressing this scalability issue is one of our research perspectives.

7 Conclusion and Research Perspectives

The goal of our work is to give a user equipped with an RFID-enabled mobile handset (mobile phone, PDA, laptop...) the ability to know the contents of distant passive RFID tags, without physically moving to them and without using a WAN (which leads to installation and/or operational costs).

---

[3] During our experiment, we observed that players were getting in a hurry at the end of a session. At that moment, they were spending their remaining time walking from tag to tag, putting their phone on the tag and immediately taking it off (in half of a second).

This paper studies how standard RFID architectures do not meet our requirements. Then it presents how we associate vector clocks to RFID tags to get an RFID-based distributed memory. This association may seem peculiar: we associate a vector clock (which is usually associated to active system elements like processors) to a piece of memory (which is intrinsically passive). This association is actually fruitful in two ways. On one hand, tags and mobile handsets can now be used to propagate information concerning the evolution of the RFID-based distributed memory; thus handsets are able to read the (cached) value of any piece of this distributed memory without being physically near the tag hosting that particular piece. On the other hand, this association turns users into network elements responsible to carry information between the tags; in other words, we create a Wireless Human-based Area Network (WHAN) playing the role of a WAN, but without installation and operational costs of a WAN. The tests we have made on an application example show that hints based on the analysis of the local view of this distributed memory are correct in 88% of the cases. The acceptability of this rate is application dependant. It is acceptable in the context of our game.

This architecture faces two issues. The first one concerns staleness. In 94% of the cases in our experiments, the view a mobile has of the distributed memory contains stale data. On the other hand, the highest rate of stale data we observed is 69%. The second issue concerns scalability. For instance, in our application, we could handle at most 341 tags. Some applications can cope with both of these limitations. It is indeed the case of our game. But the majority of applications cannot: these issues require further research.

These are several research perspectives related to this work.

From a usage point of view, [9] analyzes some of the consequences of this WHAN on human relationships between application users. This study must be deepened.

From a technical point of view, the two first perspectives concern the issues identified previously: staleness and scalability.

A third perspective is related to the generalization of our work to a system containing any kind of processing devices and a distributed memory hosted not only by RFID tags, but also by the memory of some or all of the mobile handsets. There are three *sine qua non* conditions for this generalization. First, an element of this distributed memory can only be modified in one place (the element is the property of one specific RFID tag or processing device). Second, the set of RFID tags and devices contributing to the system is defined at the beginning of the system lifetime. Third, this set is ordered so that it is possible to associate one vector clock element to each system element. A perspective of our work is to see how it is possible to release one (or more) of these three conditions. Possible answers can be found within the Optimistic Replication community [17]. But the limited size of the RFID tags' memory requires at least adapting existing answers or bringing new ones.

Acknowledgments. The author wishes to thank Eric Gressier-Soudan, Denis Conan, Annie Gentès, Aude Guyot-Mbodji, Camille Jutant and all of the reviewers, for all of their fruitful comments during the writing of this article.

References

1. Armenio, F., Barthel, H., Burstein, L., Dietrich, P., Duker, J., Garrett, J., Hogan, B., Ryaboy, O., Sarma, S., Schmidt, J., Suen, K., Traub, K., Williams, J.: The EPCglobal architecture framework. Technical Report Version 1.2, GS1 EPCglobal (September 2007)
2. Birman, K.: The promise, and limitations, of gossip protocols. SIGOPS Oper. Syst. Rev. 41(5), 8–13 (2007)
3. CÉDRIC, L3i, Musée des arts et métiers, NET Innovations, Orange, Institut Télécom—Télécom & Management SudParis, Institut Télécom—Télécom Paris-Tech., and Tetraedge. PLUG: PLay Ubiquitous Games and play more (January 2009), http://cedric.cnam.fr/PLUG/
4. Chandy, K.M., Lamport, L.: Distributed snapshots: determining global states of distributed systems. ACM Trans. Comput. Syst. 3(1), 63–75 (1985)
5. Christian, D.: 700 million of users of NFC mobiles in 5 years (September 2008) (in French),
http://www.generation-nt.com/juniper-etude-technologie-nfc-mobile-utilisateurs-actualite-151831.html
6. Couderc, P., Banâtre, M.: Beyond RFID: The Ubiquitous Near-Field Distributed Memory. ERCIM news (76), 35–36 (2009)
7. Demers, A., Greene, D., Hauser, C., Irish, W., Larson, J., Shenker, S., Sturgis, H., Swinehart, D., Terry, D.: Epidemic algorithms for replicated database maintenance. In: PODC 1987: Proceedings of the sixth annual ACM Symposium on Principles of distributed computing, pp. 1–12. ACM, New York (1987)
8. Fidge, C.J.: Timestamps in message-passing systems that preserve the partial ordering. In: Raymond, K. (ed.) Proc. of the 11th Australian Computer Science Conference (ACSC 1988), February 1988, pp. 56–66 (1988)
9. Gentes, A., Jutant, C., Guyot, A., Simatic, M.: RFID technology: Fostering human interactions. In: Blashki, K. (ed.) Proceedings of IADIS International Conference Game and Entertainment Technologies 2009, International Association for Development of the Information Society (IADIS), June 2009, pp. 67–74. IADIS Press (2009)
10. Ghellal, S., Holopainen, J., Honkakorpi, M., Waern, A.: Deliverable D4.7: Final business guidelines. Technical Report D4.7, Integrated Project on Pervasive Gaming (IPerG) (April 2008)
11. Golding, R.A.: Weak-consistency group communication and membership. PhD thesis, University of California Santa Cruz (December 1992)
12. Haselsteiner, E., Breitfuß, K.: Security in near field communication (NFC). In: Printed handout of Workshop on RFID Security RFIDSec 2006 (July 2006)
13. ITR Manager.com. City of Paris is taking care of its trees with RFID tags (December 2006) (in French), http://www.itrmanager.com/articles/59758/59758.html
14. Juang, P., Oki, H., Wang, Y., Martonosi, M., Peh, L.S., Rubenstein, D.: Energy-efficient computing for wildlife tracking: design tradeoffs and early experiences with zebranet. In: ASPLOS-X: Proceedings of the 10th international conference on Architectural support for programming languages and operating systems, pp. 96–107. ACM, New York (2002)

15. Mattern, F.: Virtual time and global states of distributed systems. In: Proc. Workshop on Parallel and Distributed Algorithms, Chateau de Bonas, France, pp. 215–226. Elsevier, Amsterdam (1988)
16. Roussos, G., Kostakos, V.: RFID in pervasive computing: State-of-the-art and outlook. Pervasive Mob. Comput. 5(1), 110–131 (2009)
17. Saito, Y., Shapiro, M.: Optimistic replication. ACM Comput. Surv. 37(1), 42–81 (2005)
18. Simatic, M., Astic, I., Aunis, C., Gentes, A., Guyot-Mbodji, A., Jutant, C., Zaza, E.: "Plug: Secrets of the Museum": A pervasive game taking place in a museum. In: Natkin, S., Dupire, J. (eds.) Entertainment Computing – ICEC 2009. LNCS, vol. 5709, pp. 302–303. Springer, Heidelberg (2009)

A Mobile Application to Detect Abnormal Patterns of Activity

Omar Abdul Baki, Joy Zhang, Martin Griss, and Tony Lin

Carnegie Mellon Silicon Valley,
NASA Research Park, Bldg. 23, Moffet Field, CA 94035
{omar.abdulbaki,joy.zhang,martin.griss,tony.lin}@sv.cmu.edu

Abstract. In this paper we introduce an unsupervised online clustering algorithm to detect abnormal activities using mobile devices. This algorithm constantly monitors a user's daily routine and builds his/her personal behavior model through online clustering. When the system observes activities that do not belong to any known normal activities, it immediately generates alert signals so that incidents can be handled in time. In the proposed algorithm, activities are characterized by users' postures, movements, and their indoor location. Experimental results show that the behavior models are indeed user-specific. Our current system achieves 90% precision and 40% recall for anomalous activity detection.

Keywords: activity monitoring; context-aware; abnormality detection; unsupervised learning; online clustering; senior care.

1 Introduction

As the baby boomer generation ages, the US census predicts a dramatic increase in persons older than 65 from 44 million in 2005 to 74 million by 2020 [1]. To cope with the massive care needs of an aging population, there must be a shift in the paradigm of care in the US. On the one hand, as people age, they would like to maintain a certain degree of independence to maintain their quality of life [2]. On the other hand, in order to allow for quality care for each person, care professionals must be able to keep track of multiple patients' daily life routines and abnormalities. Abnormal activities such as falling often result in severe medical situations. In fact, accidental falls caused more deaths than all other medical situations among seniors [3].

Existing activity monitoring systems are costly and invasive. Some solutions use wearable devices such as bracelets, pendants, or devices mounted on the waist to detect falls and monitor vital signs. Other solutions use sensors such as cameras, motion sensors, or infrared cameras to monitor vital signs, detect falls, and even recognize certain high level activities.

Most existing monitoring systems are designed to detect only certain classes of events. For example a device which detects falls may not be able to detect if an individual is still in bed at an unusual time. As a result these other types of accidents or problems may end up undetected.

In this paper, we describe the Mobile Activity Monitoring System (MAMS), a novel approach to home care monitoring based on detecting anomalous activities using mobile devices. We assume that accidents and other critical events are "abnormal" compared to a user's "normal" activities. We learn each user's normal patterns of activities and flag anomalous behaviors which are significantly different from a user's daily routine. Using a mobile device, we build a user's behavior model. Behavior is characterized by a user's posture, movement, and location. For example, an activity can be represented as "climbing" on a "stairway". In this paper, we use the built-in tri-axial accelerometer in the Nokia N95 mobile phone to infer user posture and movement, and Wi-Fi based positioning to estimate a user's indoor location. These elements are used as features for the clustering K-nearest neighbor classifier.

2 Unsupervised Anomaly Detection

Most activity recognition systems are supervised. They are trained using a set of labeled activity examples. Supervised methods have three problems in activity monitoring. Firstly, creating the labeled training data is time-consuming. This is especially the case for applications such as monitoring the elderly. It is impossible to enumerate all activities and collect data for each activity, especially for abnormal activities such as "falling down a flight of stairs."

Secondly, supervised learning solutions are limited to detecting activities defined in the training data set. Activities that are not defined in the training data will not be detected.

Thirdly, activities are user-dependent and context-dependent. People have different life styles and different daily routines; thus an activity considered to be normal for one individual may not necessarily be so for another person. Activities also depend on context. "Falling down onto a bed" could be normal but "falling down in the bathroom" should flag an alert message. Context information such as location, movement and posture are user-dependent information and can not be applied directly to other users. Thus, user-independent behavior models obtained from supervised training might not generalize well for all users.

K-nearest neighbor clustering provides for scalable activity monitoring since no labeling is required for any of the data points collected. This allows MAMS to collect data at a higher rate and larger volume than any supervised algorithm. It also achieves our desired goal of creating a personalized system capable of learning user-specific activities. More specifically users' behavior models will be completely trained using their own training data. Finally, it supports incremental learning, which we believe is essential for any usable system for activity monitoring. This is because any practical activity monitoring system should be allowed to adapt to slow changes in patterns of activity, brought about by gradual changes in the environment or activities performed.

2.1 Online Clustering Classifier

2.1.1 Learning

To avoid problems in supervised systems, we use an unsupervised clustering algorithm to detect abnormal activities in MAMS. As illustrated in Figure 1, each activity is represented by a data point and activities are grouped into *clusters* based on a *distance* measure. The *centroid* of a cluster represents an average data point amongst all member data points in the cluster. In MAMS, clusters are not labeled because of the unsupervised nature of the learning algorithm.

MAMS constantly monitors a user's activity, as new data points are generated continuously from the sensors. For each new data point, MAMS calculates its distance to each centroid of the existing clusters. If the distance to the closest centroid is smaller than a predefined threshold, then the data point is clustered and merged into this nearest cluster. Otherwise, this activity is considered novel with respect to the current observed activities. MAMS checks with the user whether this novel activity is abnormal. If the user indicates this is actually a normal activity, MAMS forms a new cluster for this activity. Otherwise, if the user confirms something abnormal happened, or if the user fails to respond within a reasonable time interval, MAMS will then quickly contact the home care giver or emergency responder to look into the situation (see Figure 2).

In our implementation we only keep track of the cluster centroid in order to optimize memory usage. In other words, once a new data point is added to a cluster, the cluster centroid is recalculated to incorporate the data point, and the data point is then discarded. Figure 2 illustrates this process in greater detail.

To allow for scalable and continuous learning, the system sets an upper limit for the number of data points stored. Without this bounded system size, real time clustering may become impossible if the number of clusters becomes too large. Therefore, to maintain this mass, a data point is removed from each of the systems k clusters for every k data points added after the system has reached its specified critical mass. In this fading process, clusters are discarded if their mass eventually fades completely. This effectively means that normal behavior becomes increasingly abnormal if it becomes a less recurrent pattern of activity. On the other hand, an abnormal activity can gradually become normal as its associated cluster grows with more and more observed data points.

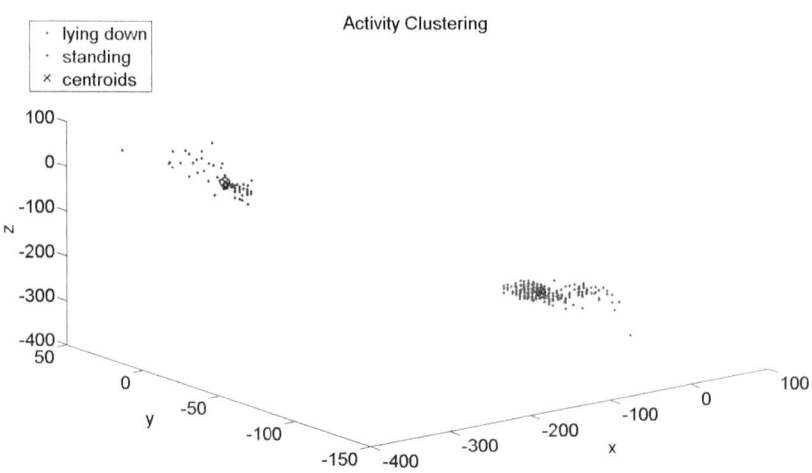

Fig. 1. Similar data points form dense regions or clusters

```
READ data_point a
FOR each existing cluster c_i:
    IF distance (a, c_i ) < min_distance
        min_distance = distance(a, c_i )
        min_cluster = cluster
IF min_distance < clustering_threshold:
    ADD data_point to min_cluster
    RECOMPUTE min_cluster.centroid
ELSE:
    CHECK with user if a is abnormal
    IF 'abnormal' or 'no user response'
        Contact home care worker or emergency
        responders;
    ELSE
        Create new_cluster for a
        Append new_cluster to cluster_array
```

Fig. 2. Pseudo-code for the online clustering process in MAMS

2.2 Activity Feature Vector

The MAMS application considers three characteristics of an activity: posture, movement, and location. Each activity data point is represented by a vector of nine feature values (see Figure 4).

Posture: We compute the mean values of the readings of the three accelerometer axes over a window of 1.8 seconds. In the absence of motion, the mean component is essentially the constant forces of gravity on each of the accelerometer axes.

Movement: We compute the variance of accelerometer axes readings over the same window of 1.8 seconds and relate the variance to the user's current movement.

Location: We use the Wi-Fi based positioning algorithm Redpin [4] to locate a user's indoor position. Redpin uses Wi-Fi fingerprinting to output a user's room level location, e.g. living room, master bedroom, bathroom etc. Fingerprinting is based on the assumption that the Wi-Fi RSS fingerprints seen over time at one location are reasonably stable. Given a database of labeled RSS fingerprints, Redpin uses the K-nearest neighbor (KNN) algorithm to predict the most plausible location of the user. It outputs a single room as the user's current location.

One of the shortcomings of Redpin is that it only outputs a single room where the user might be located. Therefore when computing the distance between two locations, the distance can take on only one of two values: 1 if two locations are from the same room or 0 otherwise. To make the location range smoother, we develop a *pseudo continuous location system* for indoor location. In this system, we consider a user's location as a unit vector in n-dimensional space, where each dimension corresponds to a room in which the user can be. Since this vector is a unit vector, each room component represents the probability of the user being in that room. Figure 3 illustrates the three dimensional case.

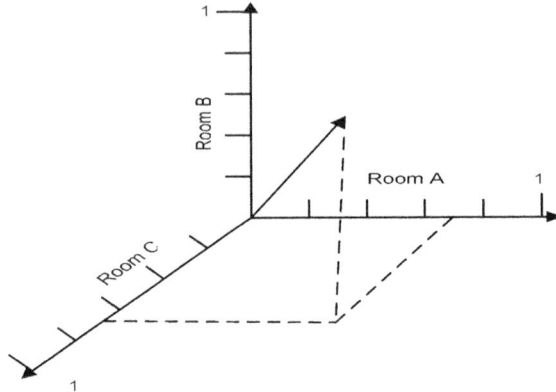

Fig. 3. Room-wise probabilities for three different rooms map a user's location to a point in 3-D space

Accelerometer Mean	Accelerometer Variance	Indoor Location
X = 150	X = 12,150	Room A: 95 %
Y = 120	Y = 5,010	Room B: 3%
Z = 130	Z = 9,800	Room C: 2%

Fig. 4. A sample activity feature vector

2.3 Distance Measurement

As each data point is generated, we attempt to assign it to the closest existing cluster. Closeness is computed as weighted distance of the posture, movement, and location components. For a new data point, a, the distance to a cluster centroid, c, is calculated using equation 1 below. Here, the net distance is calculated to be the weighted sum of the distances between a and c for each of the three feature components. The scaling factors assigned to each of the components (α, β, δ) are computed such that, on average, all components contribute equally to the distance measure.

$$\text{dist}(a,c) = \alpha \times \text{posture\_dist} + \beta \times \text{movement\_dist} + \delta \times \text{location\_dist}. \qquad (1)$$

The posture and movement distances between a and c are calculated as the Euclidean distances between their accelerometer axes mean and variance vectors respectively. To compute the location distance between the two points, we compute the Euclidean distance between their n-dimensional unit vectors of room-wise probabilities. Equations 2, 3 and 4 capture the formulas for computing the distance between two points for each of the three feature components: posture, movement and location.

$$\text{posture\_dist}(a,c) = \sqrt{\begin{array}{l}(\text{mean}(x_c) - \text{mean}(x_a))^2 + \\ (\text{mean}(y_c) - \text{mean}(y_a))^2 + \\ (\text{mean}(z_c) - \text{mean}(z_a))^2\end{array}} \qquad (2)$$

$$\text{movement\_dist}(a,c) = \sqrt{\begin{array}{l}(\text{variance}(x_c) - \text{variance}(x_a))^2 + \\ (\text{variance}(y_c) - \text{variance}(y_a))^2 + \\ (\text{variance}(z_c) - \text{variance}(z_a))^2\end{array}} \qquad (3)$$

$$\text{location\_dist}(a,c) = \sqrt{\sum_{\exists \text{room in building}} (\text{prob}_c(\text{room}) - \text{prob}_a(\text{room}))^2} \qquad (4)$$

In our study we assume that users holster or place their mobile device in their pockets in a consistent manner. In other words, when the phone is not in use, we assume its orientation with respect to the test subject to be non-changing.

3 Experiments

We test MAMS on 5 test subjects. Test subjects are full time graduate students studying and working in the same building. All students attend the same courses, and for the most part spend their time studying in their cubicles which are all located in the same room.

During the experiments, each test subject was asked to carry a Nokia N95 mobile phone for three days while a logger collected sensor data from the Wi-Fi and accelerometer sensors on each phone. Participants were encouraged to keep the phone with them at all times and to continue using it as they normally would. In other words, they were allowed to continue using the phone for making and receiving phone calls and other functions.

To ensure enough data are collected before the monitoring starts, monitoring only begins after the mass of the system has reached a specified critical level. After this point, both learning and monitoring continue simultaneously. In other words, MAMS does not check with the user if any abnormal activities occur at the beginning of the experiment.

3.1 Anomaly Detection Accuracy

In order to estimate the accuracy of MAMS in detecting normal/abnormal activities, we asked test subjects to perform 5 *abnormal* activities. These activities test various permutations of a user being in an abnormal location, assuming an abnormal posture, or moving around in an abnormal manner. Table 1 shows the list of abnormal activities participants performed and the abnormalities associated with each activity.

In this experiment a mobile application was installed on each subject's mobile phone. This application requests test subjects to perform a series of tasks. At random times, the device plays a distinct beep sound to instruct the test subject to start performing the abnormal task. Tasks were 10 seconds in length, and during this time interval, the subjects were expected to continue performing the abnormal activity. The purpose of this was to clearly mark the beginning and end of an abnormal task for evaluation.

Table 1. Abnormal Activities

	Abnormal Location	Abnormal Movement	Abnormal Posture
Jumping Jacks Student Cubicle		✓	✓
Lying Down Next to Student Cubicle			✓
Seated in Conference Room	✓		
Lying down in conference room	✓		✓
Jumping Jacks Conference Room	✓	✓	✓

We study how the system's precision and recall vary with respect to two parameters in MAMS: the clustering threshold and the behavior model size. We define precision and recall as follows.

$$\text{precision} = \frac{\text{Number of labeled abnormal activities correctly classified by MAMS}}{\text{Number of abnormal activities classified by MAMS}} \quad (5)$$

$$\text{recall} = \frac{\text{Number of labeled abnormal activities correctly classified by MAMS}}{\text{Total number of labeled abnormal activities}} \quad (6)$$

The clustering threshold is the maximum allowable distance a data point can be from its closest cluster centroid for it to be successfully assigned to that cluster. Data points further than this threshold are considered to pertain to novel activities. In the context of anomaly detection, novel activities are flagged as abnormal ones. Figure 5 shows the precision, recall and F1 score (harmonic mean of precision and recall) as the clustering threshold is varied. In this experiment, we use 2 days of unlabeled normal activity logs to generate the behavior models. We test the system's precision and recall using the third day of normal activity logs for each user and their 5 labeled abnormal activity logs.

As shown in Figure 5, increasing the threshold in the very beginning increases the F1 score, since the precision value is increasing at a larger rate than the recall is decreasing. However, after a certain point, the rate of increase in precision is exceeded by the drop in recall as the threshold is further increased.

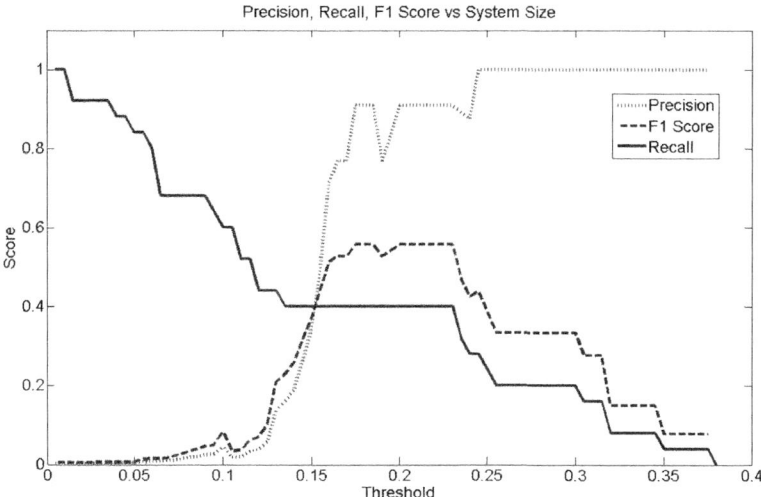

Fig. 5. Precision, recall, and F1 score vs. clustering threshold

Figure 6 illustrates the effects of increasing the model size on cases of reported novel activities. The behavior model size determines the number of historical data points kept in the MAMS system among all clusters. In Figure 6, model size is represented as log-hours, which is a measure of the time it takes the system to reach its critical mass. As the graph shows, at the early stage, MAMS reports a lot of "novel" activities since the behavior model has not covered enough normal activities yet. As the system mass increases, i.e. more user behaviors have been observed by the system, MAMS reports fewer activities to be "novel" or "abnormal". At around 4 log-hours, the system's sensitivity to abnormal data points levels off. Choosing a model

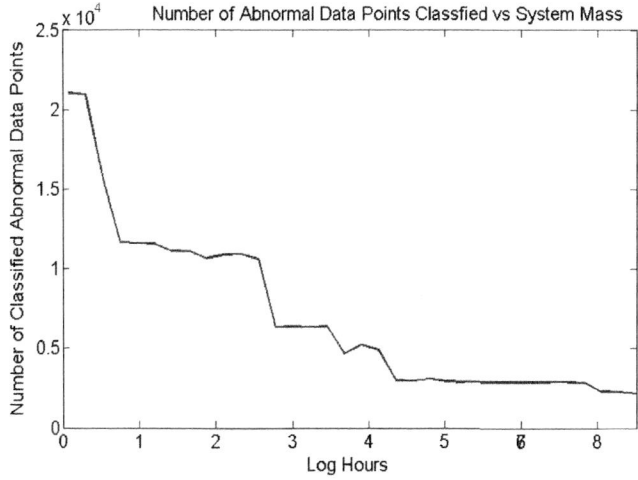

Fig. 6. As the system mass increases, the number of classified abnormal data points decreases

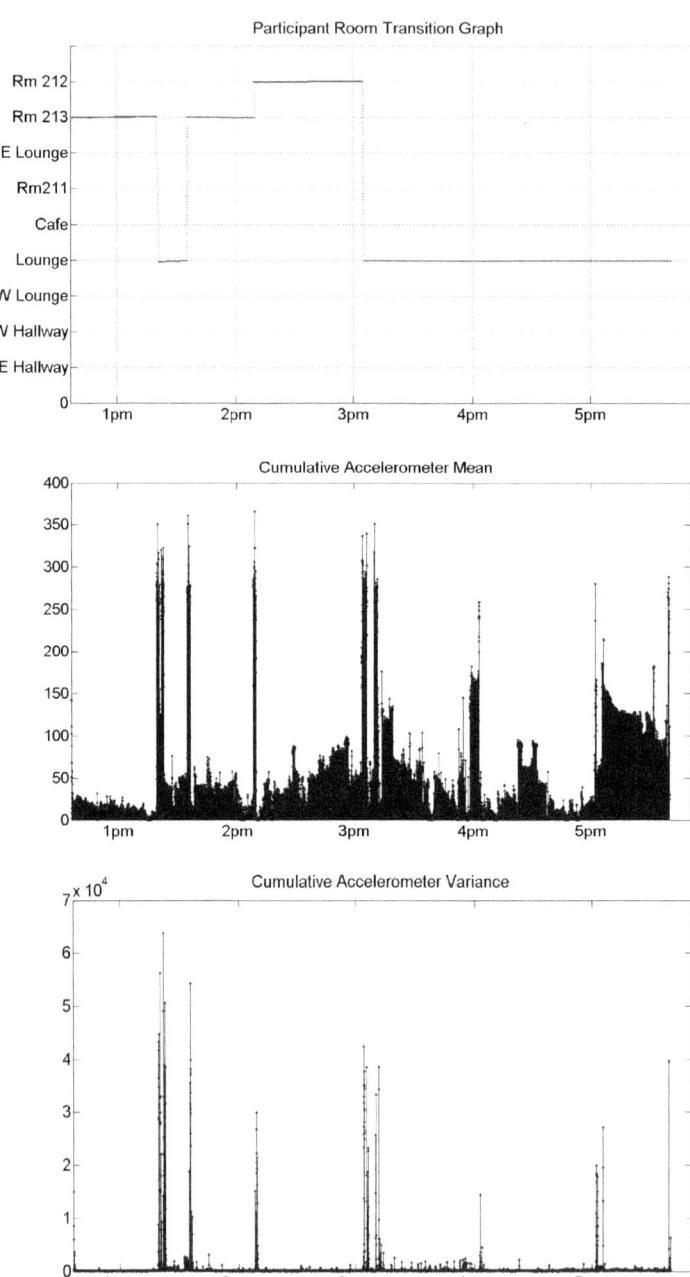

Fig. 7. Correlation among location, posture, and movement of a subject over a 5 hour period

size larger than 4 log-hours yields only marginal improvements in the decrease in false positives when identifying anomalous activities. Ultimately, choosing an appropriate model size is a trade-off between the performance constraints of the mobile device, and the level of false positives. Further experiments will be needed to determine the real-time performance of the system as the model size is increased.

Choosing a model size of 4 hours and the optimal clustering threshold, MAMS achieves 90% precision and 40% recall.

The three graphs in Figure 7 show the correlation between a user's location, posture and movement. The room location is indicated in the uppermost room transition graph. Also shown are figures illustrating the sum of the means and variance values of the three accelerometer axes. The results show that transitions between rooms bring about temporally aligned abrupt changes in the accelerometer mean and variance. However, while the user is within a room, both the accelerometer mean and variance fluctuate very little. These small fluctuations may be due either to jitter in the axes readings or to very small movements subjects make while seated or standing. This result supports our hypothesis that activities are location-dependent.

We find that removing the location feature has little effect on the system performance in our experiments (Table 2). This may be due to the fact that student participants spend most of their time seated in classrooms and at their cubicles. Future experiments may be needed to test the performance of the location feature on elderly subjects in their home environments.

Table 2. MAMS performance with the location feature enabled and disabled

Active MAMS Features	Precision	Recall	F1
Movement, Posture, Location	0.901	0.400	0.556
Movement, Posture	0.909	0.400	0.556

3.2 Cross-User Model Portability

MAMS learns behavior models for each individual user by monitoring his/her activities. We conducted experiments to test how well one user's behavior model works for other users, i.e., the portability of behavior model across different users.

Table 3 shows the results of cross-user model portability experiments. Each row in Table 3 represents the behavior model of a particular user and the columns are test subjects. Behavior models are learned from normal activity data collected over a period of 2 days. We use normal activity data from a third day as testing data. The entries in each row represent the percentage of activities which were recognized as repeated activities. With the exception of one user (user 4), MAMS performs best when the behavior model is trained from the same user to which it is being applied. These results imply that users tend to exhibit repeated patterns of activities which are unique from one person to another, and this underpins our approach of learning user-specific activities for activity recognition.

Table 3. Normal activity recall rate for behavior models applied to different users

	User1	User2	User3	User4	User5
User1 Cluster	**0.537**	0.336	0.030	0.489	0.142
User2 Cluster	0.648	**0.710**	0.229	0.463	0.229
User3 Cluster	0.422	0.390	**0.631**	0.356	0.562
User4 Cluster	0.204	0.270	0.033	0.350	**0.672**
User5 Cluster	0.171	0.124	0.000	0.110	**0.642**

4 Related Work

Much research into context and activity classification using accelerometers focuses on learning a predefined set of activities. Fabian et al. make use of wearable devices consisting of a gyroscope and a tri-axial accelerometer [5]. Three of these devices are placed on a subject's hip, dominant ankle and wrist. Using intensity values from the accelerometer axes as features, and feed forward neural networks, they report a 79.76% recognition rate for a set of 7 activities. Bao and Intille, experiment with various classifiers to classify various static and dynamic activities from accelerometer data [6]. They conclude that decision trees, which yielded 89.3% accuracy rates, perform the best. Similar studies with labeled activity recognition using accelerometer data also report accuracies in the 80%-90% range [7][8][9]. Although these approaches typically perform well for a small set of activities, they generally don't span the entire set of activities a person performs. Hence they are of limited use in real-time monitoring applications.

Studies have shown that unsupervised learning can indeed be applied to activity recognition with limited success. Krause et al. present an online learning algorithm which uses self-organizing maps and clustering to identify a set of unlabelled activities [10]. Transient or infrequent activities are filtered out to leave a set of activities which are considered to be the most recurring ones. The disadvantage of this scheme is that the number of contexts or activities generated is less than the number of activities participants identified. Hein and Kirste present a hybrid k-means clustering and Hidden Markov Model (HMM) approach to detect high-level activities [11]. Applying their classifier to data from a tri-axial accelerometer, gyroscope, and magnetometer, they report 75% recognition accuracy for high level unlabeled activities. Nguyen et al. present two unsupervised event segmentation techniques: coherent event clustering and unusual event clustering [12]. Coherent event clustering iteratively segments the raw accelerometer data into sequences and coherently clusters the segments into event classes. In unusual event clustering, one starts with a usual event model and iteratively updates the usual event model based on outlier events identified by the current model. Both methods model the event/activity using HMM and use Gaussian Mixture Models to represent the emission between the state transitions. Such models are quite complex and non-trivial to train. The large number of parameters needs to be estimated by iterative Expectation Maximization training, making it difficult to use such algorithms on mobile devices especially when online training is desired.

5 Conclusion

We describe the MAMS system, an activity monitoring system that classifies abnormal activity based on deviations from learned normal behavior. We find that our system can recognize anomalous activities based on different permutations of novel location, posture, or movement. This system learns incrementally and adapts to slow changes in activity patterns. As a result, recurrent patterns of activity gradually become normal, while less recurrent ones become anomalous. On this basis, we believe this system can be used on mobile phones to monitor the homebound elderly. On a short time scale, it may be used to detect critical events such as falls, while on a longer time scale, the system can be used to track changes in quality of life based on increasing or decreasing levels of anomalous activities.

Our results show that generated activity clusters tend to be user specific. This strengthens our approach of generating user-specific models as opposed to using one model for all users. The system recognizes abnormal activities with 90% precision and detects 40% of abnormal activities. This low recall rate may be due to the small dataset size and the pilot nature of our study. In future studies we plan to test with a wider range of labeled abnormal activities and use more training days and users. Additionally, we will evaluate several other classifiers to find one which works best for the purpose of real-time anomalous activity detection.

Although our results show that the location feature appears to bear little effect on the performance of our anomaly classifier, we will re-evaluate the performance of this feature in a larger study.

Acknowledgements

This research was supported by a grant from Nokia Research center, by CyLab at Carnegie Mellon under grant DAAD19-02-1-0389 from the Army Research Office, and by Panasonic. The views and conclusions contained here are those of the authors and should not be interpreted as necessarily representing the official policies or endorsements, either express or implied, of ARO, CMU, or the U.S. Government or any of its agencies.

We also acknowledge the efforts of Patricia Collins, Pei Zhang and the anonymous referees for their valuable feedback and suggestions on earlier drafts of this paper.

References

1. Projections of the Population by Age and Sex for the United States: 2010 to 2050, http://www.census.gov/population/www/projections/summarytables.html
2. Western Maine Community Action, Keeping Seniors Home, http://www.wmca.org/Keeping_seniors_home.htm

3. Falls Among Older Adults: Summary of Research Findings, http://www.cdc.gov/ncipc/pub-res/toolkit/SummaryOfFalls.htm
4. Bolliger, P.: Redpin - adaptive, zero-configuration indoor localization through user collaboration. In: Proceedings of the first ACM international workshop on Mobile entity localization and tracking in GPS-less environments, pp. 55–60. ACM, New York (2008)
5. Fabian, A., Gyorbiro, N., Homanyi, G.: Activity recognition system for mobile phones using the MotionBand device. In: Proceedings of the 1st International Conference on MOBILe Wireless MiddleWARE, Operating Systems, and Applications, Article No. 41, ICST, Brussels Belgium (2008)
6. Bao, L., Intille, S.: Activity Recognition from User-Annotated Acceleration Data. In: Ferscha, A., Mattern, F. (eds.) PERVASIVE 2004. LNCS, vol. 3001, pp. 1–17. Springer, Heidelberg (2004)
7. Aminian, K., Robert, P., Buchser, E.E., Rutschmann, B., Hayoz, D., Depairon, M.: Physical activity monitoring based on accelerometry: validation and comparison with video observation. Identification of Common Molecular Subsequences 37(3), 304–308 (1999)
8. Mantyjarvi, J., Himberg, J., Seppanen, T.: Recognizing Human Motion with Multiple Acceleration Sensors. In: 2001 IEEE International Conference on Systems, Man, and Cybernetics, Tucson, AZ, vol. 2, pp. 747–752. IEEE, Los Alamitos (2001)
9. Randell, C., Muller, H.: Context Awareness by Analysing Accelerometer Data. In: The Fourth International Symposium on Wearable Computers, Atlanta, GA, pp. 175–176. IEEE, Los Alamitos (2000)
10. Krause, A., Sieworik, D., Smailagic, A., Farringdon, J.: Unsupervised, Dynamic Identification of Physiological and Activity Context in Wearable Computing. In: Proceedings of the 7th IEEE International Symposium on Wearable Computers, p. 88. IEEE Computer Society, Washington (2003)
11. Hein A., Kirste T.: Towards Recognizing Abstract Activities: An Unsupervised Approach. In: Proceedings of the 2nd Workshop on Behaviour Monitoring and Interpretation. pp 102-114. Universitat Bremen, Bremen Germany (2008).
12. Nguyen, A., Moore, D., McCowan, I.: Unsupervised Clustering of Free-Living Human Activities using Ambulatory Accelerometry. Engineering in Medicine and Biology Society 22, 4895–4898 (2007)

Energy-Efficient Localization via Personal Mobility Profiling

Ionut Constandache[1], Shravan Gaonkar[2], Matt Sayler[1], Romit Roy Choudhury[1], and Landon Cox[1]

[1] Duke University, Durham, NC, USA
{ionut,sayler,lpcox}@cs.duke.edu, romit@ee.duke.edu
[2] University of Illinois at Urbana Champaign, IL, USA
gaonkar@ieee.org

Abstract. Location based services are on the rise, many of which assume GPS based localization. Unfortunately, GPS incurs an unacceptable energy cost that can reduce the phone's battery life to less than ten hours. Alternate localization technology, based on WiFi or GSM, improve battery life at the expense of localization accuracy. This paper quantifies this important tradeoff that underlies a wide range of emerging applications. To address this tradeoff, we show that humans can be profiled based on their mobility patterns, and such profiles can be effective for location prediction. Prediction reduces the energy consumption due to continuous localization. Driven by measurements from Nokia N95 phones, we develop an energy-efficient localization framework called *EnLoc*. Evaluation on real user traces demonstrates the possibility of achieving good localization accuracy for a realistic energy budget.

1 Introduction

Mobile phones are becoming a powerful platform for sensing, sharing, and querying people-centric information. A variety of applications are on the rise, many of which utilize *location* to express the context of information. Examples include geotagging, geocasting, asset tracking, context-aware search, location-specific advertisements, and visualization [1,2,3]. Most of these location based applications (LBAs) assume GPS capabilities. While GPS offers good location accuracy of around 10m, it incurs a serious energy cost that can drain a fully charged phone battery in 8.5 hours [1]. We make a few observations in light of this energy-accuracy profile.

1. In real life, the phone battery must be shared with several other needs, including voice calls, emails, pictures, SMS, and an emergency reserve. The energy budget available for localization alone is a small fraction of the battery capacity. If this fraction is assumed to be 25%, *continuous* GPS localization will sustain for less than 2.5 hours.
2. One may argue that continuous GPS localization is not necessary – GPS can be activated only on demand. While this may suffice for some services

(such as geo-tagging a photo), a large number of emerging applications are demanding frequent location updates from users. For example, GeoLife [4] plans to track a phone and display shopping lists when the phone-user is near a grocery store. Micro-Blog [1] proposes to query users in a desired region for participatory sensing applications [3]. TrafficSense [5], Pothole Patrol [6] and Nericell [7] require phones to continuously report accelerometer readings to better estimate traffic/road conditions. All these applications rely on the feasibility of continuous localization over reasonably long time scales.

3. Continuous localization over long time scales results in higher *average* error. Given an energy budget of K GPS readings, and a duration of $T > K$ time units, $(T - K)$ time units cannot be assigned location readings. Locations can only be estimated at these unassigned times, and hence, are likely to be more erroneous than actual GPS readings. When averaged over all actual and estimated readings, the *average localization error* proves to be higher than the instantaneous error (around 10m for GPS). As the energy budget decreases, the difference between average and instantaneous error increases.

4. Alternate WiFi and GSM based localizations are not obvious replacements to GPS. While these schemes are less energy-hungry, they incur higher (instantaneous) localization error (around 40m and 400m respectively). This permits greater number of location readings, each of which is less accurate. Whether this results in lower *average error* than that of a few, but accurate, GPS readings, is an open question. Whether an optimal combination of GSP/WiFi/GSM based localization meets application-demands, is also an unexplored research direction.

This paper investigates the space of energy-efficient localization for mobile phones. We quantify the energy-accuracy tradeoff through measurements on Nokia N95 phones. We formulate a theoretical framework using dynamic programming (DP), that derives the optimal localization accuracy under a given energy budget. When results show that the theoretical optimal with basic GPS/WiFi/GSM technologies may not suffice for high-accuracy applications, we explore usefulness of predictions. Habitual movements of humans, and their similarity to statistical behavior of large populations, are potential opportunities. We incorporate predictions in our offline DP and design online heuristics for use on mobile phones. Performance results, comparing the optimal to the online heuristics, confirm the feasibility of achieving localization in the order of 25m, over a day's energy budget. Certain limitations will have to be addressed before our proposed system, EnLoc, can be deployed for public use – we discuss these limitations, and indicate directions of future work. Our main contributions can be summarized as follows.

Identifying the Space of Energy-Accuracy Tradeoff. We quantify the tradeoff with measurements on Nokia N95 phones.

Analysis of the Optimal Localization Accuracy for a Given Energy Budget. For a given mobility trace, an offline dynamic program computes the maximum accuracy achievable through GPS, WiFi, GSM, or combinations thereof.

Exploiting Habitual Activity of Individuals, and Behavior of Populations, to Predict Location. Predictions incorporated into the DP offer optimal solutions. Online heuristics permit energy-efficient localization in real time.

Evaluation of Heuristics in Real Life Situations. A wireless map of the UIUC campus is created through war-driving, and real mobility traces derived from student phones. Performance of heuristics is compared with the optimal using a custom trace-based simulator.

The rest of the paper expands on each of the contributions.

2 Motivation

We used the Nokia Energy Profiler [8] to measure fine-grained power consumption in Nokia N95 phones. With this monitor running in the background, we performed GPS, WiFi, and GSM based localization. This section reports the results recorded, and highlights key observations that motivate EnLoc. However, we first provide a brief background on WiFi and GSM localization techniques, originally proposed in [9].

Localization Using WiFi and GSM
As a benchmark, we measured localization error with on-phone GPS hardware, and found an average accuracy of 10m in clear-sky, outdoor conditions. GPS was mostly unavailable indoors. To circumvent GPS hardware and its problems indoors, project Place Lab [9,10] proposed interesting alternates to GPS localization. Specifically, authors create a wireless map of a region by war-driving in the area. The wireless map is composed of sampled GPS locations, WiFi access points and GSM towers audible at these locations, and their corresponding signal strengths. This wireless map is then distributed to phones. When a phone travels through the mapped area, it estimates its own location by matching its list of audible WiFi APs to the wireless map. Several matching schemes have been proposed, including *Centroid* Computation, *Signal Strength* based estimation, and *Fingerprinting*. Place Lab experiments in Seattle region exhibit a median positioning error of 13 to 40m with WiFi, and around 94 to 196m with GSM. When performed in Champaign, IL, and Durham, NC, the errors were higher. Specifically, WiFi accuracy ranged between 25 to 40m, while GSM ranged between 300 to 400m. The increase in GSM error was due to a smaller number of cells and using a single GSM provider.

Energy Measurements for GPS, WiFi, and GSM
We measured the energy consumption on Nokia N95 phones for each localization scheme. For this, we charged the phone to its full battery capacity, and turned off all applications and communication technologies. Then, we invoked the energy monitoring program and turned on only the location sensor we aimed to measure. Our program probed the location sensor at a chosen interval, T_{probe}, that varied from 30 to 300 seconds.

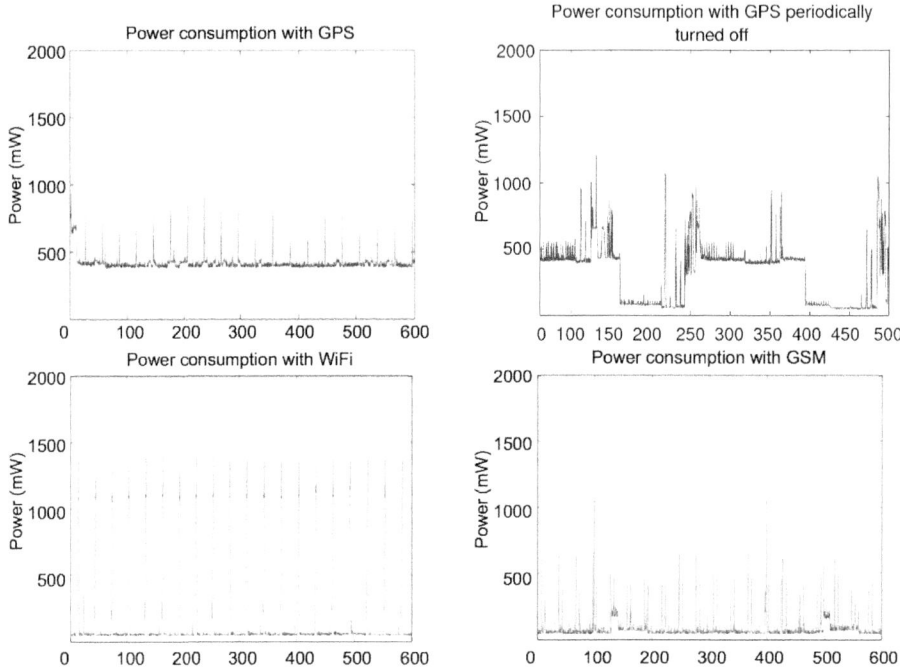

Fig. 1. Power consumption in mW: (a) GPS localization every 30 seconds. (b) GPS periodically turned off. (c) WiFi localization every 30 seconds. (d) GSM localization every 30 seconds.

Figure 1(a) shows GPS power consumptions as a function of time, for T_{probe} = 30 seconds. We see periodic spikes on top of a baseline power consumption at approximately 400 mW. The spikes correspond to the periodic probes which trigger a sensor read and a data write operation into the phone's file system. The baseline corresponds to the power consumed by the GPS receiver. To prove this, we periodically turned off all sensors, including GPS. Results in Figure 1(b) show a baseline power consumption of around 55 mW when all sensors are turned off.

Similar measurements for WiFi and GSM are reported in Figure 1(c) and (d). Observe that while the baseline power consumption for WiFi is low (55 mW), it exhibits a high spike (of around 1300 mW) at every probe. GSM based samples also exhibit similar characteristics, however, their power consumption is less.

Based on these power measurements, we computed the total battery lifetime for each sensor as if the entire battery capacity is used for localization alone. The results are presented in table 1 together with the localization accuracy of each sensor. We record battery lifetimes of around 10, 40, and 60 hours, for GPS, WiFi and GSM, respectively. When viewed against corresponding localization accuracies of 10, 40, and 400 meters, the energy-accuracy tradeoff is evident.

Table 1. Energy-Accuracy Tradeoff

Sensor	GPS	WiFi	GSM
Lifetime (h)	10	40	60
Loc. error (m)	10	40	400

3 EnLoc: Framework Design

3.1 Average Location Error (ALE)

The energy-efficient localization problem can be defined as follows. Given an energy budget E_B, time duration T, and a set of localization schemes S, design a strategy that will minimize the average localization error over the entire time, T. Formally, denote $L_{reported}(t_i)$ and $L_{actual}(t_i)$ to be the reported and actual locations of the phone at time t_i. Assuming T discrete time-points in the entire time duration, the average localization error δ_{avg}, is defined as

$$\delta_{avg} = \sum_{i=1}^{T} \left(\frac{L_{reported}(t_i) - L_{actual}(t_i)}{T} \right)$$

where $L_{reported}(t_i) - L_{actual}(t_i)$ is the distance between the reported location and the actual location at time t_i.

Assuming GPS to be the ground truth, a GPS reading at time t_j implies that $L_{reported}(t_j)$ is same as $L_{actual}(t_j)$. Similarly, a WiFi reading at t_j implies that $L_{reported}(t_j) - L_{actual}(t_j)$ is in the order of 40m. The problem then is to minimize δ_{avg} for a given energy budget. We model this energy-accuracy tradeoff as an optimization problem.

3.2 Problem Formulation

Our goal is to determine a schedule with which the location sensors should be triggered such that the ALE is minimized under a given energy budget. We envision a schedule to be a set of time instants, $\{t_1, t_2, t_3, ..\}$, and corresponding sensors $\{s_1, s_2, s_3, ..\}$, where $s_i \in [GPS, WiFi, GSM]$. The optimal schedule triggers a reading of sensor s_i at time t_i to minimize ALE. We elaborate with an example.

Consider the case in which a person begins moving at time t_0, from an initially sampled location, $L(t_0)$. Assume that until the next location reading, this location is periodically reported as the location of the phone. As the person moves away from the original location, the location error increases with time. When the person stops moving, the error remains static; when the person moves towards the $L_(t_0)$, the error starts decreasing. Without loss of generalization, Figure 2(a) and (b) show an example trajectory in discrete time, and the corresponding error versus time graph. The phone moves away for 2 time steps, pauses for 2 time steps, moves further away for another two time steps, and then pauses again for

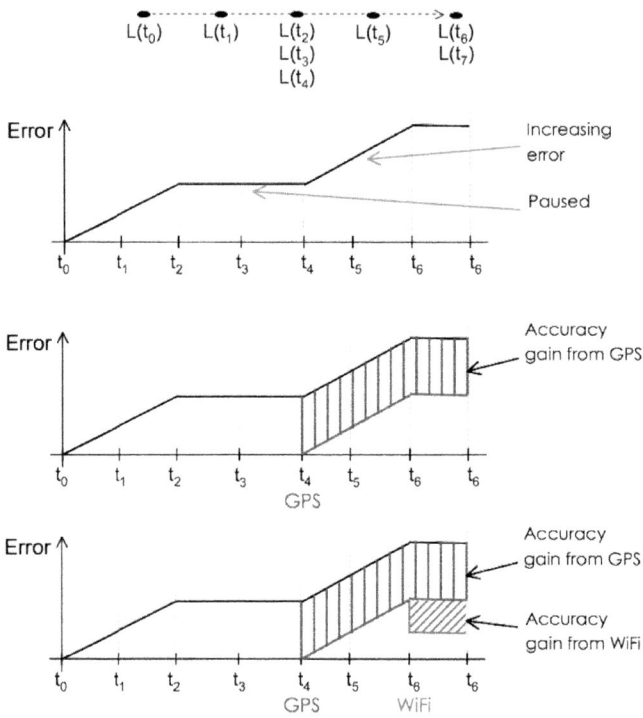

Fig. 2. Error vs time model of physical movement

one time step. The total error due to this movement can be computed as the sum of the 7 vertical lines, dropped from the error curve to the x axis. For easier explanation, let us assume this is the area under the curve.

Now, if the user takes a GPS reading at time t_4, the corresponding error graph is shown in Figure 2(c). Time t_4 onwards, the error increases as if location $L(t_4)$ is the new (known) starting location. If no other readings are taken, the reduction in total error due to the GPS reading is the shaded area labeled "Accuracy gain from GPS". If a WiFi reading is scheduled at time t_6, the instantaneous error, unlike GPS, *does not reach zero* – this results in error reduction shown by the diagonally striped rectangle. In general, the reduction depends on the time-point at which the reading is taken, as well as the sensor used. Additional location readings at subsequent time-points offer further error reductions, at the cost of expended energy. Hence, the problem of energy-efficient localization translates to finding the appropriate time-points and sensors that must be scheduled to minimize the localization error. More formally, for a given energy budget, E_B, energy consumption per-sensor, e_{s_i}, and a mobility trace, τ, the algorithm must produce a $< time_i, sensor_i >$ schedule, that minimizes the total area under the error curve. We approach this problem using dynamic programming. Related problems in the context of satellite communication and user state recognition have been investigated in [11] and [12].

3.3 Dynamic Programming

The dynamic program computes the minimum error accumulated till the end of a given trace, and the corresponding $< t_i, s_i >$ schedule that achieves it. For this, the accumulated error is expressed as a sum of (1) error accumulated until some time j when the last location reading was taken, and (2) error accumulated from j till the final time instant, T. Now, any accumulated error between two consecutive location readings at time i and j can be computed as $\sum_{k=i+1}^{j-1}(L_{reported}(k) - L_{actual}(k))$, where $L_{reported}(k) - L_{actual}(k)$ is the distance between the estimated/reported location, and the actual location. Without loss of generality, assume that $L_{reported}$ returns the last known location. Thus, the accumulated error between i and j is $\sum_{k=i+1}^{j-1}(L_{actual}(i) - L_{actual}(k))$. Since this accumulated error is a function of the sensor used at i (recall Figure 2), we define three matrices $E_{GPS}[i][j]$, $E_{GSM}[i][j]$, $E_{WiFi}[i][j]$. Each matrix corresponds to the localization sensor in the subscript, and denotes the total error from i to j, when that sensor is used at time i. The last column of these matrices gives us the error from some last reading at time j till the end of the trace at time T. Table 2 presents the notations used in the dynamic program.

For the first term, we need to minimize the accumulated error till the (last) reading at time j. This problem can in turn be formulated as finding a second last reading, say at time i, which minimizes the error until j. This can be repeated under the constraints of a maximum number of feasible readings.

Let us consider the problem of finding the minimum error accumulated till time j, assuming a GPS reading is executed at j. There must exist a previous time $i < j$, at which the previous reading was taken. At time i, any one of GPS, WiFi or GSM sensors could have been used. Since there is no other reading

Table 2. Summary of notation

Notation	Explanation
T	Trace length in time units
$I_{WiFi}[j]$, $I_{GSM}[j]$	Instantaneous error at time unit j when the WiFi/GSM sensor is used.
$E_{GPS}[i][j]$, $E_{WiFi}[i][j]$, $E_{GSM}[i][j]$	Accumulated error between i and j or sum of instantaneous errors at intermediate time units, if at i GPS/WiFi/GSM sensor was used and until j there is no other reading.
$\min_{GPS}[i][n1][n2][n3]$, $\min_{WiFi}[i][n1][n2][n3]$, $\min_{GSM}[i][n1][n2][n3]$	Minimum accumulated error till time unit i, if i was the last scheduled reading and used the GPS/WiFi/GSM sensor. The minimum is achieved with n1 GPS, n2 WiFi and n3 GSM readings
$totalError_{GPS}[j][n1][n2][n3]$, $totalError_{WiFi}[j][n1][n2][n3]$, $totalError_{GSM}[j][n1][n2][n3]$	Total accumulated error till the end of the trace if at time unit j the GPS/WiFi/GSM sensor was used and there are no further readings till the end of the trace. In total there are n1 GPS, n2 WiFi, n3 GSM readings

between i and j, some error accumulates from i to j. This accumulated error is one of $E_{GPS}[i][j]$, $E_{WiFi}[i][j]$ or $E_{GSM}[i][j]$, depending on the location sensor used at time i. The problem of finding the minimum error till time j translates into solving the minimum sum of two terms – (i) the minimum accumulated error till some time $i < j$ using some sensor (GPS, WiFi or GSM), and (ii) the error accumulated between i and j. However, the minimum accumulated error till time i has 1 GPS reading less (because one GPS reading is invested in estimating the location at time j). Since, there are three possible sensors that can be used at time i, the optimization can be formulated as:

$$\min_{GPS}[j][n1][n2][n3] = \min_i \begin{cases} \min_{GPS}[i][n1-1][n2][n3] + E_{GPS}[i][j], \\ \min_{WiFi}[i][n1-1][n2][n3] + E_{WiFi}[i][j], \\ \min_{GSM}[i][n1-1][n2][n3] + E_{GSM}[i][j] \end{cases}$$

where $j > i$, $j >= n1 + n2 + n3$, $T > j$.

The total error till the end of the trace (if the last reading is at time j), is the minimum error achieved till j plus the accumulated error between j and T.

$$totalError_{GPS}[j][n1][n2][n3] = \min_{GPS}[j][n1][n2][n3] + E_{GPS}[j][T]$$

Similarly, we can write the above equations for the case in which WiFi or GSM is used for reading the location at time j. However, we must add an additional term to denote the instantaneous location error at time j (recall that WiFi and GSM sensors have an instantaneous error represented as $I_{WiFi}[j]$ and $I_{GSM}[j]$). The dynamic program for a WiFi reading at time j is

$$\min_{WiFi}[j][n1][n2][n3] =$$
$$\min_i \begin{cases} \min_{GPS}[i][n1][n2-1][n3] + E_{GPS}[i][j] + I_{WiFi}[j], \\ \min_{WiFi}[i][n1][n2-1][n3] + E_{WiFi}[i][j] + I_{WiFi}[j], \\ \min_{GSM}[i][n1][n2-1][n3] + E_{GSM}[i][j] + I_{WiFi}[j] \end{cases}$$

and that for a GSM reading at j is

$$\min_{GSM}[j][n1][n2][n3] =$$
$$\min_i \begin{cases} \min_{GPS}[i][n1][n2][n3-1] + E_{GPS}[i][j] + I_{GSM}[j], \\ \min_{WiFi}[i][n1][n2][n3-1] + E_{WiFi}[i][j] + I_{GSM}[j], \\ \min_{GSM}[i][n1][n2][n3-1] + E_{GSM}[i][j] + I_{GSM}[j] \end{cases}$$

The total error till the end of the trace, if the last reading at time unit j uses the WiFi sensor, is:

$$totalError_{WiFi}[j][n1][n2][n3] = \min_{WiFi}[j][n1][n2][n3] + E_{WiFi}[j][T]$$

while for the GSM sensor, total error can be expressed as:

$$totalError_{GSM}[j][n1][n2][n3] = \min_{GSM}[j][n1][n2][n3] + E_{GSM}[j][T]$$

The final solution for the minimum accumulated error till the end of the trace, given a maximum of n1 GPS readings, n2 WiFi readings and n3 GSM readings, is:

$$Sol = \min_{j,n1,n2,n3} \begin{cases} totalError_{GPS}[j][n1][n2][n3] \\ totalError_{WiFi}[j][n1][n2][n3] \\ totalError_{GSM}[j][n1][n2][n3] \end{cases}$$

The $< t_i, s_i >$ schedule can be retrieved by keeping track of timing and sensor readings for each minimization sub-problem. In this way, the entire solution sequence can be retrieved and the optimal schedule computed.

3.4 Optimal Localization Error

To obtain the best localization accuracy, we executed the DP on mobility traces from the UIUC campus. We first war-drove the campus [9] and generated a wireless map of the area (as described in Section 2). Then, we distributed phones to students to gather mobility traces. A custom simulator integrated the traces with the wireless map, and executed the dynamic program. We briefly present a few relevant details.

Collecting Mobility Traces
Our tracing software (installed on mobile phones) sampled the GPS location, as well as the WiFi and GSM base station IDs every 30 seconds. For purposes of anonymity, we requested the users to turn off their mobile phones when they did not wish to be tracked. The traces were analyzed offline. For each sample, the set of WiFi APs were obtained and the corresponding locations extracted from the (war-driven) wireless map. A centroid of all the WiFi AP locations was computed and declared to be the instantaneous WiFi-based location of the phone. Since the GPS location for this time instant is known, the instantaneous error from WiFi localization can be computed. This instantaneous error (say at time, t_i) populates the $I_{WiFi}[i]$ value in the dynamic program formulation. The GSM localization error is identically computed, and used to populate the $I_{GSM}[.]$ matrix. In summary, for a given mobility trace, the exact errors due to WiFi and GSM localization is known for each time instance[1]. These errors are fed into the DP for error computation. The energy budget is specified as 25% of the phone battery, while the required duration of operation is 24 hours.

We evaluate four schemes, namely, *Optimal GPS*, *Optimal WiFi*, *Optimal GSM*, and *Optimal Combined*. As the name suggests, *Optimal GPS* corresponds to the minimum *average localization error* achieved when only GPS readings are used. In the interest of space, Table 3 reports results from three representative mobility traces. Observe that *Opt WiFi* outperforms *Opt GPS* indicating that greater number of less accurate readings is better for mobile phone localization. Also, *Opt Combined* outperforms the others, and is close to *Opt WiFi* in many of the traces. However, it is surprising that even the offline optimal error (with

---
[1] We assume that GPS is the ground truth.

Table 3. Optimal performance for different traces

	OptGPS	OptWiFi	OptGSM	OptComb
Trace 1	164.999	78.52	352.909	78.52
Trace 2	105.35	75.16	327.116	58.66
Trace 3	125.848	62.134	370.621	62.134

knowledge of the entire trace) was typically more than 60m. Online versions of these schemes, that do not have the entire trace, will naturally perform worse. This warrants enrichments that would improve the limits of energy-constrained localization. We observe that "reporting the last known location between location readings" is a source of inefficiency, and seek solutions that exploit mobility prediction.

3.5 Prediction Opportunity

The error model discussed thus far assumes that between two successive location readings, $L(t_i)$ and $L(t_j)$, the location reported is the last known location, $L(t_i)$. In reality, human behavior/mobility is amenable to prediction [13,14,15,16]. Driving on straight highways, turns on one way streets, habitual office hours, are examples of prediction opportunities. EnLoc aims to take advantage of them.

Simple Linear Predictor: We begin by considering a basic linear predictor. The location of a phone at time t_k, denoted $L(t_k)$, can be a linear extrapolation of the two previously sampled locations, $L(t_i)$ and $L(t_j)$. This can be effective when a phone moves on a straight road. However, if the phone's movement is not straight, or if $L(t_i)$ and $L(t_j)$ were highly erroneous, linear prediction may be unsuitable. Therefore, the performance of linear predictors may not be consistent, and needs evaluation. We have modified the DP to incorporate linear prediction; we omit the formal details in the interest of space.

Human Mobility Patterns: While linear prediction is a generalized approach, capturing individual human behavior may facilitate better predictions [14,15]. The intuition is that humans have habitual activity patterns, and sampling the activity at a few uncertainty points may be sufficient for predicting the rest. For instance, given that a person goes to lunch at either 12:00pm, 12:50pm, or 1:00pm, the phone may trigger GPS readings just after these times. Learning that the person has started out for lunch, her subsequent locations can be predicted (i.e., locations along the habitual path from office to the cafe). In the next section, we augment EnLoc with such human-tailored predictions. Problems arise when the person deviates from her habitual pattern.

Deviations: To cope with deviations from habitual paths, we hypothesize that statistical behavior of large populations provide useful hints. At a traffic intersection, knowing that most of the vehicles take a left turn can be valuable for prediction. In general, if a " probability map" can be generated for a given area, an individual's mobility in that area can be partially predicted. We describe the design of such a probability map later. However, assuming we have such a probability map, we are able to extend our DP to extract the optimal localization schedule for a given trace. Intuitively, the DP is expected to schedule location readings at points where the individual's actions disagree with the population behavior. Where the behavior agrees with the majority of the population, the DP can save energy.

4 EnLoc: System Solution

This section describes a complete system solution, EnLoc, for performing energy-efficient localization on mobile phones. The solution exploits both habitual mobility patterns and population-driven probability maps. EnLoc is an online solution, and unlike the DP, does not assume knowledge of the user's entire trace. Results presented in Section 5 compare EnLoc with the offline optimal schemes.

Exploiting Habitual Mobility

A study with 100,000 people has shown that individuals exhibit habitual space-time movements, with reasonably small variation [13]. If designed carefully, a few location samples may be sufficient to track the location/movement of the person over long time scales. To understand this better, we collected GPS-based mobility traces of several people over a month, and plotted them over Google Maps. Figure 3(a) shows a simplified example.

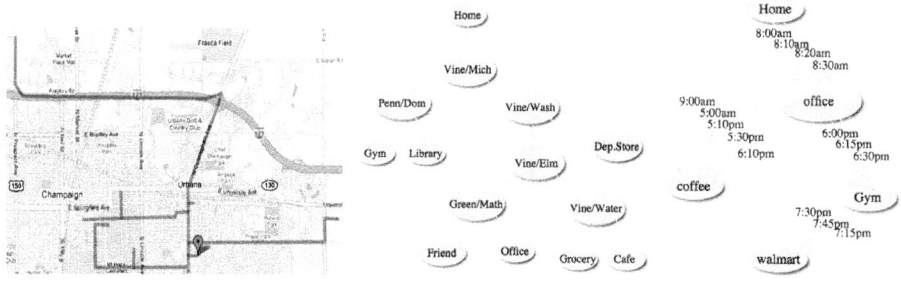

Fig. 3. (a) An anonymous user's movements over two weeks. (b) A spatial logical mobility tree. (c) A space-time logical mobility tree.

One may envision the Google maps plot as a tree, with branches at certain points – we call this the *logical mobility tree* (LMT)[2]. The vertices of this tree are the branching points on the person's actual mobility paths. While tracking a phone along a path on the tree, uncertainty arises at these branching points, and

---

[2] We are aware that the traces can very well form a graph, however, envisioning this as a tree facilitates easier explanation. The ideas we propose are general to *logical mobility graphs*.

hence, the nodes of the LMT are also called *uncertainty points*. The edges of the LMT represent physical traces that connect consecutive uncertainty points. Each edge is associated with (1) the starting time of that physical movement, (2) the average velocity on that path, and (3) the duration of travel on that physical path. Figure 3(b) shows the LMT corresponding to the physical mobility in Figure 3(a).

Our key idea is to schedule location readings right after the uncertainty points on the LMT. Such a location reading will resolve the uncertainty since the phone will be placed in one of the paths emanating from that uncertainty point. Thus, the phone's location can be reasonably predicted until it encounters the next uncertainty point, at which time, another location reading will be necessary. For instance, a location reading at the Vine/Mich intersection can tell whether the person is headed towards the Penn/Dom intersection, or the Vine/Wash intersection. Of course, problems arise due to time variations.

Observe that the LMT in Figure 3(b) is a spatial representation of a person's mobility profile. In reality, a person traverses the same edge on the LMT at many different times (e.g., one may leave for office at different times between 8:00 and 9:00am). Each of these possibilities translates into a distinct edge in the LMT. Figure 3(c) shows a hand-constructed example of such a space-time LMT representation. To accurately know when the phone leaves a particular node of the LMT, EnLoc will need to sample all the edges emanating from that node. Since a person can start moving across a large number of time points (i.e., a large number of edges), the energy budget may not permit sampling at all these times. Only a fraction of the emanating edges will need to be sampled. EnLoc designs a heuristic to sample a subset of the edges branching out of a node. We explain this with the example of Figure 3(c), which is not derived from the actual mobility trace.

Assume that current time is 8:00am, and the phone is located at home, H. Also assume that the remaining energy budget is $B_{remaining}$. The heuristic begins by identifying all the paths from H to the leaves of the tree, i.e., $P_1 = home \Rightarrow coffee$, and $P_2 = home \Rightarrow walmart$. Then, the number of location readings N_i, necessary to track the phone with certainty, is computed for each path P_i. Thus $N_1 = 4 + 6 = 10$ readings. Observe that 4 readings are necessary to track the phone leaving Home, and 6 readings for going from Office to Coffee. These 6 readings include the 5 edges from Office to Coffee (the latest being 6:10pm), as well as the 6:00pm edge from Office to Gym. If the 6:00pm edge is not included, EnLoc may not know if the phone has started moving towards the Gym. Similarly, $N_2 = 4 + 8 + 3 = 15$ readings. Now, the heuristic computes $M = max(N_i)$, a pessimistic estimate of the number of readings necessary in the future; $M = 15$ for this example. Then, the heuristic computes $F = \frac{e_H}{M}$, where e_H is the number of emanating edges from Home. In this example, $F = \frac{4}{15}$. The phone is allocated $F \times B_{remaining}$ amount of energy for detecting its departure from home. Assuming $B_{remaining}$ is 10, there are approximately 2 GPS readings available. The heuristic randomly chooses 2 time-points out of the 4, and samples the phone's location. Once that phone is found to be on one of the paths going out of Home, the heuristic predicts the phone's location based on the habitual velocity on that edge. At the next uncertainty point, the phone recomputes F using the same scheme above. The overall system is reset at night when the phone is plugged into the power outlet.

Addressing Deviation from Habits

Users may deviate from their habitual paths, and the above scheme will not be able to predict their locations. Even though deviations are not the common case, they are important because several applications may be triggered due to deviation. For instance, micro-blogging [1] may be more active when people go for vacations; location-specific information may be necessary when people are driving down untraveled paths. EnLoc addresses the case of deviation – the main idea is to exploit mobility of large populations as an indicator of the individual's mobility. Consider a person approaching a traffic intersection from Street A. Since the person has not visited this street in the past, it is difficult to predict how she will behave at the imminent intersection. Now, if a large fraction of the population is known to take a left turn onto Street B, then the person's movement can be guessed accordingly. EnLoc creates, and takes advantage of, population-driven mobility maps. The details follow.

EnLoc detects a deviation when a scheduled location reading discovers the phone in an unexpected location. At this time, EnLoc switches to the *Deviation Mode* of operation. In this mode, the residual energy budget is divided into equally spaced GPS readings across the remainder of the day. If the current time is 5:00pm, the remaining energy budget is 48 GPS readings, and there are 6 hours left before a habitual battery recharge (say at 11:00pm), EnLoc schedules 8 equally spaced readings per hour. Now, once the first location sample has been obtained, say $L(t_1)$, EnLoc uses the population activity map to predict the phone's movement. The velocity is estimated from the activity map, as well as turns at different intersections. Incorrect predictions obviously incur location error. The error accumulates until the next reading at time t_2, at which point EnLoc starts a new prediction using location $L(t_2)$ as the starting point. Of course, this heuristic needs the population-driven activity map for a given area. We created such a map of the UIUC campus, described next.

Without loss of generalization, let us consider 4-way traffic intersections. EnLoc computes 4 probabilities for each intersection, i.e., an user entering the intersection from Street A, either turns left, turns right, continues straight, or takes a U-turn. One may envision this as a 4×4 matrix, where element ij denotes the probability that the user entering street i exits through street j. Our first approach towards creating this matrix was to deploy sound/vibration sensors at traffic intersections, and count passing vehicles through simple signal processing. When this proved complicated, we adopted a much simpler approach at the expense of some inaccuracy.

On Google maps, we identified all segments of roads that border the UIUC campus. Further, we identified roads that intersect the bordering roads, and enter the campus. We call these feeder roads. Observe that all vehicles must enter the university campus through one of these feeder roads. We also identified

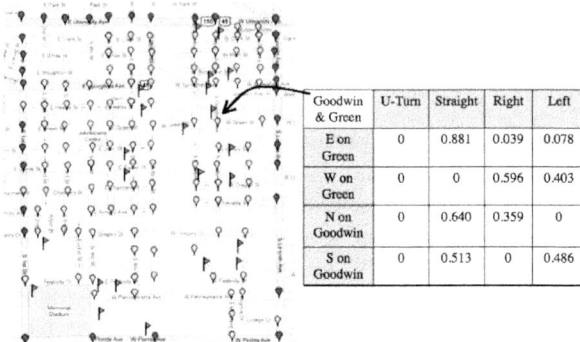

Fig. 4. UIUC campus intersection with associated probability matrix

all parking lots within the campus, their capacities, and the total number of active parking passes in the university. Now, we simulated vehicles that enter the campus through a feeder road, and drives to a pre-specified parking lot. The pre-specified parking lot is randomly chosen from the distribution of parking lot sizes (i.e., proportionally more cars are destined to bigger lots). For each vehicle, we obtained its driving direction through Google Maps APIs [17], and parsed it to extract the vehicle's movement at each traffic intersection (i.e., left/right/forward/backward). Simulations of thousands of vehicles produces the probability matrix for each intersection. Figure 4 shows the intersections on the UIUC campus, and a probability matrix for one of them. We expect the probability map to be installed in phones, and thereby used for online prediction and localization.

5 Evaluation

We evaluate EnLoc using realistic traces collected from mobile phones of UIUC students. Traces were processed offline to extract their time durations, GPS locations (every 30 seconds), as well as the WiFi APs and GSM towers overheard along the paths. Trace segments located within the war-driven campus were processed to obtain the WiFi and GSM localization errors for every 30 second time points (we assume GPS is ground truth). Further, at each traffic intersection, the mobility of the phone was predicted based on the maximum probability at the intersection. The error was computed whenever the prediction was inconsistent with the actual user's movement. All the errors were systematically incorporated into our custom simulator (which includes the dynamic program module). The energy budget for localization was assumed to be 25% of the battery capacity, while the required duration of operation was 24 hours.

An ideal evaluation of EnLoc should characterize the average localization error (ALE) over a person's complete mobility pattern (i.e., habitual and deviant paths). However, since the deviations in our traces extended far beyond the (wardriven, probability-mapped) UIUC campus, we were unable to perform an ideal evaluation. Instead, we first evaluated several trace segments confined within the UIUC campus (we pretend these are the deviant paths). Then, we evaluated the *habitual mobility profile*-based approach by pruning the deviations from a user's LMT. We believe that in reality, the localization error will be close to the average of these two cases.

Deviant Paths

Table 4 and 3 describe the different schemes examined in this section. Figure 5(a) reports the optimal localization error averaged over all mobility traces within the UIUC campus. Figure 5(c) and (e) show optimal results from two representative traces. We make the following observations from these graphs.

Table 4. Optimal and heuristic schemes

Opt-LP-GPS	Optimal with GPS + Linear Prediction
Opt-LP-WiFi	Optimal with WiFi + Linear Prediction
Opt-LP-GSM	Optimal with GSM + Linear Prediction
Opt-LP-Comb	Optimal with Combined + Linear Prediction
Opt-Map	Optimal using map predictor
Heu-Eq-GPS	Heuristic with Equally spaced GPS
Heu-Eq-WiFi	Heuristic with Equally spaced WiFi
Heu-Eq-GSM	Heuristic with Equally spaced GSM
EnLoc-Eq-Map	Heuristic with Equally spaced GPS on Map

(1) As mentioned earlier, OptWiFi consistently outperforms OptGPS. Evidently, this trend holds for the case of linear prediction as well. Opt-LP-WiFi averages an error of 28.94m, in comparison to 36.06m from Opt-LP-GPS. We conclude that, when scheduled carefully, WiFi offers better energy-efficient localization than GPS/GSM.

(2) Linear prediction performs well even for mobility traces that take frequent turns. This may appear surprising because linear prediction is expected to yield higher errors when an user has taken a right-angle turn. Examination of the optimal schedule for "manhattan type" traces offered useful insights. When the distances between consecutive turns were short, the linear predictor approximated the movement with a straight line cutting through the trace. The error is not large, as evident from Figure 6(a). On the contrary, if the trace has long stretches of straight lines, the Opt-LP scheme naturally predicts correctly (see Figure 6(b)).

(3) When using probability maps, the optimal ALE proves to be small. This is because the number of mis-predictions (at the intersections) are typically fewer than the number of location readings permitted by the budget. As a result, Opt-Map schedules a location reading wherever there is a mis-prediction. Of course, we assumed that the velocity of the phone can be perfectly predicted, and hence, errors arise only after mis-predictions. In reality, some errors will accumulate as

a result of inaccurate velocity prediction. We plan to address this issue as a part of our future work, potentially using phone accelerometers to predict the velocity.

Figure 5(b)(d)(e) present the performance of online heuristics. Consistent with our earlier observation, Heu-Eq-WiFi outperforms both Heu-Eq-GPS and Heu-Eq-GSM. However, EnLoc-Eq-Map outperforms Heu-Eq-WiFi, except in rare occasions (such as in Trace B). The reason is as follows. In trace B, the phone encounters a road block, and takes a detour from its natural path towards the destination. As a result, it takes turns that are inconsistent with the map predictions. Since this detour causes many mis-predictions, the budgeted number of readings are not sufficient to correct them. The probability map can be updated

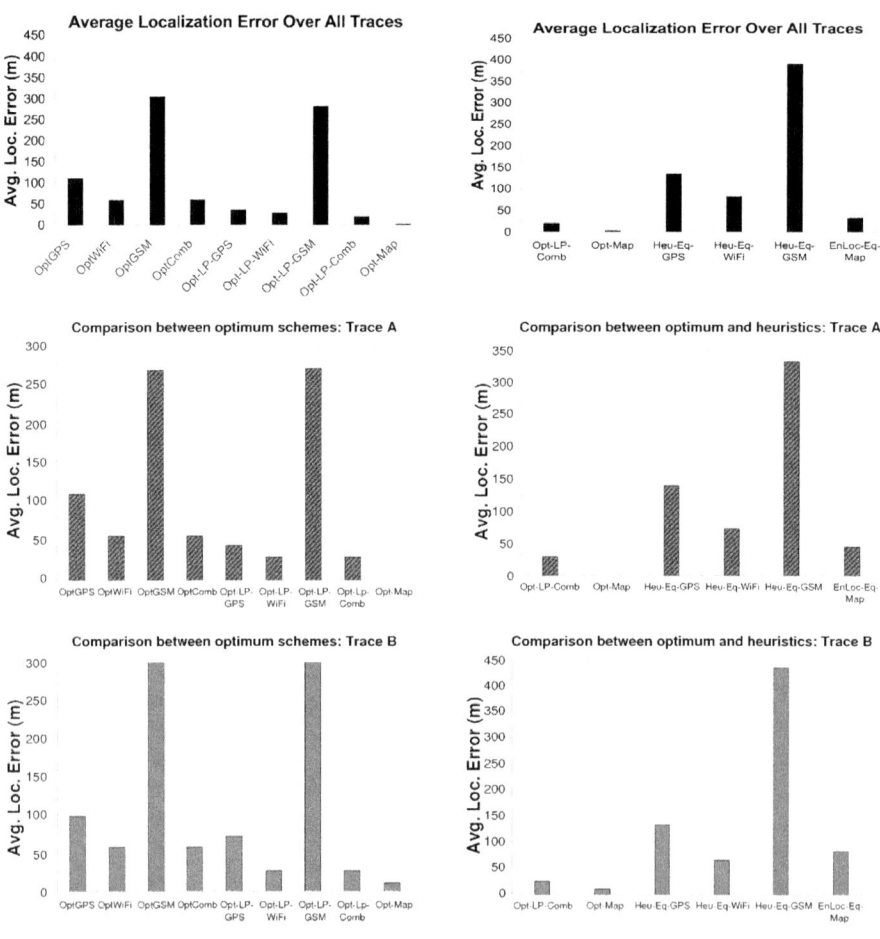

Fig. 5. Average localization error using optimal schemes: (a) Optimal over all traces, (b) Heuristic over all traces, (c) Optimal on Trace A, (d) Heuristic on Trace A, (e) Optimal on Trace B, (f) Heuristic on Trace B.

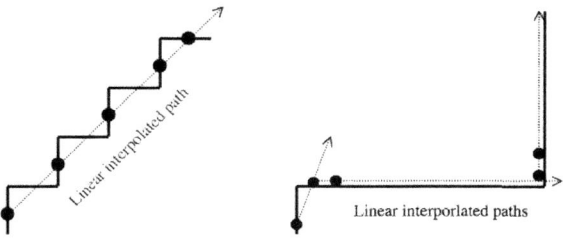

Fig. 6. Linear prediction performs well on (a) Manhattan, (b) Highway movement.

when there are road blocks in the neighborhood, and a phone can periodically download a fresh copy of the map from a web service. In general, however, results indicate that heuristics based on probability maps are reasonably effective for achieving energy-efficient localization. We are unsure if similar results may hold for places unlike university campuses – we discuss these issues in Section 6.

Figure 7 shows the variation of ALE (averaged over all traces) for increasing energy budget. The error from optimal schemes decreases monotonically with increasing energy. The EnLoc-Eq-Map scheme follows a similar trend.

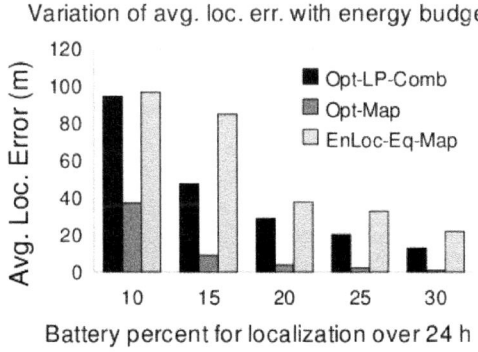

Fig. 7. ALE variation with increasing energy

Habitual Mobility

Next, we present the localization error when a person's mobility profile is utilized for prediction. We use an anonymous student's mobility profile derived from 30 days of traces. We processed her mobility traces and generated the logical mobility tree (LMT). Then, for each of the days, we executed the *Mobility Profile Heuristic*. Figure 8(a) shows that for the allocated energy budget of 25% for the entire day, the average localization error averages around 12m. This error is fairly small, particularly in view of its scalability for 24 hours. The variation of the average error with increasing energy budget is shown in Figure 8(b).

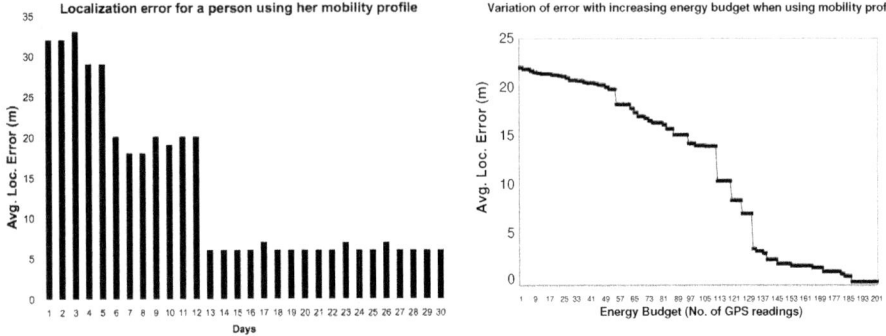

Fig. 8. (a) ALE using individual mobility profile. (b) Variation of ALE with increasing energy budget

6 Limitations and Future Work

Several limitations need to be addressed before EnLoc is ready for realistic deployment. We discuss the key limitations here, and allude to potential approaches to address them.

(1) We assumed that while moving along a predicted path, the location of the phone can be accurately tracked. In reality, varying speeds, pauses, and other random fluctuations in movement will cause this prediction to become imprecise. Our evaluation results do not account for these errors. Nonetheless, with accelerometers available on modern phones, some of these speed variations can be estimated, and used for prediction. We have investigated this in our early work in [18], and we plan to exploit it further in EnLoc.

(2) EnLoc does not proactively identify deviations from habitual paths. A person may leave office 4 hours before her habitual departure time; EnLoc will discover this after 4 hours. Techniques are necessary to quickly detect departures without investing excessive energy. A simple way could be to schedule periodic GSM readings – when the phone identifies that the GSM tower IDs have changed at an usual time, it can trigger the *deviation mode*. Accelerometers may also be effective. If the phone moves for more than a threshold duration at an unexpected time, EnLoc may suspect deviation, and schedule a GPS reading. We are investigating these possibilities in our ongoing work.

(3) Probability maps may be harder to generate for places unlike university campuses (say, a town, city, downtowns). Location updates gathered over time, from many mobile phones, can help in bootstrapping. Statistics from transportation departments, if available, can also be useful. We plan to investigate general ways of creating probability maps.

(4) The mobility profile used in EnLoc are those of graduate students, and may be less diverse (more predictable) than that of an average person. Thus, the reported errors in this paper may be optimistic. We plan to collect more (anonymous) traces, and evaluate EnLoc extensively. However, we believe that

the algorithms/results reported in this paper validate the intuition that mobility profiling per-individual can be an effective tool for energy-efficient localization.

7 Conclusion

Emerging mobile phone applications will rely heavily on location technology. We show that energy can be a critical bottleneck for continuous localization, and that attempts to increase battery-life can degrade location accuracy. This paper presents a framework, EnLoc, that quantifies this tradeoff, derives optimal performance bounds, and develops practical schemes to achieve energy-efficient localization. Proposed schemes exploit prediction opportunities inherent in human behavior. Evaluations on traces from the UIUC campus show promising results. We believe that additional use of phone sensors, such as accelerometers, compasses, cameras or microphones [19,20], may offer further benefits in localization accuracy for small energy costs.

References

1. Gaonkar, S., Li, J., Choudhury, R.R., Cox, L., Schmidt, A.: Micro-blog: Sharing and querying content through mobile phones and social participation. In: ACM MobiSys (2008)
2. Eisenman, S.B., Lane, N.D., Miluzzo, E., Peterson, R.A., Ahn, G.S., Campbell, A.T.: Metrosense project: People-centric sensing at scale. In: Workshop on World-Sensor-Web (2006)
3. Burke, J., Estrin, D., Hansen, M., Parker, A., Rmanathan, N., Reddy, S., Srivastava, M.B.: Participatory sensing. In: Workshop on World-Sensor-Web (2006)
4. Sohn, T., Li, K.A., Lee, G., Smith, I.E., Scott, J., Griswold, W.G.: Place-its: A study of location-based reminders on mobile phones. In: Beigl, M., Intille, S.S., Rekimoto, J., Tokuda, H. (eds.) UbiComp 2005. LNCS, vol. 3660, pp. 232–250. Springer, Heidelberg (2005)
5. Yoon, J., Noble, B., Liu, M.: Surface street traffic estimation. In: ACM MobiSys (2007)
6. Eriksson, J., Girod, L., Hull, B., Newton, R., Balakrishnan, H., Madden, S.: The pothole patrol: Using a mobile sensor network for road surface monitoring. In: ACM MobiSys (2008)
7. Mohan, P., Padmanabhan, V., Ramjee, R.: Nericell: Rich monitoring of road and traffic conditions using mobile smartphones. In: ACM Sensys (2008)
8. Forum. Nokia. Com, Nokia Energy Profiler, http://www.forum.nokia.com/info/sw.nokia.com/id/324866e9-0460-4fa4-ac53-01f0c392d40f/Nokia_Energy_Profiler.html
9. Cheng, Y.-C., Chawathe, Y., LaMarca, A., Krumm, J.: Accuracy characterization for metropolitan-scale wi-fi localization. In: ACM MobiSys (2005)
10. Chen, M., Soh, T., Chmelev, D., Haehnel, D., Hightower, J., Hughes, J., LaMarca, A., Potter, F., Smith, I., Varshavsky, A.: Practical metropolitan-scale positioning for gsm phones. In: Dourish, P., Friday, A. (eds.) UbiComp 2006. LNCS, vol. 4206, pp. 225–242. Springer, Heidelberg (2006)
11. Fu, A.C., Modiano, E., Tsitsiklis, J.N.: Optimal energy allocation and admission control for communications satellites. IEEE/ACM Trans. Netw. (2003)

12. Wang, Y., Lin, J., Annavaram, M., Jacobson, Q.A., Hong, J., Krishnamachari, B., Sadeh, N.: A framework of energy efficient mobile sensing for automatic user state recognition. In: MobiSys (2009)
13. Gonzalez, M.C., Hidalgo, C.A., Barabasi, A.-L.: Understanding individual human mobility patterns. Nature (2008)
14. Lee, H., Wicke, M., Kusy, B., Guibas, L.: Localization of mobile users using trajectory matching. In: ACM MELT (2008)
15. Burbey, I., Martin, T.: Predicting future locations using prediction-by-partial-match. In: ACM MELT (2008)
16. Lee, K., Hong, S., Kim, S.J., Rhee, I., Chong, S.: Slaw: A mobility model for human walks. In: IEEE INFOCOM (2009)
17. Google maps api, http://code.google.com/apis/maps/
18. Ofstad, A., Nicholas, E., Szcodronski, R., Choudhury, R.R.: Aampl: Accelerometer augmented mobile phone localization. In: ACM MELT (2008)
19. Miluzzo, E., Lane, N.D., Fodor, K., Peterson, R., Lu, H., Musolesi, M., Eisenman, S.B., Zheng, X., Campbell, A.T.: Sensing meets mobile social networks: The design, implementation and evaluation of cenceme application. In: ACM Sensys (2008)
20. Azizyan, M., Constandache, I., Choudhury, R.R.: Surroundsense: Localizing mobile phones via ambience fingerprinting. In: ACM MobiCom (2009)

Multi-agent Meeting Scheduling Using Mobile Context

Kathleen Yang, Neha Pattan, Alejandro Rivera, and Martin Griss

Carnegie Mellon University, Silicon Valley Campus
NASA Research Park, Bldg. 23, Moffet Field, CA 94035
{kathleen.yang,neha.patttan,alejandro.rivera,
martin.griss}@sv.cmu.edu

Abstract. Despite the use of newer and more powerful calendar and collaboration tools, the task of scheduling and rescheduling meetings is very time consuming for busy professionals, especially for highly mobile people. Research projects and commercial calendar products have worked since the early 1990s on implementing intelligent meeting organizers to automate meeting scheduling. However, automated schedulers have not been widely accepted or used by professionals around the world. Our research shows that the task of organizing meeting can be improved by reducing the scheduling workload and making meeting logistics more efficient. For example, new tools can decide meeting venues and dynamically handle exceptions, such as one participant not being able to arrive at the meeting location on time. In this paper, we discuss a solution to effectively employ the mobile user's context in making more intelligent decisions on behalf of the user. Business Meeting Organizer (BMO) is a multi-agent meeting scheduling system, designed to automate time and venue decisions and to handle exceptions. Rather than using a traditional multi-user calendar-based scheduler, BMO has representatives ("software secretaries" implemented as software agents) for each user to negotiate a best time for a particular meeting, taking into account availability, context and preferences. Several technologies are used by BMO to provide secure and intelligent meeting scheduling functionality. By using agent technology, BMO keeps private calendar information invisible to other meeting participants and allows diverse intelligent negotiation and scheduling policies to be employed. Through the use of a rules engine, BMO can consider meeting participants' personal preferences as to when to schedule the meeting. By means of mobile devices, BMO can get the user's context information, such as the user's physical location, which may be helpful in deciding the meeting venue and handling other meeting issues. In this paper, we discuss a solution to effectively employ the user's mobile context information in making more intelligent decisions on behalf of the user.

Keywords: context-aware; intelligent meeting scheduling; multi-agent system.

1 Introduction

In this age of constant change, one of the most effective methods to stay in touch, maintain business connections, share information, and make decisions is through meetings. It is, therefore, not surprising that many professionals spend significant amounts of time in attending meetings [1]. Much time is expended in negotiating different factors like time, venue, attendees and resources when scheduling and rescheduling meetings. The current methods of solving this problem, namely use of email, SMS or phone calls are time consuming and inefficient. It is believed that autonomous agents can significantly reduce this workload and the information overload by automating these tasks for users [3]. The problem is exacerbated for highly mobile users who may change locations frequently and are typically pressed for time in setting up meetings.

In this paper, we propose a distributed meeting scheduler for mobile devices – Business Meeting Organizer (BMO), which largely automates the process of meeting scheduling through communication between multiple software agents to negotiate and decide on the specifics of meetings. Each user has one software agent that acts as their personal secretary. This software secretary resides on their personal computer or on a secure company server and acts on the user's behalf even when the user is travelling or inaccessible. The user communicates with the software secretary using their mobile device. The secretary has access to all of the user's personal information such as calendar, contacts, current context, and preferences. The software secretaries obtain requests from their respective users and communicate with each other to decide when and where to arrange meetings. These secretaries also understand the users' preferences and priorities and make decisions accordingly.

In addition to accessing static information about the user through his/her contacts or calendar, the software secretaries also stay informed of the mobile user's context such as location, time of day or current activity. In addition to the current context, the software secretaries are also aware of how certain parameters of the context will change over time. For example, by using the user's calendar, the secretary can predict the future location of the user and may even give hints regarding activity like future travel, upcoming projects, and deadlines. The context of the user plays a very important role in managing the user's schedule for the day and in communicating important decisions back to the user. For example, if the user is going to be in San Francisco until 10 am in the morning on Friday, the secretary will not arrange any meetings for him/her at San Jose any time before 11 am (because it typically takes at least one hour to drive between these two locations). After scheduling the meeting, the secretary will send the user a confirmation with a link to the map of the exact location. If one of the users attending the meeting is not going to be able to make it to the meeting on time due to traffic congestion or if (s)he gets caught up in some other meeting, the software secretary will dynamically reschedule the meeting, notify the other attendees and inform of the revised meeting details. The software secretaries are modeled to enhance user experience by allowing soft and hard constraints on the meeting scheduling decisions. An example of a soft constraint would be: "I would rather not have meetings on Fridays, but if no other time is available, then it's ok". A hard constraint would be: "From 1 to 2 pm on weekdays, I'm not available since I am in a class".

In addition to location, we have considered other sources of context such as the time of day and related activity. Based on the user's context and preferences, the software secretary will be able to make intelligent and informed decisions and communicate them back to the user.

In the rest of the paper we first discuss background and related work in Section 2, then describe the scheduling model and architecture of BMO in Sections 3 and 4, our prototype implementation in Section 5, our initial user tests in Section 6, and finally conclusions and future work in Section 7.

2 Background and Related Work

Since the 1990s, several groups have worked on intelligent meeting organizers, such as Ameetzer [3] and CoolAgent [4]. Generally these systems focus on negotiating and deciding a suitable timeslot for meetings based on the participants' availability and personal preferences. Many of the proposed meeting organizers also focus on the use of user preferences [2] and adaptation to the user environment [5]. Some of these proposals focus on learning opportunities [6] for the user's preferences and for preferences of other users within the system based on their agents' responses. They help to reduce people's workload of scheduling meetings. Our work builds on these ideas, but specifically considers the context of the mobile user in order to make intelligent and informed decisions on behalf of the user. For example, it would be helpful for meeting organizers to decide the venue for a meeting in a way that's more convenient for all participants, based on their current location or planned location at the proposed meeting time. It would also be of great help if the meeting organizer could find a facility for a participant to join a meeting remotely if he/she is not able to arrive to the meeting on time (for example a coffee shop with Wi-Fi). It is also helpful if a meeting organizer can help (or automatically) reschedule or cancel a meeting if some attendees become unavailable for a meeting because their previous meeting got delayed or other similar reasons. As such, there are several decisions that can be made on behalf of the meeting attendees that can greatly enhance the usability of a meeting organizer. Many of these decisions are tightly coupled with the context of the user and can be easily modeled using a mobile device to obtain contextual information.

The issue of user context has not been fully addressed in designing a multi-agent personal assistant based meeting scheduling system. While similar ideas have appeared in [7], his solution consists of a broker architecture in which the broker is responsible for providing knowledge sharing service between agents and maintaining a shared model of the context. Chen proposes that the system should support policy-driven privacy protection by allowing users to define how much information is shared and with whom it is shared. In our approach, we focus on user privacy and security needs by only allowing user information to be accessible to the user's personal software secretary. Since there is no central model of the context of all users, no single entity has access to all of the information. The preferences specified for the software secretary will define what information can and cannot be shared with other software secretaries or users.

3 BMO Meeting Scheduling Model

In order to make a meeting decision, several factors need to be taken into consideration. These factors can be summarized in a scheduling tree as depicted in Figure 1.

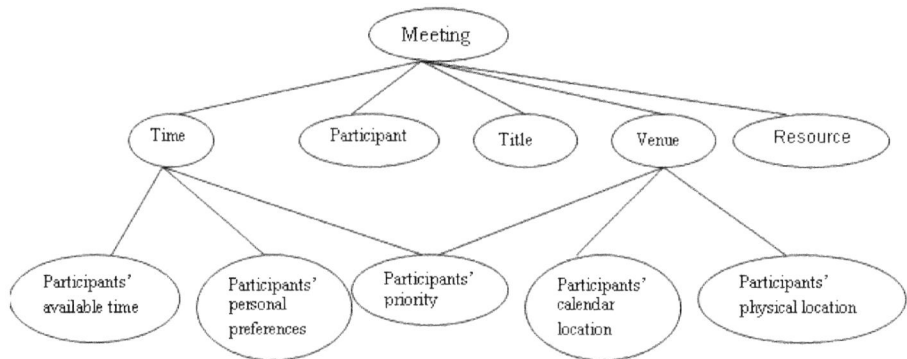

Fig. 1. Meeting Scheduling Tree

Some of the nodes in figure 1 are explained as below.

Participants' personal preferences
This node means people's time and venue preferences, for example, some people prefer to have meetings in the morning while some people prefer to have meetings in the afternoon.

Participants' priority
This node indicates how important participants are for the meeting. When a low priority person cannot attend a meeting, the meeting might not need to be re-scheduled.

Participants' physical location
This node indicates where the participants physically stay at one specific time point. For example, when a person is sitting in Mountain View library at 8:00pm, his physical location is Mountain View library. The basic location information is GPS coordinates. The outdoor GPS coordinates are retrieved from the smart phone GPS module and the indoor GPS coordinates are retrieved from GSM based locationing services. Based on the GPS coordinates, a location label based system is maintained. Each label represents a meaningful place for one particular user, such as home, office, or a geocode address. A group of GPS coordinates are associated with one label according to the user's explicit decision or some machine learning rules. For example, when the system detects a pair of GPS coordinates without associated any label, it can either

ask the user to manually input the label or assign an existing label to it if it's close enough, for example, within 10 meters.

Participants' calendar location
This node indicates where the participants stay at one specific time point according to their calendar information. For example, if the calendar shows that a person has a meeting at NASA Research Park at 10:00am; his calendar location at that time is NASA Research Park. A participant's calendar location is important because we cannot predict one's location based on his physical location but can predict based on his calendar location. Then, we can use the predicted location to decide meeting venue which can minimize the travel cost from previous meeting to next meeting. Different from the participant's physical location, his/her calendar location doesn't have any GPS coordinates. Only the label-based location system is used and the labels are subtracted from the calendar system.

Resource
This node indicates meeting resources such as teleconference system, projector and the size of the meeting room.

The other nodes are pretty self-explained, so no description is given here.

When a software secretary in the BMO system wants to schedule a meeting, it needs to establish the values of all child nodes of the *Meeting* node. If a child node has further children nodes, BMO needs to collect the value of those children nodes until it has reached the leaves. Thus, a meeting can only be decided when all the nodes in the scheduling tree have valid values. The decision on whether to schedule the meeting may be affirmative or negative depending on the exact values of the nodes.

4 BMO Architecture

The architecture of the system is shown in the Figure 2. The external services, the Map Service, Locationing Service, Calendar Service and Contacts Service, are used by the software secretary and the client application. The client application resides on the user's mobile device. Thus, the client application has easy access to all of the user's personal context information from the mobile device; in addition, some other public parts of the user's context are accessed directly by the server (e.g., calendar, history of meeting preferences, etc.) This information may vary from the user's location and activity to specific physical or medical conditions. As more hard and soft sensors get integrated with mobile devices, increased context information will become available to the client application through the mobile device.

The Software Secretary is running continuously on the user's personal computer or on a secure company server, either at home or in the office. Currently, the secretary processes context information of the user's location.

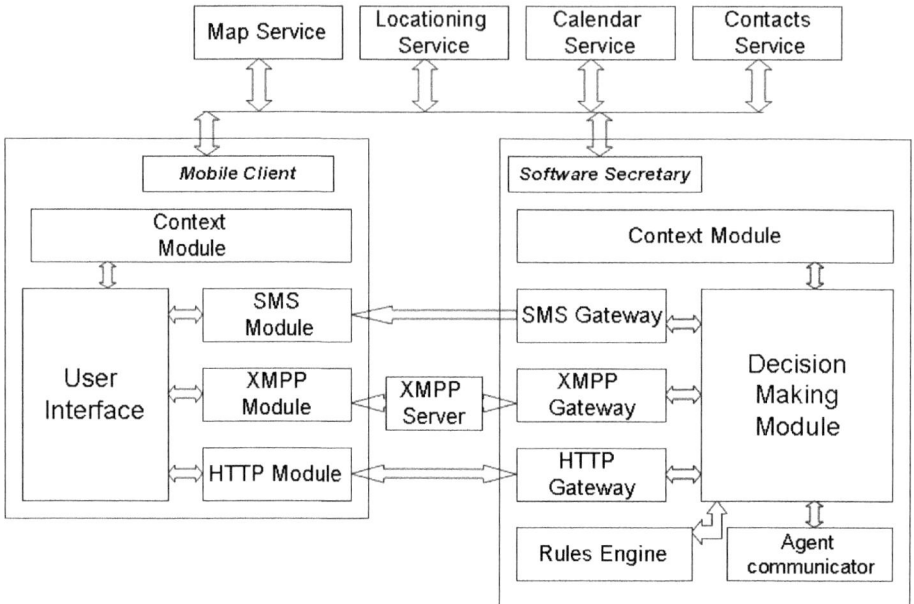

Fig. 2. Architecture of proposed system

4.1 Secretary Agent Side

The Software Secretary is a system consisting of the following components:

Context Module: This module contains the most recent context information obtained for the user. This module is triggered every time the client application sends a context update to the secretary. The context information is used by the decision-making module to make intelligent decisions based on the user's context. This module has a plug-in design, allowing us to add or remove several context sub-components that are relevant to the user.

SMS Gateway: This gateway is used to send updates to the user in case the message has a high priority and urgency. It may also be used if the user's mobile device does not support features like Wi-Fi access or data packets.

HTTP Gateway: This is used to communicate with the client application over HTTP. This may be used if the user prefers to communicate directly with the software secretary agent without relying on an intermediary service. (For example, as will be discussed next, XMPP can also be used for communication between the client and the agents. This would, however, require the messages to be routed through an external service). The HTTP gateway consists of a light-weight HTTP server running on the secretary listening for requests sent by the client application.

XMPP Gateway: This gateway is another means of transportation for messages between the client application and the secretary agent. If XMPP is used as the transport, all messages will be relayed through a third party XMPP server (also known as Jabber server). Using XMPP has the main advantage of messages being stored by the server for delayed delivery in case the destination is not currently accessible or online.

Decision-Making Module: This is the most important component of the secretary agent side system. It is responsible for combining all information from the user's context, external services and user's preferences to make decisions about meeting scheduling.

Rules Engine: This module contains all user preferences in the working memory. The rules are triggered whenever a decision needs to be made by the *Decision Making Module*.

Agent Communicator: This is used by each secretary agent to communicate with other secretaries. JADE [9], being a distributed agent platform, takes care of the communication between secretaries located in the same computer or even in completely independent hosts. The communication is handled transparently in the background, no matter what transport protocol is in use (e.g. Java-RMI, HTTP, IIOP). JADE also relies heavily on the FIPA (Foundation for Intelligent Physical Agents) and its ACL (Agent Communication Language) standards to ensure cross-compatibility with other agent systems.

4.2 Client Side

The Mobile Client software consists of the following components:

Context Module: The context module on the client side is responsible for gathering all context information of the user and will always contain the most recent information. This information is updated by periodically querying different sensors on the device, such as sensors that support location detection or movement.

SMS Module: This module is useful in intercepting the incoming messages sent by the secretary agent. These messages will be processed by the application and deleted from the user's SMS inbox. So, the messages are invisible to the BMO users. This design can avoid large volume of annoying messages in the SMS box.

XMPP Module: This module is used in interacting with the XMPP server to send and receive messages to and from the secretary agent.

HTTP Module: This module is used if the transport used for communication between the client and secretary agent is HTTP. It is used to make requests to the HTTP server running on the secretary agent side.

User Interface: This module contains the interface displayed to the user and is responsible for input and output of information to the user.

5 Implementation

We use Mobile Python [8] (referred as Python hence forth) running on Symbian S60 to implement the mobile client on Nokia N95 smart phones. Users use the mobile client to input the meeting title, preferred meeting time range and expected meeting participants. The advantages of using Python are its high coding productivity and strong support by the Symbian S60 developers and community. Also Python has access to low-level information on the system (the mobile device) that's not typically accessible by some other high-level programming languages (such as J2ME).

The Java Agent Development Framework, or JADE [9], is used to implement the software secretary agents. One software secretary is implemented as one agent in JADE. By using agent technology, the secretary of the meeting initiator doesn't need to access the other meeting participants' calendars. What the secretary needs to do is to ask the other secretaries questions such as "When will you be available today or tomorrow?" This solution helps to protect the privacy of the meeting participants, who may not want to expose their calendar information to others.

The software secretary in BMO makes use of several Google Data APIs (GData for short), such as the Calendar, Contacts and Maps API. As an instance of the Internet service API, Google APIs have several advantages compared to other traditional library APIs. First, the APIs provide functionality as well as abundant data, such as users' contact data and calendar data. Second, the APIs provide high coding productivity. Third, the library deployment is significantly minimized.

Extensible Messaging and Presence Protocol (XMPP) is used for the communication between the mobile client and the secretary. The mobile client and the secretary work as the XMPP client, sending messages to and receiving messages from the XMPP server. In this way, the mobile client can send meeting request to the secretary and the secretary can send meeting confirmation to the mobile client. The secretaries of all the participants can talk with each other to negotiate the meeting scheduling. The advantage of using XMPP is its offline capacity. If an XMPP client is offline, XMPP messages sent to it can be queued in the XMPP server. When the XMPP client connects to the Internet, it can retrieve all the queued messages from the XMPP server. For users who don't have Internet access, SMS can be used instead. For those users who are not using BMO, email can be used as the default method of communication. For instance, the email could contain a simple HTML form the user would fill in and submit it back. Making use of this technique would simplify the data exchange while maintaining a standard communication format.

As illustrated in the following screen shots (Figure 3), when a user wants to schedule a meeting using BMO s/he needs to follow four steps. After that, the meeting request is transmitted to the virtual secretary using XMPP (if connected to the Internet) or is stored in the device's memory while waiting for the Internet connection. Using these five steps, users need less than one minute to schedule a meeting on average.

Multi-agent Meeting Scheduling Using Mobile Context   231

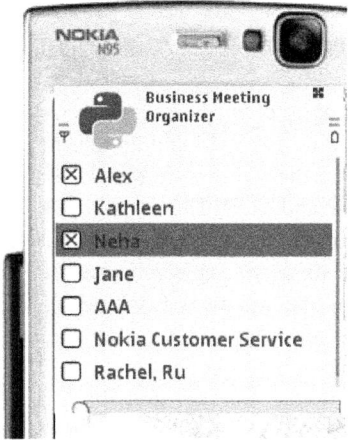

Step 1. Select the expected participants

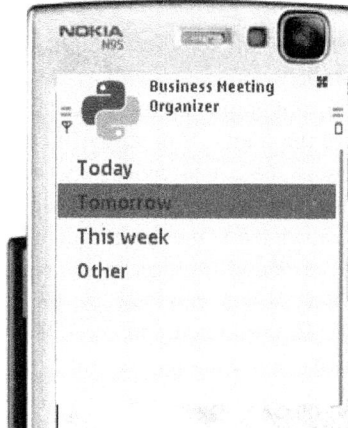

Step 2. Select the preferred time range

Step 3. Input the meeting subject and minimum required duration of the meeting

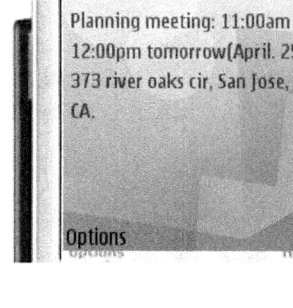

Step 4. Get SMS reminder about the meeting

Fig. 3. Mobile App Screenshots

6 Technology Trade-Offs

During the implementation, we made decisions on several technological trade-off points.

The communication protocol between the mobile client and the secretary agent is one of the trade-off points. Although we decided to use SMS and an Internet protocol to assure the connectivity between the two parts, there are more than one alternative for the Internet protocol, such as JASON, HTTP and TCP. The most important factor we considered is the accessibility of the mobile client to the Internet. Because the data plans provided by most of the carriers are still expensive, it's reasonable to assume that many users cannot access the Internet when they are away from any WIFI access point. So, we decided to use XMPP as the Internet protocol because it offers off-line capacity. For example, the meeting confirmation can be queued on the XMPP server until the participant's mobile client connects to the Internet and thus triggers the XMPP server to send the confirmation to the client.

The programming language on the mobile phone is another trade-off point. Java, C++ and Python are three primary used programming languages on the smart phones. We decided to use Python because it's a powerful, fast and secure prototyping language on Symbian platform. We can use two to three lines of codes to access many capacities, such as GPS capacity, SMS capacity, and Calendar capacity. It doesn't provide any mechanism to explore the layers under the mobile device middle ware or other applications.

7 Experiments and Data Analysis

We observed and measured people scheduling meetings in five ways: a) using BMO, b) using "Friend's calendar" function of Google calendar, c) exchanging emails, d) using the telephone and e) face-to-face negotiation.

We also decided to observe meetings with four participants, because in our survey people stated that when meeting has four participants or more, scheduling the meeting becomes painful.

Table 1. Time for meeting scheduling

	Tools use for scheduling	Number of participants	Negotiation time (minutes)	Request pending time (minutes)
Meeting #1	BMO	4	1	0
Meeting #2	Google calendar	4	2	0
Meeting #3	Email	4	7	50
Meeting #4	Telephone	4	11	0
Meeting #5	Face-to-face	4	5	0

In table 1, the time used for scheduling meetings is recorded for the five different approaches. The negotiation time is the time that people spend to talk with each other, discussion via email and check calendar. The request pending time is the period of time during which people wait for the email response. This request pending time is specific to scheduling meeting via emails.

From the data, we found that using email is the least efficient approach because the negotiating uses seven minutes and the time waiting for email response is 50 minutes.

Telephone and face-to-face are better than email because there is no pending time between the time when the meeting is requested and when the confirmation is responded. However, they still cost five and eleven minuets respectively. It's still painful for meeting organizers.

The Google Calendar can help people organize such a meeting similarly efficient to BMO. However, it has several disadvantages compared to BMO.

First, all meeting participants must expose their calendar to others. Many people don't feel comfortable about this privacy exposure. Second, Google Calendar doesn't consider context information, such as people's personal preferences, location and priorities. Third, people need to check others' availability manually. It's unpleasant work when a meeting has four or more participants.

In contrast, BMO can protect people's privacy, consider context information and automatically negotiate and decide a meeting time slot and venue.

Figure 4 displays the communications between three different secretaries (Kathleen, Neha and Alex's) while arranging one meeting.

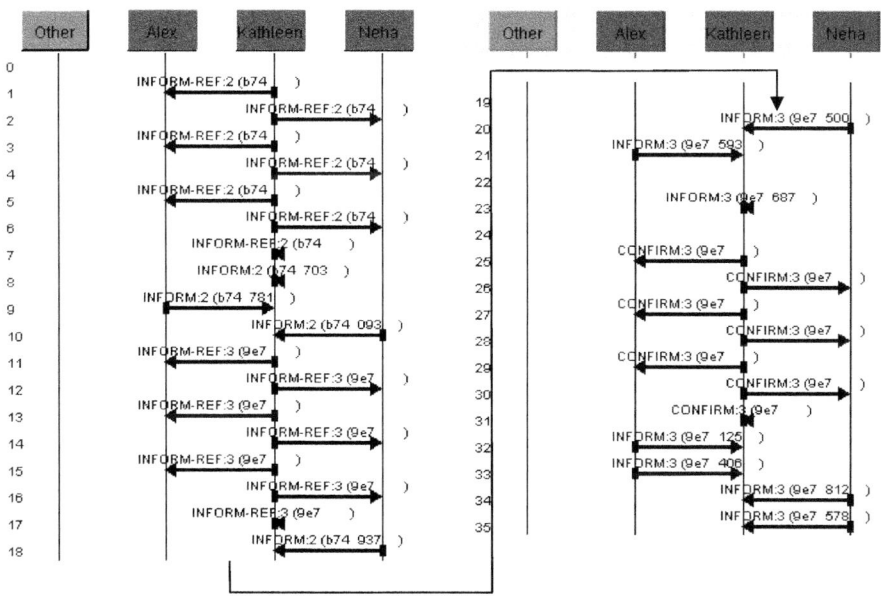

Fig. 4. Agent communication

In Figure 4, it is easy to see that there are many messages being exchanged, yet the time it takes these software secretaries to organize the meeting is in order of a few seconds. If the meeting participants were to employ traditional means of coordinating a meeting, such as discussing the options by telephone, the number of distinct communications among the participants would potentially be much higher and the results might not be as optimal.

8 Conclusion and Next Steps

As discussed in Section 6, we observed through several tests that the time taken to schedule and reschedule meetings with the intelligent meeting organizer is a fraction of the time spent in performing the same tasks manually by using email or calling all attendees. In addition, our solution also scales well when meetings require many more attendees for whom location and time constraints need to be considered. The next step for us is to take the system out to test in the field with additional users and to analyze data on usability, reliability and effectiveness of using the tool.

Currently, the system requires all users to have an installed and running software secretary. BMO is built to require communication between each of the secretary agents representing the various participants to negotiate the meeting time and venue details. We would like to make a provision for secretaries to be able to communicate with other users who may not have software secretaries. In this case, if a secretary agent is requesting a meeting to a user who does not have a secretary agent, the requester agent sends an email or text message to the user directly using standard meeting requests (such as iCal attachments) or by providing a web form the user could fill in that the software secretaries would be able to understand to continue the negotiation process.

These secretary agents can be implemented as part of the client application on the mobile device rather than on the user's personal workstation. We plan to prototype such a standalone intelligent meeting organizer on the Android platform using lightweight agents provided by the Light Extensible Agent Platform (LEAP) for mobile platforms.

We would also like to improve the preference module to allow more interesting scheduling patterns. For instance, allowing the user to decide whether or not to schedule adjacent meetings (with no breaks). Another example is to inform the secretary that instead of selecting the next available timeslot for the meeting, it should try to balance all meetings throughout the week. Perhaps users would also like their secretaries to know that it's best for them to have meetings early in the day or to respect lunch hours at all costs, etc. Another kind of preference can be related to the progress of particular projects. For example, if the project is running late and a bottleneck has been detected, it's of extreme importance to schedule a meeting to address the issue and all participants should be aware that this meeting has priority over all others.

Another area we will investigate is to associate some form of weighting or economic value to decide which meetings are more important or urgent than others; for example, a meeting with the CEO is sufficiently important that even canceling or rescheduling other meetings or a planned trip is appropriate. Similarly, user preferences and soft constraints can associate weights with alternative meeting slots

In addition to understanding user preferences and context, the tool can be made more powerful by learning the behavior, preferences and context of the user. This can be achieved by using appropriate machine learning algorithms within the secretary agent to achieve a high level of intelligence within the agent.

We believe that, in the future, automated software agents will play a very important role in simplifying the lives of business professionals and organizing meetings will be one of the several functions carried out by them.

Acknowledgements

This research was supported by grants from Nokia Research Center and the CyLab at Carnegie Mellon under grant DAAD19-02-1-0389 from the Army Research Office. The views and conclusions contained here are those of the authors and should not be interpreted as necessarily representing the official policies or endorsements, either expressed or implied, of ARO, CMU or the U.S. Government or any of its agencies.

We also acknowledge the efforts of Patricia Collins and the anonymous referees for their valuable feedback and suggestions on earlier drafts of this paper.

References

1. Maes, P.: Agents that reduce work and information overload. Communications of the ACM 37(7), 30–40 (1994)
2. Haynes, T., Sen, S., Arora, N., Nadela, R.: An automated meeting scheduling system that utilizes user preferences. In: Proceedings of the First International Conference on Autonomous Agents (1997)
3. Lin, J.H., Wu, J., Lai, S.L.: A Multi-agent meeting scheduler that satisfies soft constraints. In: Proceedings of the Second International Conference on Multiagent Systems (1997)
4. Griss, M., Letsinger, R., Cowan, D., Vanhilst, M., Kessler, R.: CoolAgent: Intelligent Digital Assistants for Mobile, Professionals – Phase 1 Retrospective, HPL Technical Report (November 2001)
5. Sen, S., Durfee, E.H.: On the design of an adaptive meeting scheduler. In: Proceedings of the Tenth Conference on Artificial Intelligence for Applications (March 1994)
6. Crawford, E., Veloso, M.: Opportunities for learning in multi-agent meeting scheduling. In: Proceedings of the AAAI 2004 Symposium on Artificial Multiagent Learning, Washington, DC (2004)
7. Chen, H.: An intelligent broker architecture for context aware systems (2002)
8. Scheible, J., Tuulos, V.: Mobile Python: Rapid Prototyping of Applications on the Mobile Platform. Wiley, Chichester (2007), ISBN: 978-0-470-51505-1
9. Bellifemine, F.L., Caire, G., Greenwood, D.: Developing Multi-Agent Systems with JADE. Wiley, Chichester (2009), ISBN: 978-0-470-05840-4

Bayesian Networks-Based Interval Training Guidance System for Cancer Rehabilitation

Myung-kyung Suh[1], Kyujoong Lee[1], Alfred Heu[1], Ani Nahapetian[1,2], and Majid Sarrafzadeh[1,2]

[1] Computer Science Department, [2] Wireless Health Institute
University of California, Los Angeles
{dmksuh,lkjrsy,alfredheu,ani,majid}@cs.ucla.edu

Abstract. The number of cancer patients who live more than 5 years after surgery exceeds 53.9% over the period of 1974 and 1990; Treatments for cancer patients are important during the recovery period, as physical pain and cancer fatigue affect cancer patients' psychological and social functions. Researchers have shown that interval training improves the physical performance in terms of fatigue level, cardiovascular build-up, and hemoglobin concentration, the feelings of control, independence, self-esteem, and social relationship during cancer rehabilitation and chemotherapy periods. The lack of proper individual motivation levels and the difficulty in following given interval training protocols results in patients stopping interval training sessions before reaching proper exhaustion levels.

In this work, we use behavioral cueing using music and performance feedback, combined with a social network interface, to provide motivation during interval training exercise sessions. We have developed an application program on the popular lightweight iPhone platform, embedded with several leveraged sensors. By measuring the exercise accuracy of the user through sensor readings, specifically accelerometers embedded in the iPhone, we are able to play suitable songs to match the user's workout plan. A hybrid of a content-based, context-aware, and collaborative filtering methods using Bayesian networks incorporates the user's music preferences and the exercise speed that will enhance performance. Additionally, exercise information such as the amount of calorie burned, exercise time, and the exercise accuracy, etc. are sent to the user's social network group by analyzing contents of the web database and contact lists in the user's iPhone.

Keywords: exercise guidance system, interval training, music recommendation, social networks, rehabilitation.

1 Introduction

In a survey of cancer patients over 1974 and 1990, over 53.9% of the cancer patients survive more than 5 years after surgeries. As these patients have a chronic illness, treatments that relieve physical pain, fatigue, psychological and social problems are complex and comprehensive. Lehmann et al. [1] identified the needs of cancer population and concluded that 438 of 805 patients surveyed had impairments and functional limitations such as psychological distress, general weakness, work-related problems, finance, housing, family support, and dependence in activities of daily life after cancer surgeries. Among those factors, physical performance and lean mass are among the most important clinical factors to predict the survival rates of cancer patients. Factors related to cardiovascular and musculoskeletal models may advance the understanding of the adaptation in cancer patients. Hence, measures of aerobic fitness and daily activity are useful to measure and determine the patients' needs and their progress in rehabilitation efforts (Gerber [2]).

For cancer patients, cancer fatigue resulting from muscle weakness, pain or sleep disruption is seen most frequently and it causes disruptions in physical, emotional, and social functions. Since emotional and social functions are highly related to physical health status, improving physical performance and recovery is significant in aiding patients regain physical, mental and social health.

Many researchers and physicians recommend interval training to improve aerobic capacity, and to restore physical functions, and cardiovascular systems. Dimeo et al. [4] showed that among the group participating in interval training exercise, fatigue and somatic complaints did not increase and psychological distress such as obsession, fear, interpersonal sensitivity, and phobic anxiety were diminished. Interval training consists of interleaving high intensity exercises with rest periods. The high intensity activity is followed by low intensity activity, referred to as the recovery period. The improvement in physical performance such as fatigue level, maximum performance, cardiovascular build-up, and hemoglobin concentration (Dimeo et al. [3]) enhanced by interval training can increase the feelings of control, independence, self-esteem, and result in better social relationships with others. In addition, interval training is a well known exercise protocol which helps strengthen and improve one's cardiovascular system ([5] – [11]). Moreover, it has been shown to help with weight loss, general fitness, and the reduction of heart and pulmonary diseases, as compared with other continuous exercise methods. During interval training, the body's energy production system is utilized, and both aerobic and anaerobic energy sources are activated. Energy from these two sources is then efficiently distributed throughout the body for the duration of the workout period.

Despite interval training's health benefits, properly following given interval training protocols is not simple. Dimeo et al. [3] assumed that less motivated patients would stop the interval training session before reaching a certain proper exhaustion level.

To provide motivation and guidance in interval training exercise sessions, we present our behavioral cueing system developed for the popular iPhone platform. It uses music, sensor readings, and social networking to encourage and motivate users to follow a healthy exercise plan. As iPhones are very light, small and embedded with sensors, they can serve as a cheaper, a more convenient, and a multi-purpose alternative to traditional exercise equipment.

When recommending music to the user, our system uses three different filtering methods: content-based, context-aware, and collaborative filtering. Especially for collaborative filtering, Bayesian networks are applied to the system to select the appropriate music, based on music preferences of others and a probabilistic model. Additionally, competitive group exercise methods incorporated into the system may motivate users to participate in interval training type exercises. By using social networks, such as sending emails to friends and uploading rankings on a shared website, etc., users can be effectively motivated to continue interval training for their rehabilitation.

2 Interval Training Motivations

2.1 Music Motivation

Terry et al. [12], Karageorghis et al. [13], and Mohammadzadeh et al. [14] examine how the rhythm of music related to personal factors and situational factors promotes more exercise with less stress. Athletes respond to the rhythmical qualities of music by synchronizing movement patterns to tempo. Rhythm related to the user's preference and situational conditions affect exercise and stress level. Works in time to music increases the likelihood of harder exercise for longer periods of time in a broad range of level of fitness. Listening to music provides benefits such as improvement in mood, pre-event activation or relaxation, dissociation from unpleasant feelings such as pain and fatigue, reduced rating of perceived exertion (RPE), especially during aerobic training. The exercise output can be extended through synchronization of music with movement, it can be enhanced when rhythm or association is matched with required movement patterns, and the music increases likelihood of athletes achieving flow states. Thus, music can be a good source of motivation during interval training, especially when the exercise method is tedious and repetitious.

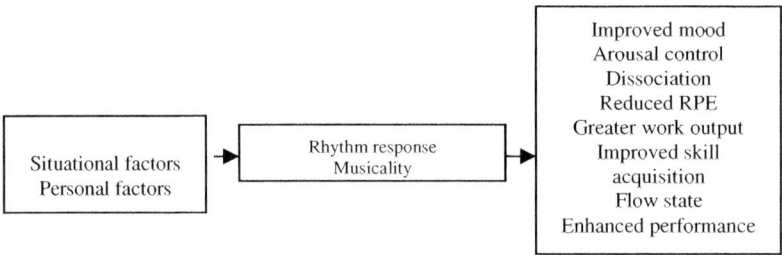

Fig. 1. Benefits of listening to music in sports and exercise contexts

2.2 Competitive Group Exercise

Competitive group interval training is another form of user motivation, with known beneficial effects on physiological functions. The experimental results (Kilpatrick et al. [15], Figure 2, and Figure 3) indicate that sport participants are more motivated to engage in physical activity as a means for enjoyment and to achieve positive health benefits. Sport participation is strongly related to affiliation, competition, enjoyment, and challenge. In addition, a program of aerobic, and endurance activities such as interval training, undertaken in a group setting, stimulates and improves physiological and cognitive functions and subject wellbeing (Williams et al. [16]). Since doing exercise together can help people maintain affiliation with friends and promote to exercise more, the designed system relates to social networks.

Subscale	Sample item
Affiliation	**To spend time with friends**
Appearance	To look more attractive
Challenge	**To give me goals to work toward**
Competition	**Because I like trying to win in physical**
Enjoyment	**Because I enjoy the feeling of exerting myself**
Health pressures	Because my doctor advised me to exercise
Ill-health avoidance	To prevent health problems
Nimbleness	To stay/become more agile
Positive health	To have a healthy body
Revitalization	Because it makes me feel good
Social recognition	**To show my worth to others**
Strength and endurance	To increase my endurance
Stress management	Because it helps reduce tension
Weight management	To stay slim

Fig. 2. Several exercise motives

Subscale	Men	Women	Total
Affiliation	4	1	2
Appearance	11	12	12
Challenge	3	3	4
Competition	1	4	1
Enjoyment	2	2	3
Health pressures	14	14	14
Ill-health avoidance	13	13	13
Nimbleness	7	10	8
Positive health	8	7	7
Revitalization	6	5	5
Social recognition	9	11	9
Strength and endurance	4	6	6
Stress management	10	9	10
Weight management	12	8	11

Fig. 3. Ranking of exercise motivation: Range = 1 (most important) to 14 (least important)

2.3 Light Weight Wireless Smartphone

Without fitness equipment such as a treadmill, interval training is very difficult to follow, since individuals are unable to accurately determine or guess their current exercise intensity. On the other hand, traditional fitness equipment has significant space and cost restrictions. Therefore, the use of an inexpensive mobile handheld device can be an effective means for guiding interval training exercises. Ease of use, compatibility, and communicability are key issues related to mobile device use and adoption by individuals (Kleijnen [18]).

Sarker [17] mentions factors influencing the use of mobile handheld device by individuals, including technology, communication/task characteristics, modalities of mobility, and context (Fig. 4.). Those same features support our choice of a smart phone platform for our system.

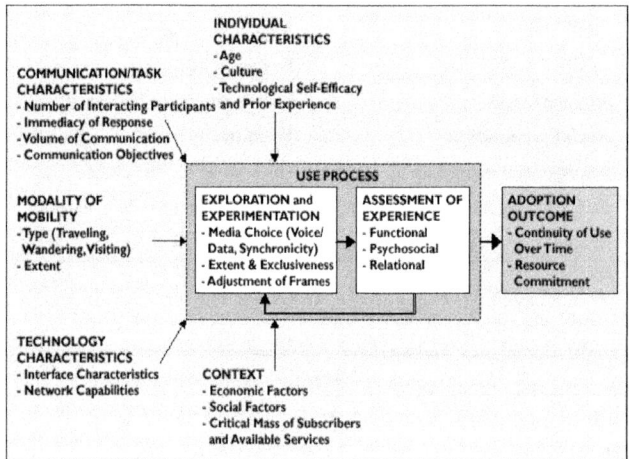

Fig. 4. Factors influencing mobile handheld device use and adoption [17]

Accelerometers are currently among the most widely studied wearable sensors for activity recognition. They are also very useful for interval training. By analyzing data obtained from three different axes, the accuracy of exercise and the caloric consumption can be calculated. This information can be used to calculate one's score to motivate the user to compete with other peoples in order to obtain better score. Accelerometers are commonly embedded in smart phones, such as Apple iPhone, Google G1, and Nokia N95. Moreover, smart phones can help individuals to follow scheduled speeds, during a specific time period, and can also give feedback using the cell phone's sound, vibration, and other graphical interfaces. Additionally, their calculation functionalities can also be leveraged.

With technology, the easy interface of a mobile device is important. An iPhone's 3.5 inch multi-touch display with 480-by-320-pixel resolution makes users navigate by touching the screen. The multi-touch display layers a protective shield over a capacitive panel that senses touches using electrical fields. It then transmits that information to the LCD screen below it, and the iPhone software enables the flick, tap, and pinch. The iPhone display also supports multiple languages and characters for users worldwide.

Poor network characteristics act as severe inhibitors to use and adoption. For example, the lack of coverage in many areas tend to reduce the sense of freedom and safety in many subject's minds. The iPhone 3G uses a technology protocol called HSDPA (High-Speed Downlink Packet Access) to download data quickly over UMTS (Universal Mobile Telecommunications System) networks. Accessing the internet to load information is twice as fast on 3G networks as on 2G EDGE networks. Since the iPhone 3G meets worldwide standards for cellular communications, a user can make calls and surf the web from practically almost anywhere. When a user is not in a 3G network area, the iPhone uses a GSM network for calls and an EDGE network for data.

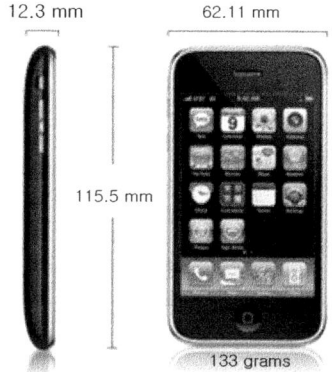

Fig. 5. The size and weight of an iPhone

According to the market research group NPD, Apple's iPhone 3G topped the sales charts in the third quarter of 2008. Therefore, many iPhone users and developers share information regarding iPhone systems and applications via several websites or magazines. In addition, the number of interactions among iPhone user groups and developer groups continues to increase.

Research has shown that, for a variety of reasons, humans are attracted to the natural environment (Knoph [18], [19], and Kaplan [20]). These investigations have shown that when encountering or presented with images of natural environments, subjects experience a variety of positive psychological, social, and physiological outcomes. Since an iPhone weighs only 133g, it is easy to carry outdoors where interval training can be performed; the only limitation is the possession of a light 3G smart phone.. In a gym, on the other hand, a person is limited to a treadmill

Sarker [17] also emphasizes on the modalities of mobility for the use of adoption in mobile devices. Traveling, wandering, and visiting are seen as three ways to qualifying the essence of mobility (Kristoffersen [21]). Traveling is defined as the process of going from one place to another in a vehicle. Wandering refers to an extensive local mobility where an individual may spend considerable time walking around. Visiting refers to stopping at some location and spending time there, before moving to another. Three different types of mobilities are associated with different motivations of a user. For instance, safety is an important concern for a person traveling frequently. Therefore, the iPhone's 3G internet connectivity, and light size can be a good source of motivation to use in this system for all three different groups.

3 System Design

In our system, we use the iPhone [22] and a web server system for our development. The iPhone is a 133g, 3.5 inch multi-touch display smart phone that supports both Wi-Fi and Bluetooth. Furthermore, the iPhone has an in-built accelerometer, a light and a proximity sensor. By using an embedded 3-axis accelerometer on the iPhone, activity patterns are detected. QuickTime, a proprietary digital media player application, is embedded in the iPhone, and it is capable of handling various formats of digital video, media clips, sound, text, animation, music, and interactive panoramic images. Also, the iPhone features the Safari web browser with 3G and Wi-Fi, which allows the user to access the Internet almost anywhere. When songs and users' information are stored on a web server, the 8 or 16 gigabytes storage of an iPhone doesn't have to be wasted. The remaining data storage can be used for other purposes. Based on given information, the system recommends suitable songs, and connects users.

3.1 Music Recommendation

As mentioned, music is a good source of motivation during exercise by synchronizing movement patterns to the rhythm of the music. In addition, personal factors such as one's age, sex, cultural background, and education level also affect one's taste in music. Therefore, the group of users who share similarity can be generated by collaborative filtering technique. A user's choice in music in the past can affect the list of recommended music by content-based filtering. Content-based recommendation systems with collaborative filtering analyze the content of the objects that a user has preferred in the past, and they recommend other relevant contents by using one's history and the nearest neighbor's information (Park [23], Adomavicius [27], Balabanović [28]).

Based on training data, a user model is induced that enables the content-based filtering technique to classify unseen items. The training set consists of the items that the user has found interesting. These items form training instances that all have an attribute. This attribute specifies the class of the item based on either the rating of the user or on implicit evidence. With the class information, the list of recommended items is determined. A content-based filtering system selects items based on the correlation between the content of the items and the user's preferences, as opposed to a collaborative filtering system that chooses items based on the correlation among people with similar preferences. Collaborative filtering is the process of filtering for information or patterns using techniques involving collaboration among multiple agents who share similarities in attributes. The LIBRA system (Mooney [29]) recommends books by using content-based and collaborative filtering. Many systems use hybrid features of content-based and collaborative filtering to recommend items. Those systems combine knowledge about users who liked a set of items with knowledge of a particular content feature associated with the item in one user's set.

User context is any information that can be used to characterize the situation of an entity. Context includes location of use, the collection of nearby people and objects, accessible devices, and changes to these objects, etc. A context-aware system in recommendation systems is to provide a user with relevant information and services based on one's current context (Van Setten [25]). Context information which characterizes the situation of the user, place or objects should be considered to recommend proper music to the user. For example, Van Setten [25] realizes the mobile tourist application which takes account the user's interests and contextual factors such as the place last time visited. The PILGRIM recommendation system (Brunato [26]) uses mobility-aware filtering techniques with a GPS system. Our system recognizes the exercise context by using user input such as exercise time, the amount of calorie to be burned and data stream from an accelerometer. Using this information, the system can recommend suitable music for the exercise context.

Our system classifies songs by genre and the speed of the music, and it recommends the songs to the user based on the user's history and background. The system classifies users based on age, gender, and residential location. Recommendations for users are in the same group are affected by the annotations of other members of the group. In addition, music which corresponds to the speed and the intensity of interval training is recommended to the user. A 3-axis accelerometer embedded in the iPhone and user input are used to select a suitable music. The system filters music recommendations using a comparison or target speed and current speed. For instance, when the user runs slower than the target speed, faster music is played. Therefore, our system realizes the collaborative, content-based, and context-aware recommendation by analyzing the user's information, the group the user belongs to, and the speed of the exercise. As user's annotation is accumulated in the database, the list of the recommended music is also updated and made more suitable for the user and the exercise context. Thus the recommended music is modified and adapted, as the number of annotation data increases.

Fig. 6. An example of the music recommendation part in our system

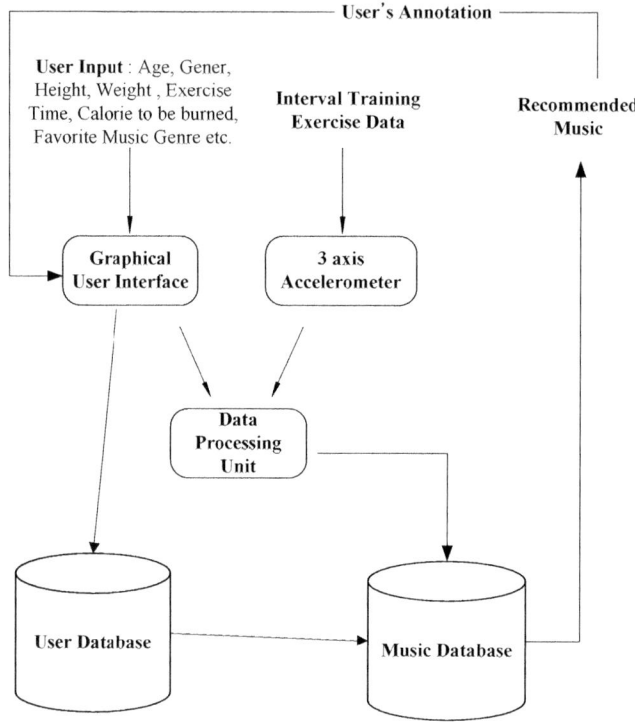

Fig. 7. The structure of context-aware, content-based, and collaborative music recommendation system

The music database currently stores 822 mp3 files, which is around 4.64 gigabytes. Since songs are not stored on the iPhone, the remaining storage can be used for other purposes.

The preference of music can be determined by a listener's age, gender, ethnic group, residential area, family and peer group, etc. LeBlanc ([30], [31]) examines that the age of a music listener exerts a strong influence on overall preferences of music style and the speed of music. Christenson [32] denotes gender is central to the ways in which popular music is used and tastes are organized. Even though the underlying structure of music preference cannot be accounted for by reference to two or three factors, there are crucial difference between males and females in terms of their mapping of music types. In addition, preferences in popular music also vary according to the neighborhood in which the music listener lives (Johnstone [33]). People can also be affected by their family and peer groups, or their backgrounds, such as music training experience and level of education. There are additional factors which affect the music preference shown in Fig. 8.

Using the structure in Fig. 8. and information which can be obtained by iPhone such as sex, age, and exercise intensity, location data obtained by GPS embedded

Bayesian Networks-Based Interval Training Guidance System for Cancer Rehabilitation 245

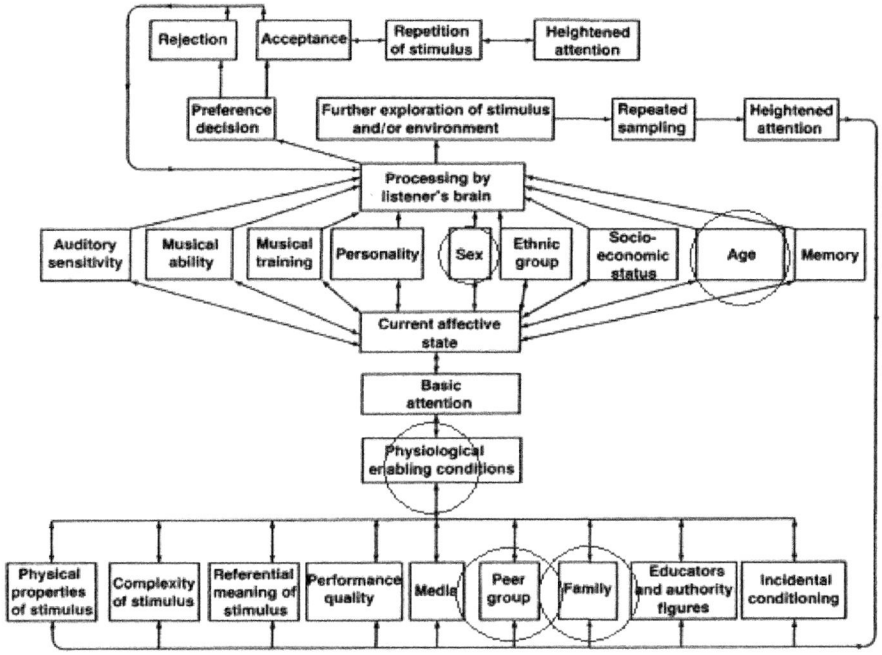

Fig. 8. Sources of variation in music preference

in the iPhone, etc. , a Bayesian network structure in Fig. 10. is designed. Peer group is assumed to be formed by age, gender, and residential district, since these factors affect perceptions of relationships (Furman [34]). Physiological enabling condition is assumed to be related to the exercise intensity information such as exercise time, and calories to be burned in the user's input.

Based on the above assumptions, a Bayesian network model is obtained and is used to calculate the probability that the given song is recommended by people sharing similarities with the user. Bayesian networks are a probabilistic model using a set of random variables and conditional relations. This probabilistic model attempts to make valid predictions based on only a sample of all possible observations.

$$\Pr(CityA, Age5-10, Speed0-5, Male \mid SelectMusicA) =$$
$$\frac{\Pr(CityA) * \Pr(Age5-10 \mid CityA) * \Pr(Speed0-5 \mid Age5-10) * \Pr(Male \mid Speed0-5) * \Pr(SelectMusicA \mid Male)}{\Pr(SelectMusicA)}$$
$$\geq Threshold$$

Fig. 9. Probabilistic model of Fig. 10

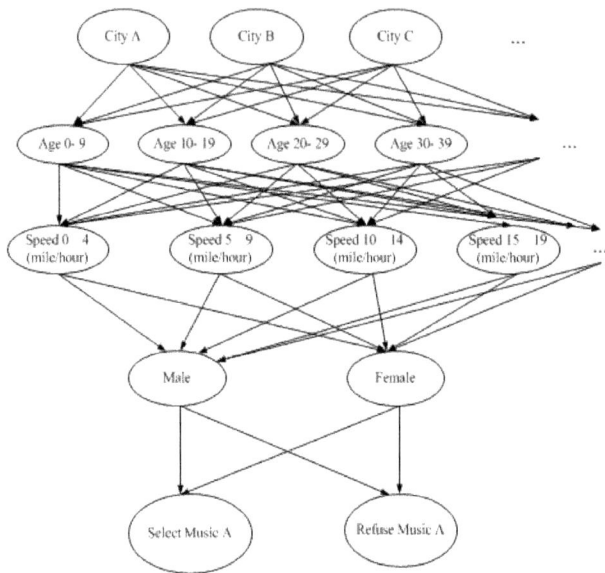

Fig. 10. Bayesian networks related to music preference and choice

3.2 Interval Training Game with Social Network

With the inserted user input (Fig 11), the system design is customized to an interval training protocol. The user input screen requires user information including weight, height, the amount of time to exercise, and the amount of calorie to be burned. By using the following equation, the system designs a customized exercise plan for the user.

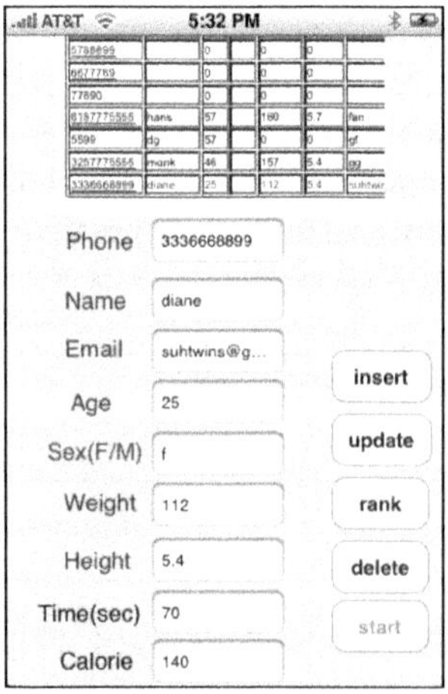

Fig. 11. User input screen

$W = m \cdot g \cdot h \cdot h\_rate \cdot v' \cdot t + \frac{1}{2} \cdot m \cdot v^2 \cdot t$ (J)

m: mass (kg)
g : the gravitational constant (kg/m2)
v : the speed of running (m/s)
v' : the number of steps per second (steps/s)
t : collapsed time (s)
h : height (m)
h_rate : the rate of height lifted up when walking/running

An iPhone provides a 3.5 inch 480-by-320-pixel resolution multi-touch display, an audio system which has a frequency rate from 20Hz to 20,000Hz, and vibration functionality. Our system gives three different feedback commands to the user through the use of sound, vibration and animation based on the developed schedule.

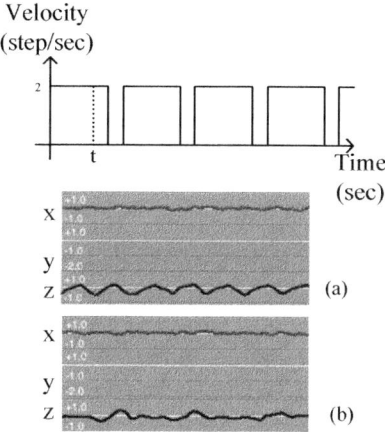

Fig. 12. Scheduled interval training and the accelerometer data for the accurate exercise (a) and the inaccurate exercise (b) at time t

By comparing the determined schedule of the interval training exercise with the exercise data collected via a 3-axis accelerometer embedded in an iPhone, the accuracy of the exercise is calculated. The system gives a lower score to the user who does not complete the interval training session accurately in order to motivate the user to exercise more and more precisely in an effort to get a better score. For instance, the accuracy of the exercise shown in Fig. 12. is 50%, since only 3 steps among 6 given commands are completed.

$$Accuracy = \frac{The\ number\ of\ correct\ steps}{The\ number\ of\ times\ commands\ are\ given}$$

The user's interactions with the iPhone are synchronized with the user's online profile registered in the web database of our system and the user's friend group, a group of users who participate in the iPhone interval training guidance system. E-mails containing the accuracy of the exercise sessions, exercise session time, and the amount of calories burned, etc. can be sent to other members in the user's social networking group after an exercise session is completed in order to increase the interaction among the user's friend group and to use the sharing of the information to increase the motivation to exercise, in the hopes of competing with friends.

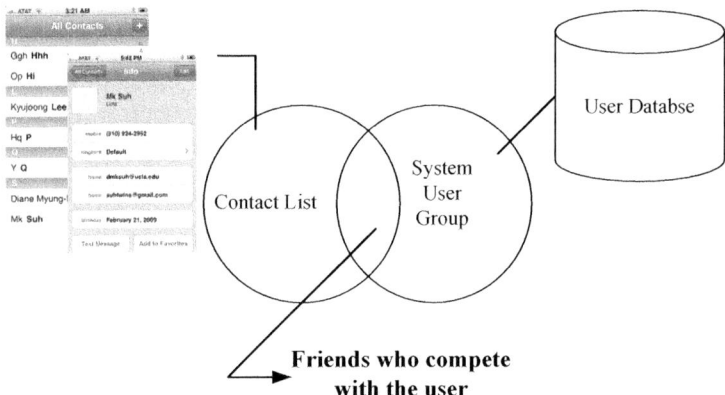

Fig. 13. The method to get the friend group who compete with the user in our system

4 Experimental Results

4.1 The Effect of Interval Training Guidance Systems

As shown in Table 1, eight different individuals participated in the experiment to test the effectiveness of our system. All of these individuals are between the ages of 20 and 30, and they all live in Los Angeles, California. Exactly half of the participants are female.

Table 1. Information about individuals who participated in the experiment

	Individual 1	Individual 2	Individual 3	Individual 4	Individual 5	Individual 6	Individual 7	Individual 8
Gender	Female	Male	Male	Female	Female	Male	Male	Female
Age	25	24	27	28	25	29	27	25
Weight (kg)	51	61.4	73	49.5	50.5	62	70	51
Height (cm)	158	170	175	163	164	172	175	158
Residential District	Los Angeles, CA	Los Angeles, CA	Los Angeles, CA	Los Angeles, CA	Los Angeles, CA	Los Angeles, CA	Los Angeles, CA	Los Angeles, CA

Compared with uncontrolled condition, the experiment (Fig 4.2.) shows that exercise commands generally help users to exercise more accurately. Especially for individual 4, the accuracy of the exercise was improved from 53.97% to 88.71%. Dimeo [3] showed, 7 weeks after discharge, improvement of maximum performance, cardiovascular system build-up, fatigue level, and hemoglobin concentration is significantly increased in the interval training exercise group.

It shows that the most effective command combination is different from each person. Hale [35] also mentions different people have different time to perform information processing task depending on the type of information. This result also can be affected by the circumstance of the experiment. For example, the sound command is not a proper exercise command to the user in the noisy area or for the user who has a hearing problem.

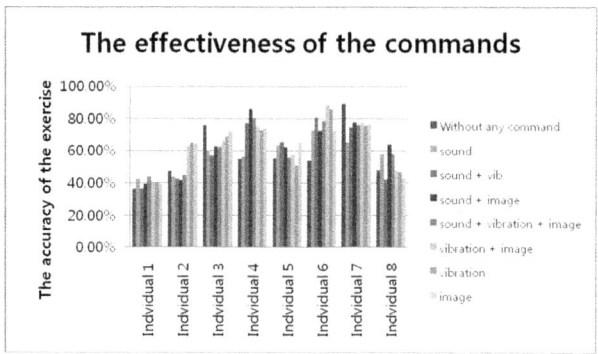

Fig. 14. The accuracy of the exercise with given commands

Since the vibration affects the reading of the 3-axis accelerometer data, it has more than 85 % of the false positive rate(Fig 4.3). When using probabilistic model(Fig 4.4) to calculated true positive rate of the movement detection with using vibration commands, the true positive detection rate of indvidual 5 is 0.44.

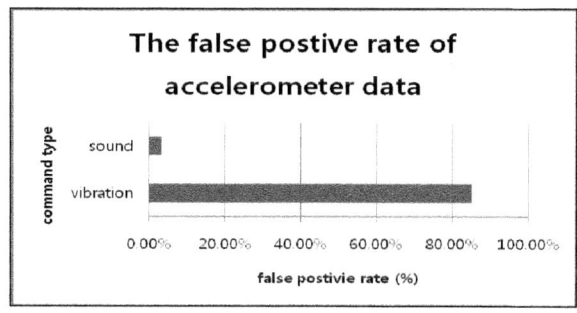

Fig. 15. The false positive rate of accelerometer data with and without vibration command

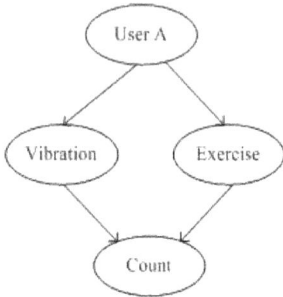

Fig. 16. Probabilistic model to calculate true positive rates using vibration commad

4.2 A Music Recommendation System Suitable for Interval Training Exercise

Each individual, as shown in Fig. 14, was requested to annotate his/her preference in music by using the music annotation user interface. Members in each individual group share gender, age, and residential area. Each song in the web database was annotated more than 8 times. Fig. 17 shows the number of refused songs among 10 recommendations for a 30 years old, 180cm, and 80kg individual living in Los Angeles, California. Compared with the method which recommends music preferred by people who share the same age range, gender, similar exercise intensity and residential area, Bayesian networks-based recommendation method is better for selecting suitable exercise music.

In addition, by using content-based filtering, if there is music that only one specific individual does not like, this song will not be recommended for that user, but still will be recommended to others until the rating is below the threshold level. Music is also selected by the intensity of the interval training. The appropriateness of music in the exercise context is more suitable than in the uncontrolled condition using random music. This content-based, context-aware, and collaborative music recommendation system will choose suitable music for the user who is engaged in the interval training.

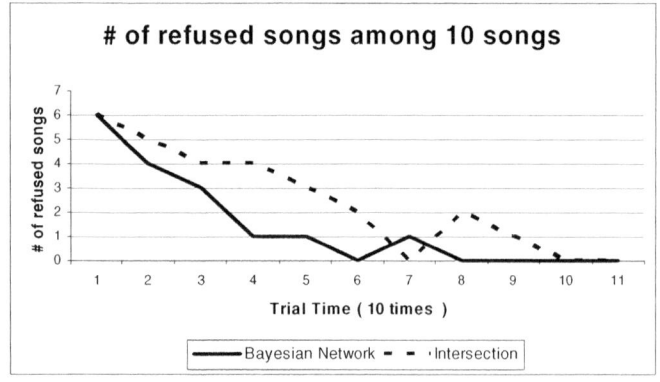

Fig. 17. The number of refused music among 10 recommended

5 Conclusions

Interval training involves a series of intensive exercises to provide proper stress to cardiovascular and musculoskeletal systems with recovery periods. Interval training exercises include establishing a period for high intensity activity, followed by a period of low intensity activity referred to as a recovery period, a frequency with which this cycle is repeated, and finally a number of cycles corresponding to one complete session after which the subject ceases activity. Interval training improves the physical performance in terms of fatigue level, cardiovascular build-up, and hemoglobin concentration, the feelings of control, independence, self-esteem, social relationship during cancer rehabilitation and chemotherapy periods. Furthermore, interval training is beneficial for weight loss, rehabilitation, general health, and cardiovascular build up. However, the lack of proper individual motivation levels and the difficulty in following given interval training protocols results in patients stopping interval training sessions before reaching proper exhaustion levels. Conventional interval training methods also have the limitation that an individual subject may not be equipped with the capability to predict the level of effort required to provide an optimal training intensity.

Our game-like and social networking application with a music recommendation system on a light and small smartphone can be used to prompt participation in the interval training without any large or expensive equipment. The performance of the user can be measured by sensor readings, specifically accelerometers embedded in the iPhone. Additionally, the exercise information of the user is sent to the user's friends group. Our experiments show that customized interval training schedules and commands generated based on the user information increase the accuracy of the interval training up to 88.71%.

A content-based, context-aware, and Bayesian networks-based collaborative filtering algorithm incorporates user music preferences and the exercise speed to play music to enhance performance. Based on user information combined with the information of other individuals in the same group and the speed of the interval training, the list of recommended music is generated and evolved. The experiment shows that the Bayesian networks-based collaborative method is better for recommending exercise music than the intersection methods. As more data related to the music is accumulated, the experiments show improvement in the selection of recommended music.

References

[1] Lehmann, J.F., DeLisa, J.A., Warren, C.G., et al.: Cancer rehabilitation: assessment of need, development, and evaluation of a model of care. In: Archives of physical medicine and rehabilitation (1978)
[2] Gerber, L.H.: Cancer rehabilitation into the future. Cancer (2001)
[3] Dimeo, F.C.: Aerobic exercise in the rehabilitation of cancer patients after high dose chemotherapy and autologous peripheral stem cell transplantation. Cancer (1997)
[4] Dimeo, F.C.: Effects of physical activity on the fatigue and psychologic status of cancer patients during chemotherapy. Cancer (1999)

[5] Gorostiaga, E.M., Walter, C.B., Foster, C., et al.: Uniqueness of interval and continuous training at the same maintained exercise intensity. Eur. J. Appl. Physiol. (1991)
[6] Astrand, I., Astrand, P.O., Christensen, E.H., et al.: Intermittent muscular work. Acta Physiol. Scand (1960)
[7] Billat, V., Slawinski, J., Bocquet, V., et al.: Intermittent runs at vV. O2max enables subjects to remain at VO2max for a longer time than submaximal runs. Eur. J. Appl. Physiol. (2000)
[8] Essen, B., Hagenfeldt, L., Kaijser, L.: Utilization of blood-born and intra-muscular substrates during continuous and intermittent exercise in man. J. Physiol, Lond (1977)
[9] Fox, E.L., et al.: Frequency and duration of interval training programs and changes in aerobic power. Journal of Applied Physiology 38(3)
[10] Reindell, H., Roskamm, H., Gerschler, W.: Das Interval training. John Ambrosius Barth Publishing, Munchen (1962)
[11] Billat, L.V.: Interval Training for Performance: A Scientific and Empirical PracticeSpecial Recommendations for Middle- and Long-Distance Running. Part I: Aerobic Interval Training. Sports Med. (2001)
[12] Terry, P.C., Karageorghis, C.I.: Psychophysical effects of music in sport and exercise: an update on theory, research and application. In: Joint Conference of the Australian Psychological Society and the New Zealand Psychological Society (2006)
[13] Karageorghis, C.I., Terry, P.C.: The magic of music in movement. Sport and Medicine Today (2001)
[14] Mohammadzadeh, H., Tartibiyan, B., Ahmadi, A.: The effects of music on the perceived exertion rate and performance of trained and untrained individuals during progressive exercise. Physical Education and Sport (2008)
[15] Kilpatrick, M.: College Students' Motivation for Physical Activity: Differentiating Men's and Women's Motives for Sport Participation and Exercise. Journal of American college health (2005)
[16] Williams, P.: Effects of group exercise on cognitive functioning and mood in older women. Australian and New Zealand Journal of Public Health (1997)
[17] Sarker, S., Wells, J.D.: Understanding mobile handheld device use and adoption. Communications of the ACM (2003)
[18] Kleijnen, M., de Ruyter, K., Wetzels, M.G.M.: Factors influencing the adoption of mobile gaming services. In: Mennecke, B.E., Strader, T.J. (eds.) Mobile Commerce, Idea (2003)
[19] Knopf, R.C.: Recreational needs and behavior in natural settings. In: Altman, I., Wohlwill, J.F. (eds.) Human behavior and environment: behavior and the natural environment, vol. 6. Plenum Press, New York (1983)
[20] Knopf, R.C.: Human behavior, cognition, and affect in the natural environment. In: Stokols, D., Altman, I. (eds.) Handbook of environmental psychology, vol. 1. Wiley, Chichester (1987)
[21] Kaplan, R., Kaplan, S.: The experience of nature: a psychological perspective. Cambridge University Press, Cambridge (1989)
[22] Kristoffersen, S., Ljungberg, F.: Mobility: From stationary to mobile work. In: Braa, K., Sorensen, C., Dahlbom, B. (eds.) Planet Internet (2000)
[23] Apple iPhone: http://www.apple.com/iphone/
[24] Park, H.: A context-aware music recommendation system using fuzzy bayesian networks with utility theory. LNCS. Springer, Heidelberg (2006)
[25] Cano, P., et al.: Content-based music audio recommendation. In: Proc. ACM Multimedia (2005)

[26] Van Setten, M.: Context-aware recommendations in the mobile tourist application COMPASS. LNCS. Springer, Heidelberg (2004)
[27] Brunato, M.: PILGRIM: A Location Broker and Mobility-Aware Recommendation System. In: Proceedings of the First IEEE International Conference on Pervasive Computing and Communications (2003)
[28] Adomavicius, G., Tuzhilin, A.: Toward the next generation of recommender systems: A survey of the state-of-the-art and possible extensions. IEEE T Knowl. Data En. (2005)
[29] Balabanović, M.: Fab: content-based, collaborative recommendation. Communications of the ACM (1997)
[30] Mooney, R.J.: Content-based book recommending using learning for text categorization. In: Proceedings of the Fifth ACM Conference on Digital Libraries (2000)
[31] LeBlanc, A.: Tempo Preferences of Different Age Music Listeners. Journal of research in music education (1988)
[32] LeBlanc, A., Sims, W.L., Siivola, C., Obert, M.: Music style preferences of different age listeners. Journal of Research in Music Education (1996)
[33] Christenson, P.G., Peterson, J.B.: Genre and gender in the structure of music preferences. Communication Research (1988)
[34] Johnstone, J., Katz, E.: Youth and popular music: A study in the sociology of taste. American Journal of Sociology (1957)
[35] Furman, W.: Age and sex differences in perceptions of networks of personal relationships. Child development (1992)
[36] Hale, S.: Global processing-time coefficients characterize individual and group differences in cognitive speed. Psychological science (1994)

Mobile Context-Aware Personal Messaging Assistant

Senaka Buthpitiya[*], Deepthi Madamanchi, Sumalatha Kommaraju,
and Martin Griss

Carnegie Mellon Silicon Valley,
NASA Research Park, Bldg. 23, Moffet Field, CA 94035
{senaka.buthpitiya,deepthi.madamanchi,sumalatha.kommaraju,
martin.griss}@sv.cmu.edu

Abstract. A previous study shows that busy professionals receive in excess of 50 emails per day of which approximately 23% require immediate attention, 13% require attention later and 64% are unimportant and typically ignored. The flood of emails impact mobile users even more heavily. Flooded inboxes cause busy professionals to spend considerable amounts of time searching for important messages, and there has been much research into automating the process using email content for classification; but we find email priority depends also on user context.

In this paper we describe the Personal Messaging Assistant (PMA), an advanced rule-based email management system which considers user context and email content. Context information is gathered from various sources including mobile phones, indoor and outdoor locationing systems, and calendars. PMA uses separate scales of importance and urgency to prioritize emails and to decide on an appropriate action, such as SMS to user, defer to later, file or forward. Initial results yield 96% recall and 88% precision in importance classification of emails; 95% recall and 92% precision in urgency classification of emails. PMA shows a 30X reduction in false-negative rates over existing systems. A key contribution of our work has been to leverage an extensible set of context information, gathered in a mobile environment, for the classification of emails and customizable decision making.

Keywords: Context-aware computing, mobile, messaging, email assistant, Jess rules, prioritization, productivity, urgency, priority, SMS.

1 Introduction

Today email is not only used as a messaging tool but also for task management, information storage and archiving, scheduling and social communication [1]. This clearly explains that email has evolved into a primary and very frequently used communication tool. Therefore, the problem of email overload is ubiquitous [2]. There is a strong need for a system which can intelligently prioritize and manage email for busy professionals. On average a professional can receive upwards of 50 email messages per day, including all the email inboxes the user owns; typically only an average

---

[*] Corresponding author.

of 23% of these emails are important at a given instant in time based on his context. These important emails either get lost in the flood of unnecessary email or are forgotten even after being marked for later processing because there is no reminder system to notify users about the unprocessed-marked emails. These deferred important emails should instead be presented at the right time. It is more helpful to have a user examine a message when needed, rather than in the order it arrives in the mail box, often interrupting important work. While this is important to all busy professionals, the problem is particularly acute for users who are increasingly mobile, with limited time and dynamically changing situations.

We envision a personal messaging system that addresses many of the concerns described above. For example, a busy anesthetist may want to receive notifications of only the important mails at night. Usually his nurse, who works from 9 a.m. to 5 p.m., has access to his inbox to take care of them. But, during the rest of the day, he may want to have a system which notifies him of important emails. During the night, he wants all urgent emails to be sent to his smart phone, perhaps summarized as SMS. The incoming urgent emails could instantly trigger an alarm. Similarly, a mobile engineer may notice oxidation on some equipment – not yet serious, but something to deal with eventually. The project manager is informed about this oxidation, not when the message is sent, but instead when she is visiting that area. If a worker sends a message to his manager about a repair issue in a manufacturing facility (requiring future attention), the manager may receive it while he is, for example, enmeshed in arranging a conference. Since the information is not time critical, he does not address it at that time, and it sits in the manager's inbox where it is likely to be forgotten; instead it would be better if this information is delivered (or a reminder issued) only when he is planning maintenance for equipment in that area.

We have developed a Personal Messaging Assistant (PMA) prototype, which primarily captures incoming (email) messages, analyzes them based on the email content and the user context and stores or processes the messages for immediate or eventual delivery. The PMA system consists of two parts: an email scanning server and a client on a mobile device. The messages coming into the inbox are scanned by the server periodically at stipulated time intervals and repeatedly prioritized by an intelligent rule-based system that makes decisions based on message content, user preferences and contextual information. Contextual information transmitted from the mobile device includes current location, topics of interest for the user, feedback regarding received messages, current activity, or social context such as who is nearby. The system also uses information from email content, the calendar, history of movement and patterns of information access. The mobile device periodically sends user context information to the server via http; and also receives selected messages delivered via SMS[1]. This system can be customized to suit an individual's preferences and social context, such as a dynamically changing list of important colleagues.

While a multitude of email-content based email sorting/classification programs exist, we believe our system outperforms these systems by taking context-information into consideration. PMA outperforms other context based email sorting programs [3]

---

[1] Even if the user's mobile phone has a built-in email client, such as Blackberry, prioritization and filtering of email will help reduce clutter and ensure that the most appropriate messages are seen first. We used SMS for convenience to avoid modifying the client email reader, and as a ubiquitous channel available on most phones.

with the use of separate scales to measure the importance and urgency of an email. This allows PMA to classify and decide on an action to take with an email with greater accuracy. The improved performance PMA offers can be seen with the large reduction in false-negatives and the high accuracy and precision results. The rest of the paper is divided into five sections which describe, respectively, other work and programs related to PMA; the design and implementation of the PMA system; performance testing done on the system and an interpretation of the results; the future improvements we hope to make with the system; and finally we conclude with a summary of our work and contributions.

2 Related Work

The Email overload has been a problem since the early 1990s [2]. Since then several techniques have been proposed to effectively manage emails. The idea of prioritizing and filtering the emails based on personal attributes is not a novel concept. We list such work and highlight their key features in this section. We also highlight relevant research related to filtering email spam.

2.1 Personal Email Assistant

The Personal Email Assistant [3] prototyped a system which can prioritize, filter, index and re-file all the emails. This system used information retrieval (Lucene) and statistical methods (WEKA, SVM) to classify incoming emails from a Microsoft Exchange email server, and then a rule system to decide on what to do with each class of mail. However, their system did not address mobile users, or changing dynamic context. The system was also envisioned as part of an ensemble of information management agents, rather than as a distinct tool.

2.2 Cool Agent Personal Assistant

The HP Laboratories Cool Agent project [4][5], continued later as the UCSC ScateAgent project [6], developed a series of context-aware personal and team assistant agents to manage information, arrange meetings, help software engineering teams and offer travel advice. Each of these used rule-based systems in conjunction with personal preferences and context, access to calendar information as well as mobile devices. In both systems, the context-aware and preference-aware notification agents play an important role. The user can set his preferences and the notification agent notifies the user whenever there is a change in the system. The notification agent used in this system is very similar to the delivery agent in PMA.

The CoolAgent Personal Assistant (PA) also uses multiple agents which can route messages and notifications. Based on the preferences and context, the notification agent routes the messages via one or more channels (email, voicemail, IM/jabber, or pager). One part of the overall PA vision is for the Personal Email Assistant (PEA) to interact as a peer or child of the other agents. In one direction, the PEA uses the calendar agents, the notification agents and the PA to find information and to communicate with the user. In the other direction, email messages concerning meetings could trigger the meeting agents, or at least monitor, prioritize and route email relevant to specific meetings.

2.3 Conventional Email Rules and Filters

Almost every email client can allow the user to set up some rules which can recognize (or filter) messages which have certain keywords or email header information. The rules can then delete, file or forward the messages. For example, Microsoft Outlook has two levels of rule settings, one on the Microsoft Exchange Server and another on the local outlook client. Microsoft Exchange Server uses rules on mailboxes and other folders to automatically execute actions on objects in the folders. One can use rules to develop applications that carry out predefined or custom actions, even at times when the client application is not running. Rules can be performed on the Exchange store (server-side) or on Microsoft Outlook (client-side). One of the main disadvantages of client-side rules is that they can only run when Outlook is running online. Also, the rules are very rigid and have only two states for any set rule: Yes or No. In our work, the PMA server can also act on the user's behalf even when the email client is not running.

Gmail provides filters to manage the flow of the incoming messages. Using these filters, one can perform only basic operations such as forwarding, deleting, or labeling based on the combinations of keywords, sender, recipients etc. These filters are very rigid and do not take into account contextual information. Also certain useful settings, such as forwarding emails as IM or SMS are not possible.

2.4 Spam Filters

In the past decade, communication through emails has grown exponentially and so has spam. As of 2002, the number of spam messages sent daily was 2.4 billion [7], whereas by mid-2007, it had reached an estimated 100 billion per day [8]. Spam filtering uses a variety of techniques such as rule-based rankings, Bayesian word distribution filters, distributed adaptive blacklists, white list verification filters, Bayesian trigram filters, etc.

Rule-based spam filters like SpamAssassin [9] evaluate a large number of patterns and match the content of an email. SpamAssassin uses techniques such as keyword filters, email header analysis, email content databases, statistical filters, and negative rules. SpamAssassin [10] mainly uses rules and weights (also known as scores). Each rule performs a test on the email and attaches a weight to the email. After an email is processed by all the rules, the final weight, which is an addition of individual weights, is compared to the threshold weight. If the score exceeds the value of the threshold, the email is tagged as spam. This process of assigning weights to the email, comparing to a threshold and tagging has influenced the PMA design. SpamAssassin also lets its users configure their email delivery system. Through this, the users can decide whether the tagged email is spam or not and perform an appropriate action on the email.

Our rule and weights based approach (described more fully below), has a number of unique features, such as the use of a wide variety of extensible context information gathered from a mobile environment, the representation of all user information as context information, the processing of context information from a rule based system, etc. Unlike the earlier systems, the PMA system separately calculates the importance and urgency of an email as two separate indicators which affect priority and therefore to achieve superior email classification results.

3 Approach – Ranking Emails

The primary objective of the Personal Messaging Assistant (PMA) is to effectively prioritize and categorize emails for immediate or delayed delivery, forwarding or filing, by taking context information into consideration. The system decides to deliver specific emails to the user as and when an email (i.e., its content and envelope) becomes important and/or urgent in the current context. It also decides on the appropriate mode of delivery (e.g., SMS, Text-to-Voice on the mobile phone, IM/XMPP) based on the current context of the user. Earlier prototypes of PMA and past work in the field [11] have shown that using a single prioritization metric to achieve this objective was ineffective, as it could not simultaneously account for both urgency and importance of an email.

In our work, we use two separate and independent scales to rate the emails being processed. The first scale represents an email's importance. The second scale represents an email's urgency. The use of these two scales is based on the observation that important emails need not necessarily be urgent and vice-versa. For example, consider: 1) An email from a user's spouse, asking to pick up their son in an hour is both important and urgent; 2) An email from the user's boss, about a deadline that is two weeks away being pushed back another month is important but possibly not urgent; 3) An email from an online auctioning system informing the user about a better bid is urgent but, in most cases, not important; 4) An email invitation from a colleagues is neither urgent or important (see Fig. 1).

Therefore by rating incoming emails using these two different scales, PMA is able to make more flexible decisions than if it were using a single scale. The importance and urgency values for each email is assigned based on the current context of the user and vary with variations in the user's context.

	Unimportant	Important
Non-Urgent	Evite for a BBQ.	From manager: Client visit pushed back by another month.
Urgent	Online auction: you were out bid.	Son missed his bus, pick him up from school.

Fig. 1. Classifications of Importance and Urgency (adapted from [12])

3.1 Context Representation

Context is the differentiating factor that makes PMA unique in the way it handles email. We classify context information into two categories: static context and dynamic context. As their names suggest, static context encapsulates the user preferences, and priorities that either remain relatively constant over time or change gradually. On the other hand, dynamic context might change almost every instant. Static context includes information such as user email contacts, mobile phone book contacts, age, social group, email address, names, and preferred emails. Dynamic context is comprised of information such as user's location and activity that are communicated at regular intervals to the server from the mobile devices. Dynamic context also includes user specific information such as calendar information, topics of interest, people nearby, etc. While static context is mainly gathered from explicit user input and the user's email inbox and managed on the servers, the dynamic context is partly updated from the mobile device at regular intervals and partly fetched from external servers.

For example, to gather location information, we use a combination of exterior and interior locationing. In our prototype PMA, we use an extended Wi-Fi Redpin service [13, 14] that sends indoor location information to the PMA server while outdoor locationing is achieved through a combination of GPS and cell ID information. Fetching context information from external servers is exemplified in the PMA prototype with the use of the Business Meeting Organizer system (also presented at this conference) to retrieve calendar information. Context information received by the PMA system from various sources is stored in the context engine of the system.

The dynamic context information and the static context is represented within the PMA system in a context data object with fixed fields for location, activity and user's email addresses and an extendable set of tag-value fields for representing additional context information. The tag is a textual representation of the type of the context data being represented and the value section contains the value of the context data type. Fixed fields for location, activity and user email addresses were created in the data object as they always represent a meaningful value.

The PMA system stores both types of context (static and dynamic) using the same set of extensible tag-value pairs. By storing both static context (which is generally considered as user information [15]) and dynamic context (which is regarded in most context-aware systems [16] as the only form of context information) the same way allows the PMA rule systems to use a uniform method to access this information. We similarly store user preferences.

The exact context information represented in a context data object at any given moment varies with the context of the user at that point in time. For example – 1) when the user is in a meeting the context data object holds information as to the topic of the meeting, the chair of the meeting, etc. (gathered from the user's calendar), 2) when the user is driving to work the context data object holds information regarding the estimated time of arrival (gathered from the GPS system).

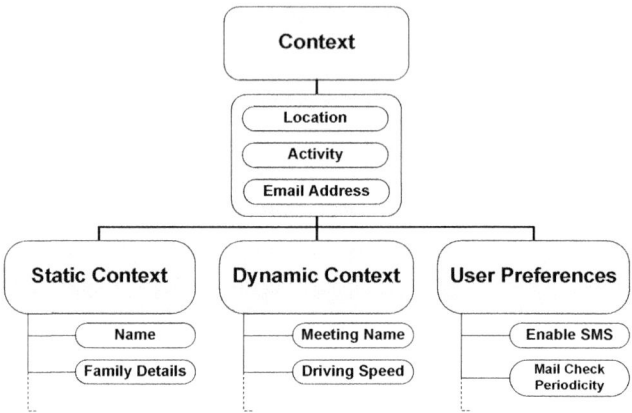

Fig. 2. Types of Context

3.2 Architecture

The PMA architecture consists of five main components, shown in Fig. 3. They are the email preprocessor, the context-generator, the importance processor, the urgency processor and the delivery agent.

PMA processes emails using rules. Furthermore, to make the rule system simpler and more adaptable, the system uses a concept of buckets to classify and prioritize emails. We use the main components of context (currently location and activity) as a tuple to select a bucket into which the emails are placed and then a set of rules specific to that bucket are invoked to process the emails. The rules themselves take various additional aspects of context into consideration. For cases where a location-activity tuple of the user's current context does not have a defined bucket for the rules, a generic bucket is used. The five components are:

- *Context Generator* – retrieves context information from various sources (location information from the user's mobile phone, user's schedule from his online calendar, etc.), and converts the raw context data into the PMA compatible representation described above.
- *Email Preprocessor* – preprocesses emails retrieved from the user's mailbox. Preprocessing involves the removal of non-textual components (images, video clips, etc.) embedded in the emails, and stemming the words in the textual part of the email's body and the email's subject. The preprocessor creates two separate stemmed word lists, one each for the subject and the body of the email which also includes the frequency of each stemmed word.
- *Importance Processor* – selects the appropriate rule-set for this bucket of emails to calculate a numerical value denoting the importance of each email.
- *Urgency Processor* – selects an appropriate rule-set to calculate a numerical value denoting the urgency value of each email.
- *Delivery Agent* – selects appropriate rule-sets for the email buckets for the current location and activity contexts. The delivery rules consider the calculated importance and urgency values and the user's current context to decide on which mails

to deliver to the user immediately, in which order and by which means. The delivery agent can also decide to forward select emails to an address in a predefined set of addresses (belonging to a peer, colleague, family member, etc.). The delivery agent can sort the emails in the inbox into separate folders that allow the user to quickly ascertain which emails have a higher priority. The delivery agent can also keep track of all the delivered emails so as to avoid "re-deliveries". This component is designed in such a way so as to allow fast and easy extensions to delivery methods and sorting options (e.g., add a desktop notification program as a new delivery target). This agent can be customized by the user to allow finer grained control over the entire system.

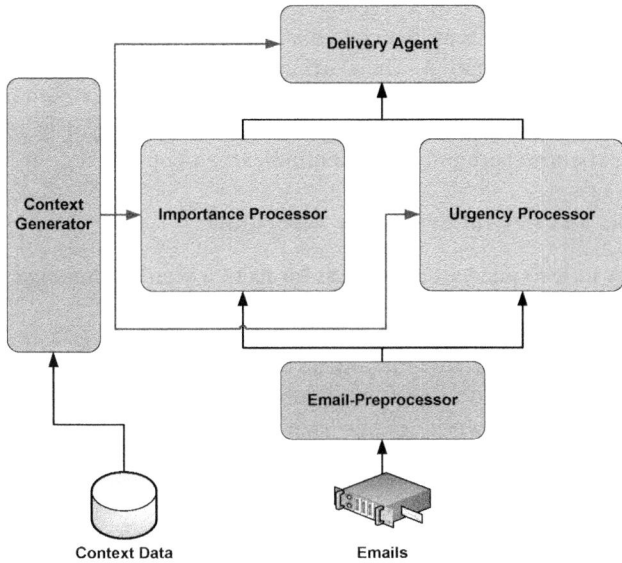

Fig. 3. Architectural overview of the PMA system

3.3 Rule System

The rule systems used in the Importance Processor, Urgency Processor and the Delivery Agent are written using the JESS rule engine for Java [17]. In the current implementation of the rules, only forward-chaining rules are used.

As the rule systems are not able to handle context information directly (i.e., – context data in its raw form is cumbersome to match against features of an email), the Java modules containing the rule engines convert context values to appropriate JESS facts for the rule engine.

The JESS rule engine matches facts to select the rules to be fired. JESS ignores rules which contain facts that have not been defined. This allows the PMA designer to dynamically add or remove (especially remove) the extensible context fields from its context representation without the need to alter the rule-sets.

3.4 Email Content Recognition

PMA uses three different techniques of content analysis to classify an email. The first type is to match the words in the subject and body of the email with lists of words (keywords) denoting varying levels of importance and urgency. The lists in which words occur add or subtract from a weight denoting the email's importance or urgency.

The second type is to compare words occurring the subject and body of an email with dynamic context information. This allows the system to identify emails which become relevant due to the user's current context (e.g., – when the user is heading to a meeting and an email related to the subject of the meeting arrives, the system is able to assign a higher importance and urgency to it).

The third type is to compare the contents and other metadata of an email with the user's static context information. The exact comparisons made vary depending on the bucket holding the email, for example the system may check if the email contains a reference to a user's family member, or it may check if the mail is addressed to the user rather than being a carbon-copy or a blind-carbon-copy.

3.5 Stemming and Lexical Frequency Algorithm

To allow PMA to recognize all equivalent forms of a word in an email subject line or body all textual input from emails to the PMA are stemmed, using the Porter Stemmer Algorithm [18]. The implementation of the algorithm processes a body of text and returns the stemmed words list with only a single instance of each stemmed word and a count of the number of occurrences of the word. For example, the words "work", "worked", "works" and "working" are stemmed to the word "work", whereas the word "worker" is stemmed to the word "worker".

3.6 Structure of the Email Rule System

Each of the components that use the JESS rule system (i.e., – the importance processor, the urgency processor and the delivery agent), employ a flat-rule-architecture.

The flat rule architecture consists of an arbitrary number of rules, each of which performs a single content recognition operation (belonging to one of the three types of content recognition categories described above). In the urgency and importance processors, the rules increment, decrement or scale the importance/urgency values, maintained by the PMA, for each email. The delivery agent employs a simple set of threshold rules (acting on the urgency and importance values) to determine the action for each email in the current context.

An alternative to this method, used in previous iterations of PMA, is to use a layered model for rules. In this type of architecture the firing of one or more rules in a lower layer creates facts that trigger a rule on a higher layer. The current PMA rule systems moved away from this type of implementation for two main reasons. I.e., – 1) to keep the rule system simple enough to allow ordinary users to edit the rules themselves 2) in a flat rule architecture it is easier to implement mechanisms to provide feedback to the user to explain why decisions were made within the system (to improve user confidence [19]).

The customization of the rule systems in the importance processor and the urgency processor can be performed in three different ways. The first involves the addition of new rules to match content of the emails with one instance of the keywords, the static context or the dynamic context. The second is to vary the impact of each rule on the final outcome of deciding whether an email is important/urgent or not. The impact of each rule on the overall decision is adjusted by changing the amount by which the rule adjusts the importance/urgency value of an email when the rule is triggered. The third is to alter the list of keywords understood by the PMA system. This could be done either by the addition or deletion of keywords or by changing the category to which a particular word belongs.

Shown below are three separate rules in the importance processor of the PMA when the user's context tuple is outdoor-driving. The first rule, "driving-rule-1", checks if an email is from an important person (important persons are a group maintained by the context generator); if so the importance of that email increases by 50 (this value can be customized). The second rule, "driving-rule-2", checks if an email was received as a carbon copy; if so the importance of that email is reduced, by 10. The third rule, "driving-rule-3", checks the subject section of an email for words defined as high importance keywords; if any are found in the subject line of an email the importance of the email is increased for each occurrence, by 20.

Examples of three simple rules in the importance processor

```
(defrule driving-rule-1
"Increase email importance if from an 'Important Person'"
 (declare ( no-loop TRUE))
 (and
  (context (othertags $?before GROUP_06_IMP_PPL$?after))
  (email (messageID ?message_id) (from ?from))
  (test (isSubstring ?from ( implode$ (first$ ?after)))))
=>
 (incrementImportancePacketValue ?message_id 50 ))

(defrule driving-rule-2
"Reduce email importance if email is received as a CC"
 (declare ( no-loop TRUE))
 (and
  (context (users_email_addrs $? ?owners_addr $?))
  (email (messageID ?message_id) (cc $? ?owners_addr$?)))
=>
 (incrementImportancePacketValue ?message_id -10 ))

( defrule driving-rule-3
"Increase email importance if highly-important keywords found in subject"
 (and
  (email (messageID ?id)(subject $? ?h_imp_wrd $?))
  (subject_keywords (h_imp_wrds $? ?hgh_imp_wrd $?)))
=>
 (incrementImportancePacketValue ?id 20 ))
```

Shown below is part of the definition of the Context class, which contains representations of context within the PMA system. The class contains a timestamp field, a location field, an activity field, a field for the user's email address(es), and an arraylist is used to hold the extensible tag-value pairs of context

Part of the context representation class in Java

```
public class Context {
      Date         date_current;
      String       location, activity, owners_email_addrs;
      ArrayList<ContextTag>   additional_tags;
      .
      .
      .
}
```

4 Testing and Results

4.1 Baseline

Initially statistical data from several user mailboxes was collected and analyzed to observe how many emails were received each day, and the number of emails that were left unread. Emails in the main inbox were counted separately from emails filed manually or automatically filed in folders or the Trash folder. The statistical data regarding the amount of emails deleted by users could not be accurately discovered, since the Trash folder is periodically cleared by the email system. Therefore the baseline is an underestimate of the actual complexity. Data was collected for a 120 day period except for Trash folder data which was collected for a 30 day period. Table 1 shows these results.

These results and previous email usage research [1] indicate users receiving high quantities of emails tend to leave larger percentages of emails unread. Based on these statistics, the PMA system was designed with the users receiving in excess of 40 emails per day in mind.

Table 1. Summary per-day baseline email statistics

	Received	Deleted	Filed	Unread
Mailbox 1	79.4	0	59.9	12.3
Mailbox 2	38.1	3.6	23	3.1
Mailbox 3	14.3	4.1	4.5	1
Mailbox 4	12.3	0.2	0.1	1.8
Mailbox 5	10.5	0	8.4	0
Mailbox 6	175.7	44.7	56.6	77.6

4.2 Effectiveness Testing

The first stage of testing was directed at discovering the effectiveness of PMA, by executing it in three different configurations.

The three configurations are: 1) as is without any customization of the system (PMA-a), 2) with customizations done using separate email inboxes as data sources for the customization (PMA-b), 3) using the target email inboxes as data sources for customization, in a 4-fold cross validation (PMA-c). In all three configurations, the static component of the user's context information (e.g. - user's name, email addresses) were provided to the PMA system. In the customized configurations the keyword collection used by the system was increased using a data source (configuration 2 - the data source was a separate inbox, configuration 3 - the data source was a section of the inbox used in the test).

This stage of testing was carried out on three users' inboxes, each with approximately 75 emails. The context was generated synthetically (for repeatability and time considerations) while the test was being carried out. Classifications done by the PMA system was checked for correctness against manual classification performed by the owner of the inbox, for each of the context scenarios.

The results of the importance prioritization in this stage of testing are shown in Fig. 4 along with the sorting provided by Gmail Labels and Gmail Rules and random assignment as benchmarks. The Gmail Rules and Labels were created manually by the PMA team analyzing 20% of the mail box as a guide. Approximately 25 rules were created.

The results of the urgency prioritization in this stage of testing are shown in Fig. 5 with random urgency prioritization as a benchmark. In each test the results were validated against manual classification of the entire inbox by the inbox owner, separately for importance and urgency for a series of contexts.

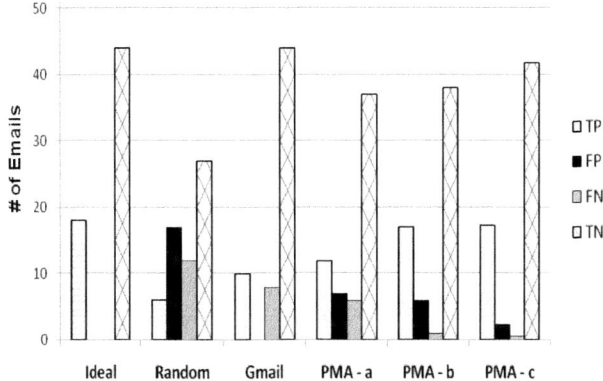

Fig. 4. Importance Prioritization

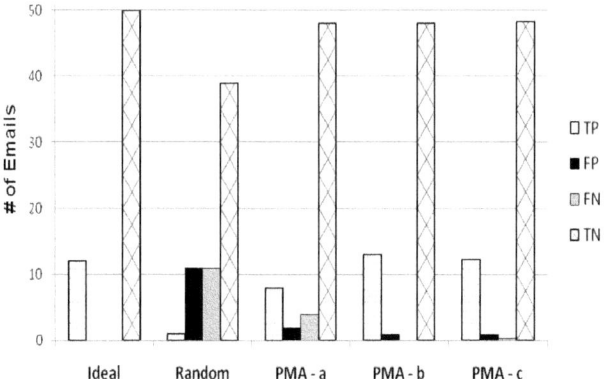

Fig. 5. Urgency Prioritization

In both figures, true-positives (i.e., – correct classifications as important and urgent respectively) are displayed in the left most (speckled) columns, false-positives (i.e., – incorrect classifications as important and urgent respectively) are displayed in the second (black) columns, false-negatives (i.e., – incorrect classifications as unimportant and non-urgent respectively) are displayed in the third (gray) columns and true-negatives (i.e., – correct classifications as unimportant and non-urgent respectively) are displayed in the fourth, right most (diagonal-pattern) columns.

The importance prioritization data shows that Gmail labels perform slightly better than the PMA system (PMA - c) at labeling unimportant emails (depicted by the high true-negative count), yet the high false-negative count shows that Gmail would classify a large number of emails as unimportant where as the PMA system tends to be more cautious and reduce the false-negatives to a minimum while allowing a low number of false positives. The reduction in false positives is approximately 30x over the Gmail system. The PMA results are advantages on the premise that it is better to receive some un-important mails along with the important ones, rather than missing some of the important mails.

The urgency classification data shows that the PMA system is very adept at urgency classification even in an un-customized state, while the effectiveness is increased (high number of true-positives/negatives and extremely low number of false-positives/negatives) when the PMA system is customized.

Table 2. Summary of precision and recall of importance classification

	Random	Gmail	PMA - c
Recall	33.3%	55.6%	96.3%
Precision	26.1%	98.2%	88.2%

Table 3. Summary of precision and recall of urgency classification

	Random	Gmail	PMA - c
Recall	8.3%	N/A	94.8%
Precision	8.3%	N/A	92.6%

4.3 Variation of Effectiveness with Customization

The second stage of testing was directed at discovering the variation of effectiveness of PMA's sorting with the increase of customization. This first level of simple customization of PMA was done solely by increasing the set of keywords understood by the system. While simple, it is quite powerful. The basic keyword set of the PMA system presently consists of approximately 125 keywords, which were manually selected by the research team. The tests were carried out on a constant set of inboxes while the customization was increased by roughly adding 15 keywords between each test. The keywords used in increasing customization were selected randomly from the keyword bank created manually by the research team.

Fig. 6 show the variation of importance classification accuracy with the number of keywords defined. The solid line in the graph shows the variation of true-positives (TP) a.k.a. emails properly classified as important, while the dashed line in the figure plots the false-positives (FP, emails improperly classified as important). Fig. 7 plots the accuracy of unimportance classification with the number of keywords defined. In this case the solid line represents the true-negatives (TN, i.e., – emails correctly classified as un-important) and the dash line represents emails incorrectly classified as un-important (false-negatives – FN).

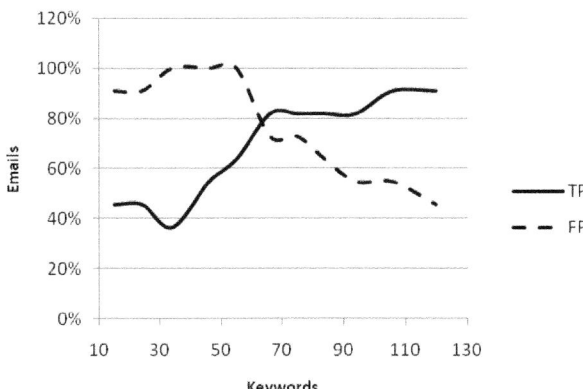

Fig. 6. Variation of importance classification accuracy with the number of keywords

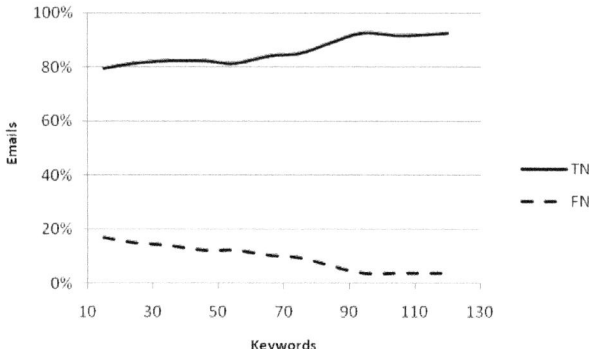

Fig. 7. Variation of unimportance classification accuracy with the number of keywords

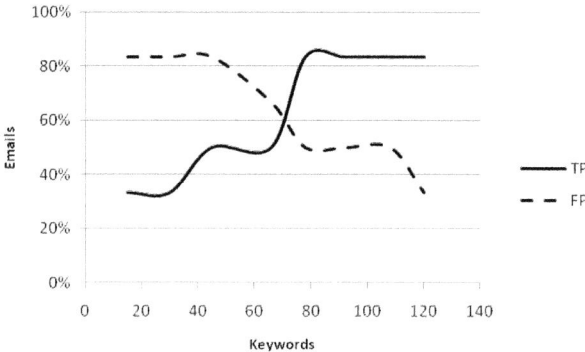

Fig. 8. Variation of urgency classification accuracy with the number of keywords

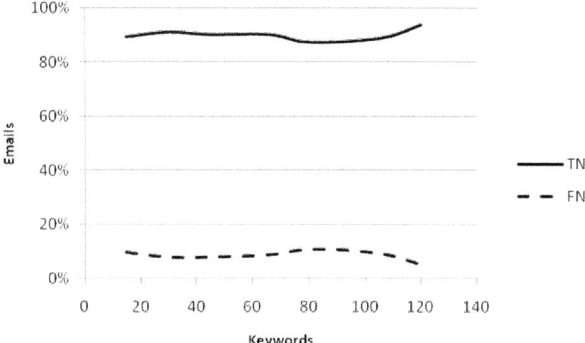

Fig. 9. Variation of non-urgency classification accuracy with the number of keywords

Fig. 8 show the variation of urgency classification accuracy with the number of keywords defined. Fig. 9 plots the accuracy of non-urgency classification with the number of keywords defined.

From the data depicted in figures 6, 7, 8 & 9 it is clear that the accuracy of importance and urgency ratings improve with the increase in keywords.

4.4 PMA Action Sampling

The third stage of testing was to evaluate the percentages of various actions taken on emails (i.e., SMSed to the user, filed for later viewing, forwarded to a peer) and the effectiveness of these actions in different context situations. The PMA makes these decisions based on the delivery agent rules. For testing, the PMA system was customized and used on an inbox with a 75 email sample. The results of the test are shown in figure 15, where the actions taken by PMA for each location-activity pair is represented by the column on the left with the distribution of ideal actions represented to the right (note: 50 mails on which no actions were taken have been omitted from the graph).

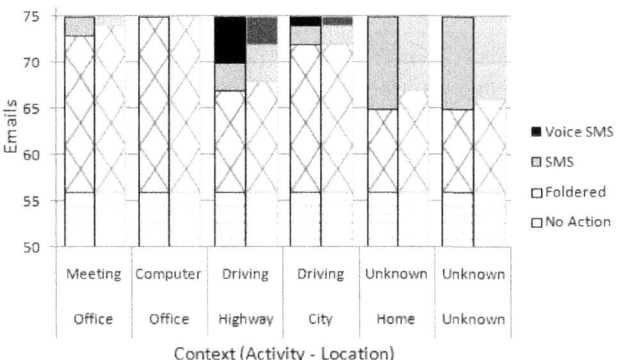

Fig. 10. Variation of actions taken on emails with change in context

The action sampling data shows that SMS as a delivery method is given less priority when the user is deemed to be in a less disturb-able state, as can be seen by the number of SMS sent to the user when he is in a meeting or driving through busy streets within the city. When the user is deemed to be in state where operation of the mobile phone might be difficult, i.e., driving, the PMA decides to use text-to-voice-SMS with the highest priority emails while some of the other emails are delivered (in summary form) as standard SMS. In cases where the PMA infers that the user has ready access to his inbox, PMA does not send SMS but rather files all high priority emails for faster viewing. These decisions are made by a set of rules that take not only the emails content but also the user's context into consideration. The context used in this test was generated synthetically for repeatability and time considerations.

5 Future Work

Performance, generalization and personalization are three dimensions that could be improved in the future versions of the system, which is the major focus for our future work.

The performance of the system is two dimensional. It can be improved in terms of scanning email content and learning the user context information. Advanced machine learning schemes could be used to automate the learning of keywords from user feedback; Naïve-Bayesian methods [20] are under consideration for this purpose. Also, during the initial iterations of the development of PMA, the system run-time was not a major concern, as more emphasis was placed on using context information for email sorting. Going forward, system performance could be improved to consume less system time on larger mail boxes. Lexical analysis or thesaurus based canonic word form analysis on email content to replace the word stemming method used in the current prototype is under consideration.

The next steps to generalize the PMA are to make the system work with other email client accounts like Yahoo! mail and Hotmail and adding handling support for additional message types like SMS, IM, RSS, HTTP and Voice.

Personalization of the system is another area that would be addressed in future research. This includes the creation of a user interface to allow users to create/edit custom rules. Also planned is a user interface on the mobile device to allow user feedback regarding actions taken on emails. Determining the costs of data transfer and processing on the mobile device in terms of power and bandwidth are areas of research under consideration for future work.

We also plan to conduct a larger scale usability test to study the effectiveness of the PMA system in comparison to human-based sorting. The study will also attempt to gather users' requirements for customization.

6 Conclusion

The key goal of PMA is to make email processing for users not just easy but more accurate. PMA considerably reduces time spent by users on filtering emails by sorting and delivering messages that are relevant to the user in his current context. Unlike other email filtering systems that depend solely on email content for filtering and sorting, PMA takes into account the content of emails and the contextual information of the user. Another unique aspect of the system is the consideration of urgency and importance of an email as separate dimensions for classification; thereby PMA is able to integrate a temporal aspect into email sorting. By combining the use of context-information and email content in classification with the idea of separate scales for importance and urgency, PMA is able to intelligently decide whether an email is to be delivered immediately via SMS, deferred for later delivery forwarded to another address, filed, etc.

It is scalable for all inbox sizes and types and offers better performance in terms of identifying email priorities. It could be easily personalized to suit the requirements of any user for better accuracy. The system is highly efficient over continuous usage compared to discontinuous usage.

This system saves ample time for users struggling with email flooding, by filtering emails and decision making on behalf of the user. It gives better performance in terms of filtering emails compared to existing rule based systems. It goes beyond simple email filtering to integrate context-awareness, in an extendable manner, to perform advanced email management.

Acknowledgements

This research was supported by a grant from Nokia Research Center and by CyLab at Carnegie Mellon under grant DAAD19-02-1-0389 from the Army Research Office. The views and conclusions contained here are those of the authors and should not be interpreted as necessarily representing the official policies or endorsements, either express or implied, of ARO, CMU, or the U.S. Government or any of its agencies.

We also acknowledge the efforts of Patricia Collins, Anind K. Dey, Pei Zhang and the anonymous referees for their valuable feedback and suggestions on earlier drafts of this paper.

References

1. Dabbish, L., Kraut, R., Fussell, S., Kiesler, S.: Understanding Email Use: Predicting Action on a Message. In: Proceedings of the SIGCHI conference on Human factors in computing systems, pp. 691–701. ACM, New York (2005)
2. Whittaker, S., Sidner, C.: Email Overload: Exploring Personal Information Management of Email. In: Proceedings of the SIGCHI conference on Human factors in computing systems, pp. 276–283. ACM, New York (1996)
3. Bergman, R., Griss, M., Staelin, C.: A Personal Email Assistant. Technical report, HPL-2002-236, ACM HP Laboratories, Palo Alto (2002)
4. Griss, M., Letsinger, R., Cowan, D., Sayers, C., VanHilst, M., Kessler, R.R.: CoolAgent: Intelligent Digital Assistants for Mobile Professionals - Phase 1 Retrospective. Technical report, HPL-2002-55(R1), HP Laboratories, Palo Alto (2002)
5. Barton, J., Kindberg, T.: The CoolTown User Experience. Technical report, HPL-2001-22, ACM HP Laboratories, Palo Alto (2001)
6. Yin, M., Griss, M.: SCATEAgent: Context-Aware Software Agents for Multi-Modal Travel. In: Applications of Agent Technology in Traffic and Transportation, pp. 69–84. Birkhäuser, Basel (2004)
7. The Big Business of Fighting Spam, http://www.ecommercetimes.com/story/20702.html
8. Anti Spam SpamUnit – Spam Statistics, http://www.spamunit.com/spam-statistics
9. Seewald, A.K.: Combining Bayesian and Rule Score Learning: Automated Tuning for SpamAssassin. Intelligent Data Analysis. Technical report, TR-2004-11 Austrian Research Institute for Artificial Intelligence, Vienna, Austria (2004)
10. McDonald, A.: SpamAssassin: A Practical Guide to Integration and Configuration. Packt Publishing, Birmingham (2004)
11. Kiritchenko, S., Matwin, S.: Email Classification with Co-Training. In: Proceedings of the 2001 Conference of the Centre for Advanced Studies on Collaborative research, pp. 8–18. IBM Press, Indianapolis (2001)

12. Covey, S.R.: The Seven Habits of Highly Effective People. Free Press, New York (1989)
13. Bolliger, P.: Redpin - adaptive, zero-configuration indoor localization through user collaboration. In: Proceedings of the first ACM international workshop on Mobile entity localization and tracking in GPS-less environments, pp. 55–60. ACM, New York (2008)
14. Lin, H., Zhang, Y., Griss, M., Landa, I.: WASP: An Enhanced Indoor Locationing Algorithm for a Congested Wi-Fi Environment. Technical report, MRC-TR-2009-04, Carnegie Mellon Silicon Valley, Moffett Field (2009)
15. Carrizo, C., Hatalkar, A., Memmott, L., Wood, M.: Design of a Context Aware Computing Engine. In: Proceedings of the IET 4th International Conference on Intelligent Environments, pp. 1–4. IEEE Press, New York (2008)
16. Salber, D., Dey, A.K., Abowd, G.D.: The Context Toolkit: Aiding the Development of Context-Enabled Applications. In: Proceedings of the SIGCHI conference on Human factors in computing systems, pp. 434–441. ACM, New York (1999)
17. Friedman-Hill, E.: Jess in Action: Java Rule-based Systems. Manning Publications, Greenwich (2003)
18. van Rijsbergen, C.J., Robertson, S.E., Porter, M.F.: New Models in Probabilistic Information Retrieval. Technical report, no. 5587, British Library Research and Development (1980)
19. Li, B.Y., Dey, A.K., Avrahami, D.: Why and Why Not Explanations Improve the Intelligibility of Context-Aware Intelligent Systems. In: Proceedings of the 27th international conference on Human factors in computing systems, pp. 2119–2128. ACM, New York (2009)
20. Witten, I.H., Frank, E.: Data Mining: Practical Machine Learning Tools and Techniques, 2nd edn. Morgan Kaufmann, San Francisco (2005)

OCRdroid: A Framework to Digitize Text Using Mobile Phones*

Mi Zhang[2], Anand Joshi[1], Ritesh Kadmawala[1], Karthik Dantu[1], Sameera Poduri[1], and Gaurav S. Sukhatme[1,2]

[1] Computer Science Department
[2] Electrical Engineering Department
University of Southern California, Los Angeles, CA 90089, USA
{mizhang,ananddjo,kadmawal,dantu,sameera,gaurav}@usc.edu

Abstract. As demand grows for mobile phone applications, research in optical character recognition, a technology well developed for scanned documents, is shifting focus to the recognition of text embedded in digital photographs. In this paper, we present OCRdroid, a generic framework for developing OCR-based applications on mobile phones. OCRdroid combines a light-weight image preprocessing suite installed inside the mobile phone and an OCR engine connected to a backend server. We demonstrate the power and functionality of this framework by implementing two applications called PocketPal and PocketReader based on OCRdroid on HTC Android G1 mobile phone. Initial evaluations of these pilot experiments demonstrate the potential of using OCRdroid framework for real-world OCR-based mobile applications.

1 Introduction

Optical character recognition (OCR) is a powerful tool for bringing information from our analog lives into the increasingly digital world. This technology has long seen use in building digital libraries, recognizing text from natural scenes, understanding hand-written office forms, and etc. By applying OCR technologies, scanned or camera-captured documents are converted into machine editable soft copies that can be edited, searched, reproduced and transported with ease [13]. Our interest is in enabling OCR on mobile phones.

Mobile phones are one of the most commonly used electronic devices today. Commodity mobile phones with powerful microprocessors (above 500MHz), high-resolution cameras (above 2megapixels), and a variety of embedded sensors (accelerometers, compass, GPS) are widely deployed and becoming ubiquitous. By fully exploiting these advantages, mobile phones are becoming powerful portable computing platforms, and therefore can process computing-intensive programs in real time.

In this paper, we explore the possibility to build a generic framework for developing OCR-based applications on mobile phones. We believe this mobile solution to extract information from physical world is a good match for future trend [16]. However, camera-captured documents have some drawbacks. They suffer a lot from focus

---
* This work was supported in part by NSF grant CCR-0120778 (CENS: Center for Embedded Networked Sensing), and by a gift from the Okawa Foundation.

loss, uneven document lighting, and geometrical distortions such as text skew, bad orientation, and text misalignment [14]. Further, since the system is running on a mobile phone, real time response is also a critical challenge. To address these problems we have developed a framework called OCRdroid. It utilizes the embedded sensors (orientation sensor, camera) combined with an image preprocessing suite to address the issues mentioned above. We have evaluated the OCRdroid framework by implementing two applications called PocketPal and PocketReader. Our experimental results demonstrate the OCRdroid framework is feasible for building real-world OCR-based mobile applications. The main contributions of this work are:

- A real time algorithm to detect text misalignment and guide users to align the text properly
- Utilizing orientation sensors to prevent users from taking pictures if the mobile phone is not properly oriented and positioned
- An auto-rotation algorithm to correct skewness of text
- A mobile application called PocketPal which extracts text and digits on receipts and keeps track of one's shopping history digitall
- A mobile application called PocketReader which provides a text-to-speech interface to read text contents extracted from any text sources (magazines, newspapers etc)

The rest of this paper is organized as follows. Section 2 discusses work related to OCR and image processing on mobile phones. Section 3 describes the OCRdroid framework and PocketPal and PocketReader applications. Design considerations of OCRdroid are described in detail in Section 4. The system architecture and implementation are presented in Section 5, with experiments and evaluations in Section 6. Section 7 discusses limitations and future work, and Section 8 summarizes our work.

2 Related Work

Mobile phones are some of the most popular wearable computers. As elucidated in the earlier section, their capability in terms of computing, sensing and communication is only increasing by the day. It is perceivable that mobile phones are used for a variety of other applications other than voice communication. One such application is recognizing and understanding characters or Optical Character Recognition (OCR). OCR has been in research for a number of years. The technoloqy is quite mature when it comes to recognizing scanned documents. However, it is still not customized for the cameras that come on modern day cellphones. We list below some related work that form pieces of work that we used or benefited from in designing OCRdroid.

Project Visual Codes [8] aims to develop an 2-dimensional visual code recognition system on camera-equipped mobile phones. Using a lightweight recognition algorithm based on planar homography, authors have shown the possibility to detecting objects in the user's immediate surroundings so as to bridge the gap between the physical and the virtual world. Project in [18] has presented a new algorithm of image reorganization for EAN/QR barcodes in mobile phones. Authors have demonstrated the robustness and little calculation cost of their algorithm that is suitable to be implemented in mobile phones for practical situations. Several other applications applied computer vision techniques to interpret images in specific scenarios. In [26], authors demonstrated a mobile sensing system that analyzed images of air sensors to extract indoor air pollution information by comparing the sensor to a calibrated color chart. In [11], a digital museum guidance system is built on camera equipped mobile phones to enable museum visitors to identify exhibits by capturing photos of them.

There has been intensive research on building mobile systems for text extraction and recognition both in the commercial space and in the academia world. In industry, ABBYY [1] offers a powerful but compact mobile OCR engine which can be integrated within various mobile platforms such as Windows Mobile, Nokia Symbian, iPhone, and Android. Another company called WINTONE [9] also claims that its mobile OCR engine can achieve 95 recognition accuracy for English documents. In academia, project in [15] implemented a business card reader in mobile device with build-in camera. Using multi-resolution analysis method, the authors have shown the possibility to reduce the memory requirement and improve the computation speed for computing-intensive image processing on the phone. Work in [13] and [12] both focus on text extraction and recognition of signs from nature scenes. In [13], authors demonstrated an image preprocessing suite on top of OCR engine with novel approaches for foreground/background detection and skew estimation using morphological edge analysis. Authors in [12] presented a hierarchical framework for sign detection. A novel local intensity normalization method was proposed to effectively handle lighting variation problem for camera-taken images.

OCRdroid is different from previous text recognition mobile systems in that it focuses on building a generic framework for OCR on mobile phones augmented with the ability to interact with users. The effort was also to guide users to use phones more efficiently.

3 OCRdroid Description and Applications

OCRdroid is a generic framework for developing OCR-based applications on mobile phones. This framework not only supports the baseline character recognition functionality, but also provides an interactive interface by taking advantage of high-resolution camera and embedded sensors. This interface enriches the interactivity between users and devices, and as a result, successfully guides users step by step to take good-quality pictures.

In this section, we describe two applications called PocketPal and PocketReader we have implemented on OCRdroid framework. Several potential applications that can be built on this framework are also covered.

3.1 Pocket Pal

PocketPal is a personal receipt management tool installed in one's mobile phone. It helps users extract information from receipts and keep track of their shopping histories digitally in a mobile environment. Imagine a user, Fiona, is shopping in a local shopping mall. Unfortunately, she forgets what she bought last weekend and hasn't brought her shopping list with her. Fiona takes out her mobile phone and starts the PocketPal application. PocketPal maintains her shopping history in a chronological order for the past 2 months. She looks over what she bought last week and then decides what to buy this time. After Fiona finishes shopping, she takes a good-quality picture of the receipt under the guidance of PocketPal via both audio and visual notifications. All the information on the receipt is translated into machine editable texts and digits. Finally, this shopping record is tagged, classified and stored inside the phone for future reference. PocketPal also checks the total and reminds users in an unobtrusive way if the spending increases one's monthly budget. We have studied a lot of receipts with different formats. Some of the most common items are:

- Date
- Time
- Shop name
- Shop location
- Contact phone number
- Shop website
- Name of each item bought
- Price of each item bought
- Amount of tax
- Total amount of the purchase

PocketPal can recognize all of these items, categorize them into different categories, and display to the users.

3.2 Pocket Reader

PocketReader is a personal mobile screen reader that combines the OCR capability and a text-to-speech interface. PocketReader allows users to take pictures of any text source (magazines, newspapers, and etc). It identifies and interprets the text contents and then reads them out. PocketReader can be applied in many situations. For example, users can quickly capture some hot news from the newspaper and let PocketReader read them out if users do not have time to read. What's more, PocketReader can act as a form of assistive technology potentially useful to people who are blind, visually impaired, or learning disabled. A visually impaired person, trying to read a newspaper or a description of a medicine, can ask PocketReader to read it loud for him.

3.3 Other Potential Applications

It is feasible to apply OCRdroid to digitize physical sticky notes [16] to build a mobile memo and reminder. If the message contains important timing information, such as a meeting schedule, the system can tag this event and set an alarm to remind user of this event. In addition, OCRDroid framework can be combined with a natural language translator to diminish the language barrier faced by tourists. As a result, tourists can take pictures of public signage and have the same access to information as locals [13].

4 Design Considerations

OCR systems have been under development in both academia and industry since the 1950s. Such systems use knowledge-based and statistical pattern recognition techniques to transform scanned or photographed images into machine-editable text files. Normally, a complete OCR process includes five main steps [10]:

- noise attenuation
- image binarization (to black and white)
- text segmentation
- character recognition
- post-processing, such as spell checking.

These steps are very effective when applied to document text collected by a scanner. Such documents are generally aligned, have clear contrast between text and uniform background. However, taking pictures from a portable camera, especially ones embedded inside a mobile device, may lead to various artifacts in the images. As a result, even the best available OCR engines fail on such images. Problems include uneven document lighting, perception distortion, text skew, and misalignment. Some of these issues are illustrated in Figure 1. In addition, since the system is installed on mobile devices, real time response is another critical issue that needs to be considered. Among the issues mentioned above, some exhibit inherent tradeoffs and must be addressed in a manner that suit our applications. This section presents a pertinent range of design issues and tradeoffs, and discusses proposed approaches applicable to our OCRdroid framework.

4.1 Lighting Condition

Issue: An embedded camera inside the mobile phone has very little control of lighting conditions. Uneven lighting is common, due to both the physical environment (shadows, reflection, fluorescents) and uneven response from the devices. Further complications occur when trying to use artificial light, i.e. flash, which results in light flooding.

Proposed Approach: Binarization has long been recognized as a standard method to solve the lightning issue. The goal of binarization process is to classify image pixels from the given input grayscale or color document into either foreground (text) or background and as a result, reduces the candidate text region to be processed by later processing steps. In general, the binarization process for grayscale documents can be grouped into two broad categories: global binarization, and local binarization [23]. Global binarization methods like Otsu's algorithm [19] try to find a single threshold value for the whole document. Each pixel is then assigned to either foreground or background based on its grey value. Global binarization methods are very fast and they give good results for typical scanned documents. However, if the illumination over the document is not uniform, global binarization methods tend to produce marginal noise along the page borders. Local binarization methods, such as Niblack's algorithm [17] and Sauvola's algorithm [21] compute thresholds individually for each pixel using information from the local neighborhood of that pixel. Local algorithms are usually able to achieve good results even on severely degraded documents with uneven lightning conditions. However, they are often slow since computation of image features from the local

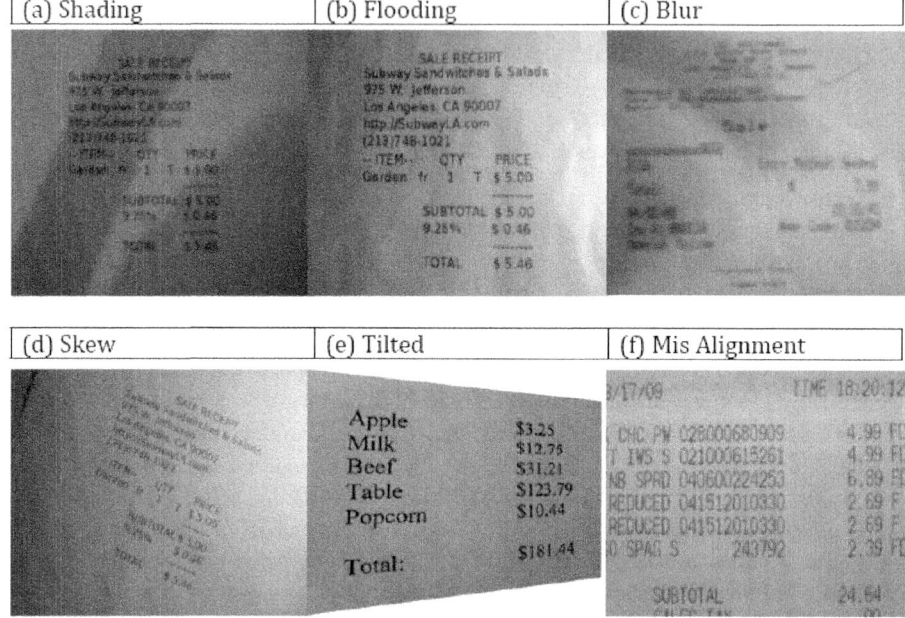

Fig. 1. Different issues arising in camera-captured documents: (a) shading (b) flooding (c) blur (d) skew (e) tilted (f) misalignment

neighborhood is to be done for each image pixel. In this work, in order to handle uneven lightning conditions for our camera-captured documents, we adopt Sauvola's local binarization algorithm. We have also tried Background Surface Thresholding algorithm in [22]. Based on our experiments, we found Sauvolas's algorithm worked better and faster.

4.2 Text Skew

Issue: When OCR input is taken from a hand-held camera or other imaging devices whose perspective is not fixed, text lines may get skewed from their original orientation [10]. Based on our experiments, feeding such a rotated image to our OCR engine produces extremely poor results.

Proposed Approach: Skew detection process is needed before calling the recognition engine. If any skew is detected, an auto-rotation procedure is performed to correct the skew before processing text further. There are many techniques in literature [10] but are based on the assumption that documents have set margins. However, this assumption does not always holds true in real world scenarios. In addition, traditional methods based on morphological operations and projection methods are extremely slow and tend to fail for camera-captured images. In this work, we choose a more robust approach based on Branch-and-Bound text line finding algorithm (RAST algorithm) [25] for skew detection and auto-rotation. The basic idea of this algorithm is to identify each line independently and use the slope of the best scoring line as the skew angle for the entire text segment. After detecting the skew angle, rotation is performed accordingly. Based on our experiments, we found this algorithm to be highly robust and extremely efficient and fast. However, it suffered from one minor limitation in the sense that it failed to detect rotation greater than 30°.

4.3 Perception Distortion (Tilt)

Issue: Perception distortion occurs when the text plane is not parallel to the imaging plane. It happens a lot if using a hand-held camera to take pictures. The effect is characters farther away look smaller and distorted, and parallel-line assumptions no longer hold in the image [14]. From our experience, small to mild perception distortion causes significant degradation in performance of our OCR engine.

Proposed Approach: Instead of applying image processing techniques to correct the distortion, we take advantage of the embedded orientation sensors to measure the tilt of the phone. Users are prevented from taking pictures if the camera is tilted. This reduces the chances of perception distortion considerably. We also consider the situation where the text source itself is titled. In this case, we have provided users a facility to calibrate the phone to any orientation so that the imaging plane is parallel to the text plane. For example, if one user wants to take a picture of a poster attached on the wall, he can first calibrate the phone with the imaging plane parallel to the wall surface and then take the picture.

4.4 Misalignment

Issue: Text misalignment happens when the camera screen covers a partial text region, in which irregular shapes of the text characters are captured and imported as inputs to the OCR engine. Figure 1 (f) shows an example of a misaligned image. Misalignment issue generally arises when people casually take their pictures. Our experiment results indicate that the OCR result is significantly affected by misalignment. Moreover, misaligned images may lead to loss of important data.

Proposed Approach: We define that a camera-captured image is misaligned if any of the four screen borders of the phone cuts through a single line of text either horizontally or vertically. Based on this definition, we set a 10-pixel wide margin along each border as depicted in Figure 2.

If any foreground pixel is detected within any of those four margins, there is a high probability that the text is being cut. Hence, we can conclude that the image is misaligned. We faced two major challenges while designing an algorithm for this task

1. Our algorithm should provide real time response
2. Our algorithm must be able to deal with random noise dots and should not classify them as part of any text

We based our alignment-checking algorithm on the fast variant of Sauvola's local binarization algorithm described in [24] so as to provide real time resposne. Further, we run

Fig. 2. Four 10-pixel wide margins

Fig. 3. The route to run Sauvola's binarization algorithm

the Sauvola's algorithm within four margins in a Round-Robin manner as depicted in Figure 3.

Specifically, we first go around the outermost circle in red. If no foreground pixel is detected, we continue on the green route. By following this approach we can detect the misalignment faster and quit this computing-intensive process as soon as possible. Once a black pixel is detected, it is necessary to verify whether it is a true pixel belonging to a text character or just some random noise. Figure 4 demonstrates a perfectly aligned receipt with some noise dots located within the top margin, which are circled by a red ellipse. To judge whether if it is noise, whenever a black dot is detected, we check all its neighbors within a local W x W box. If more than 10% of its neighboring pixels inside the local box are also black dots, then we conclude that current pixel belongs to a text character. This inference is based on the observation that text characters always have many black pixels besides each other whereas the noise is generally randomly distributed. Based on our experiments, this newly designed alignment-checking algorithm takes no more than 6 seconds and has an accuracy of 96% under normal lighting condition.

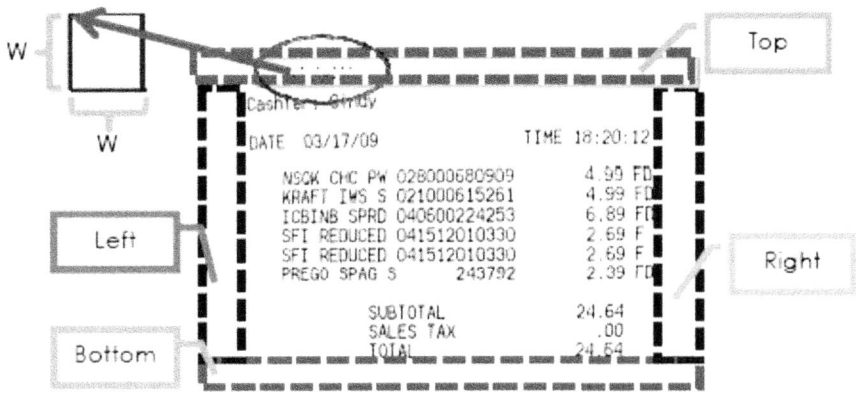

Fig. 4. A perfectly aligned receipt with noise dots located within the top margin and circled by a red ellipse. A WxW local box is defined to help filter out the noise dots.

4.5 Blur (Out Of Focus)

Issue: Since many digital cameras are designed to operate over a variety of distances, focus becomes a significant factor. Sharp edge response is required for the best character segmentation and recognition [14]. At short distances and large apertures, even slight perspective changes can cause uneven focus.

Proposed Approach: We adopted the AutoFocus API provided by Android SDK to avoid any blurring seen in out-of-focus images. Whenever we start the application, a camera autofocus handler object is initiated so that the camera itself can focus on the text sources automatically.

5 System Architecture and Implementation

Figure 5 presents the client-server architecture of OCRdroid framework. The software installed inside the phone checks the camera orientation, performs alignment checking, and guides the user in an interactive way to take a good-quality picture. The mobile phone plays the role as a client sending the picture to the back-end server. The computing-intensive image preprocessing steps and character recognition process are performed at the server. Finally, the text results are sent back to the client.

The OCRdroid client program is currently developed on HTC G1 mobile phone powered by Google's Android platform. However, it can be extended with minimal effort to any other platform powered phones with orientation sensors and an integrated embedded camera. The OCRdroid server is an integration of Tesseract OCR engine from Google [7] and Apache (2.0) web server. We have implemented 2 applications called PocketPal and PocketReader based on this framework. A series of screenshots and a demo video of these applications can be found at our project website [5].

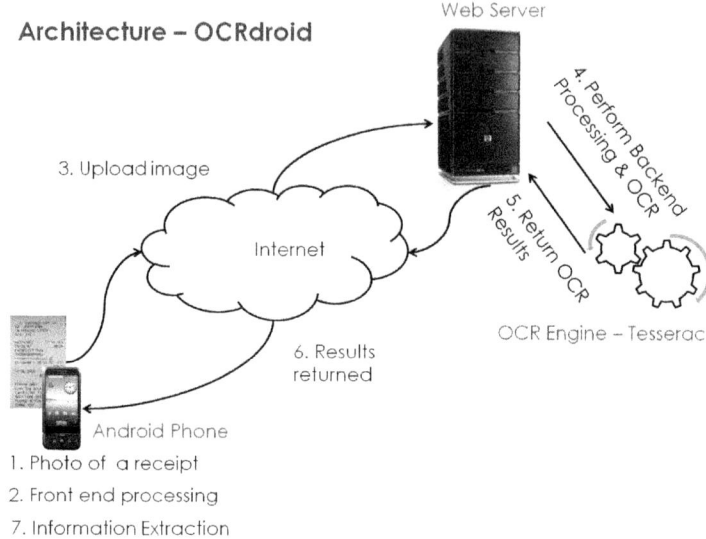

Fig. 5. Overview of OCRdroid framework architecture

The following subsections discuss the details of the implementation of both the client and the server.

5.1 Phone Client

We developed the OCRdroid phone client program using Google's Android software stack. Our client software requires access to phone camera, embedded sensors, background services, network services, relational database and file-system inside the phone. All these necessary APIs are provided by Android SDK (version 1.1).

Compared to desktop or notebook computers, mobile devices have relatively low processing power and limited storage capacity. The mobile phone users always expect instantaneous response when interacting with their handsets. OCRdroid framework is designed to minimize the processing time so as to provide real time response. We enforce real time responsiveness by adopting many strategies, including:

- Designing computationally efficient algorithms to reduce processing time
- Spawning separate threads and background services for computation-intensive workloads to keep the GUI always responsive
- Using compressed image format to store and transmit image data

Figure 6 presents an architectural diagram for the client software. The client software consists of seven functional modules. By default, the client is in idle state. When the user initiates the application, the client starts up a camera preview to take a picture. The preview object implements a callback function to retrieve the data from the embedded 3.2 mega-pixel camera, which satisfies the 300dpi resolution requirement of the OCR engine. Client also implements an orientation handler running inside a background service thread that is responsible for preventing users from tilting the camera beyond a

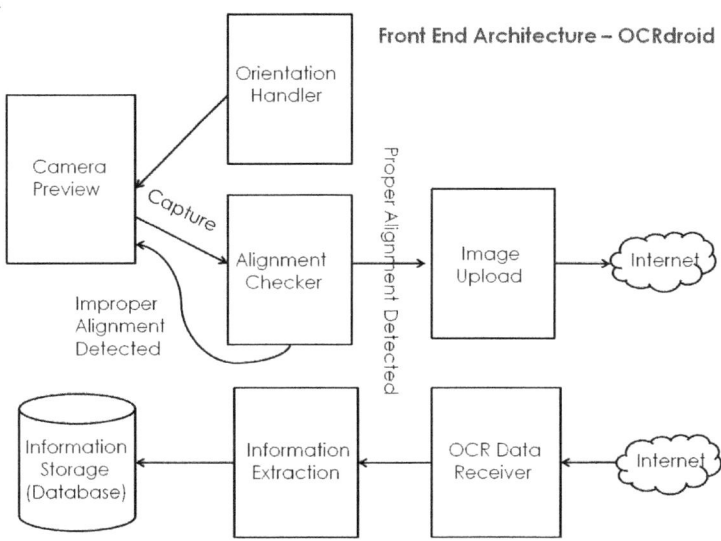

Fig. 6. OCRdroid Client(Phone) Architecture

threshold while taking a picture. It periodically polls the orientation sensors to get the pitch and roll values. A threshold range of [-10, 10] is set for both sensors. If the phone is tilted more than the allowed threshold, users are prevented from capturing any images. Figure 7 presents a screenshot showing how the tilt detection module works. The image on the right shows a case where user has tilted the phone. As a result, a red colored bounding box is displayed, indicating the camera is not properly oriented. As soon as the orientation is perfect, the box turns to green and the user is allowed to capture an image.

Once the user takes a picture, the misalignment detection module is initiated to verify if the text source is properly aligned. In case of a misaligned image, the client pops up

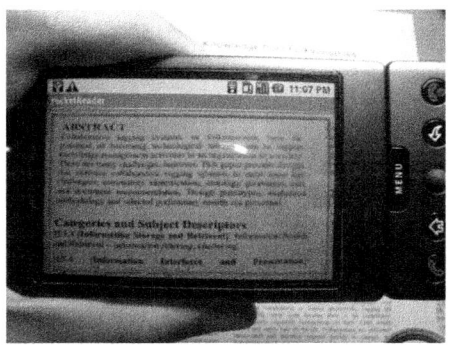
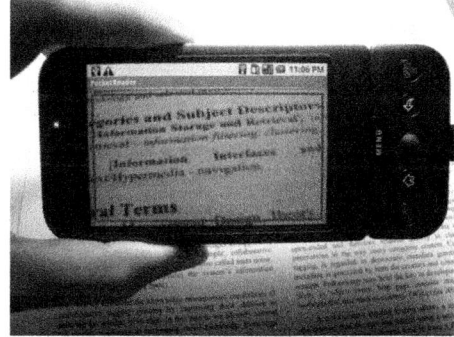

Fig. 7. Screenshot demonstrating tilt detection capability of OCRDroid using embedded orientation sensor. A green (red) rectangle indicates correct (incorrect) tilt.

an appropriate notification to instruct the user where misalignment is detected. As soon as the text source is properly aligned, the new image is transported to the server over a HTTP connection. An OCR data receiver module keeps listening on the connection and passes the OCR result sent from the server to information extraction module where corresponding information is extracted and parsed into different categories. Finally, the OCR result is displayed on the screen and automatically stored inside the local database for future references.

5.2 Back-End Server

OCRdroid backend server is an integration of an OCR engine and a web server. Figure 8 presents the architectural diagram for the backend server. Once images are received at web server, shell scripts are called in sequence to perform binarization and image rotation (if any skew is detected). Conversion of image is done using a popular open source tool called ImageMagick. Once the image preprocessing procedures are completed, the intermediate result is fed to the OCR engine to perform text segmentation and character recognition.

There are many open source as well as commercial OCR engines available, each with its own unique strengths and weaknesses. A detailed list of OCR engines is available at Wikipedia. We tested some of the open source OCR engines such as OCRAD [4], Tesseract [7], GOCR [2] and Simple OCR [6]. Based on our experiments, we found that Tesseract gave the best results. We also tested a complete document analysis tool called OCRopus [3]. It first performed document layout analysis and then used Tesseract as its OCR engine for character recognition. However, OCRopus gave rise to one additional complication in our PocketPal application because of its inherent multi-column support. The document layout analyser identified receipts to be in a 2-column format. As a result, it displayed the names of all shopping items followed by their corresponding prices.

Backend End Architecture – OCRdroid

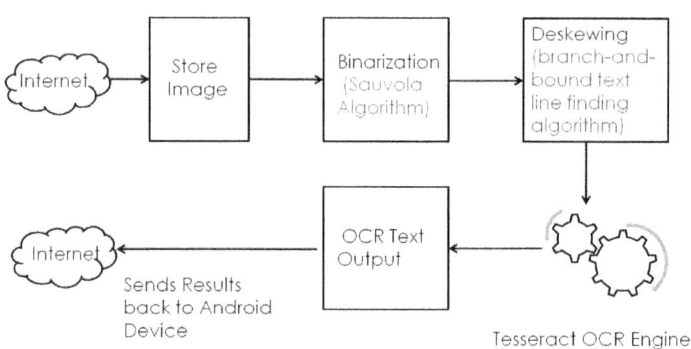

Fig. 8. Software architecture For OCRdroid Backend

This required us to carry out extra operations to match the name and the price of each item. Therefore, we choose Tesseract as our OCR engine.

Once the whole process finishes successfully, OCRdroid server responds to the client side with an OK message. On receieving this message, the client sends back a HTTP request asking for the OCR result. Finally, OCRdroid server sends the text back as soon as the request is received.

6 Experiments and Evaluation

We evaluate the OCRdroid framework by implementing PocketPal and PocketReader applications. The OCR accuracy and timing performance are of our interest. We start by describing the text input sources and defining performance metrics. Then we present the results of a set of preprocessing steps and detailed performance analysis.

6.1 Test Corpus

The system was tested on ten distinct black and white images without illustrations. Tests were performed under three distinct lighting conditions. All the test images were taken by HTC G1's embedded 3.2 megapixel camera. The images and corresponding results can found at our project website [5].

6.2 Performance Metrics

In order to measure the accuracy of OCR, we adopt two metrics proposed by the Information Science Research Institute at UNLV for the fifth annual test of OCR accuracy [20].

Character Accuracy. This metric measures the effort required by a human editor to correct the OCR-generated text. Specifically, we compute the minimum number of edit operations (character insertions, deletions, and substitutions) needed to fully correct the text. We refer to this quantity as the number of errors made by the OCR system. The *character accuracy* is defiend as:

$$\frac{\#characters - (\#errors)}{\#characters} \quad (1)$$

Word Accuracy. In a text retrieval application, the correct recognition of words is much more important than the correct recognition of numbers or punctuations. We define a word to be any sequence of one or more letters. If m out of n words are recognized correctly, the *word accuracy* is m/n. Since full-text searching is almost always performed on a case-insensitive basis, we consider a word to be correctly recognized even if one or more letters of the generated word are in the wrong case (e.g., "transPortatIon").

In addition, real time response is another very important metric for mobile applications. We evaluate timing performance in terms of processing time taken by each of the preprocessing steps.

6.3 Experimental Results and Analysis

In this section, we give both qualitative and quantitative analysis of the performance improvement brought by each of our preprocessing steps.

Binarization. Binarization plays a very important role in OCR preprocessing procedure. Figure 9 presents one test case to demonstrate the importance of binarization process. The effectiveness of binarization algorithm heavily depends upon the lighting

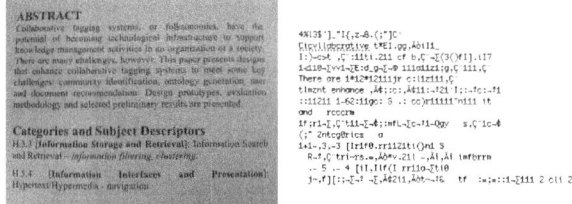

(a) An image of text source taken under normal lighting condition

(b) OCR output without binarization

(c) OCR output with binarization

Fig. 9. Comparision of OCR results with and without Binarization Module

conditions when image is captured. To measure the performance, we carried out our experiments under three distinct lighting conditions:

- *N*ormal lighting condition: This refers to situations when images are captured outdoors in the presence of sunlight or indoors in an adequately lit room.
- *P*oor lightening condition: This describes situations when users take images outdoors during night or capture images in rooms which have very dim lighting.
- *F*looding condition: This describes situations when the source of light is very much focused on a particular portion of the image, whereas the remaining portion is dark.

Table 1 lists the character accuracy of OCR output with and without binarization under three lighting conditions. As expected, OCR output with binarization achieves much higher peformance than its counterpart without binarization. Under normal lighting condition, the OCR output with binarization achieves 96.94% character accuracy. However, the performance degrades significantly if lighting condition is poor or flooded. The main reason to cause this problem is the text in the shaded regions tend to have very dark and broad boundaries. This confuses the OCR engine and leads to relatively low character accuracy.

Table 1. Comparision of OCR character accuracy with and without binarization under three lighting conditions

Type of lighting conditions	Character accuracy without binarization(%)	Character accuracy with binarization (%)
Normal	46.09	96.94
Poor	9.09	59.59
Flooding	27.74	59.33

Skew Detection and Auto-Rotation. Our Tesseract OCR engine failed to produce any meaningful results if the input image is skewed more than 5°. In order to examine the performance of our skew detection and auto-rotation algorithm, we rotated all the images in our text corpus by 5°, 10°, 15°, 25°, and 35° in both clockwise and counter-clockwise directions. These rotated images are pictured under three different lighting conditions and then passed to the OCR engine. Figure 10 and Figure 11 demonstrate the performance in terms of average character accuracy and average word accuracy.

As presented, our skew detection and auto rotation algorithms work quite well for image rotation up to 30° in both clockwise and counter-clockwise directions. On the

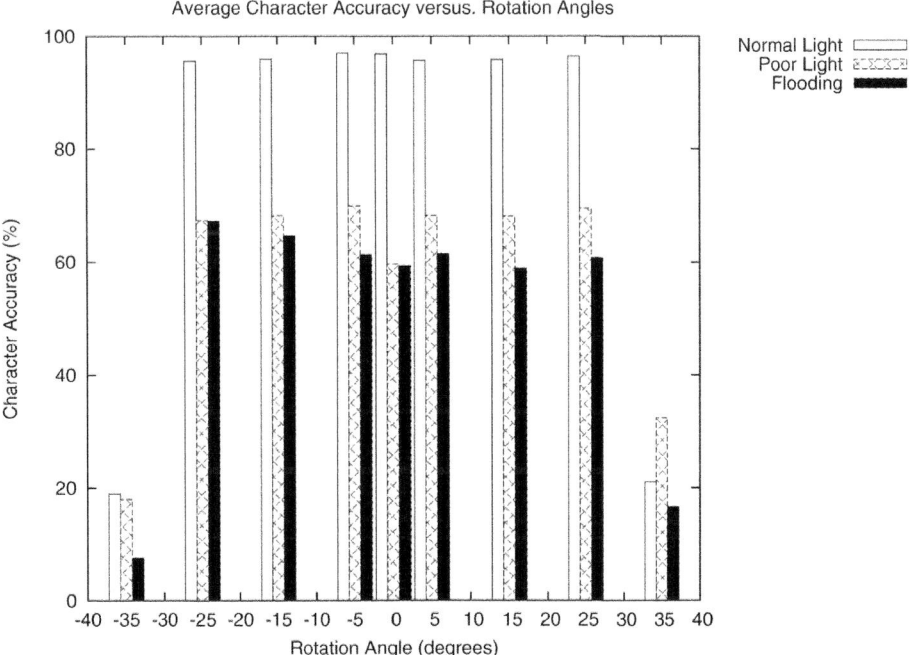

Fig. 10. Average Character Accuracy of OCR output at different rotation angles under three different lighting conditions

Table 2. Experimental results indicating accuracy for misalignment detection algorithm

Lighting conditions	Type of images	No. of images	No. of images detected misaligned	No of images detected properly aligned
Normal	Misalgined	15	14	1
	Properly Aligned	15	2	13
Poor	Misalgined	15	14	1
	Properly Aligned	15	7	8
Flooding	Misalgined	15	13	2
	Properly Aligned	15	6	9

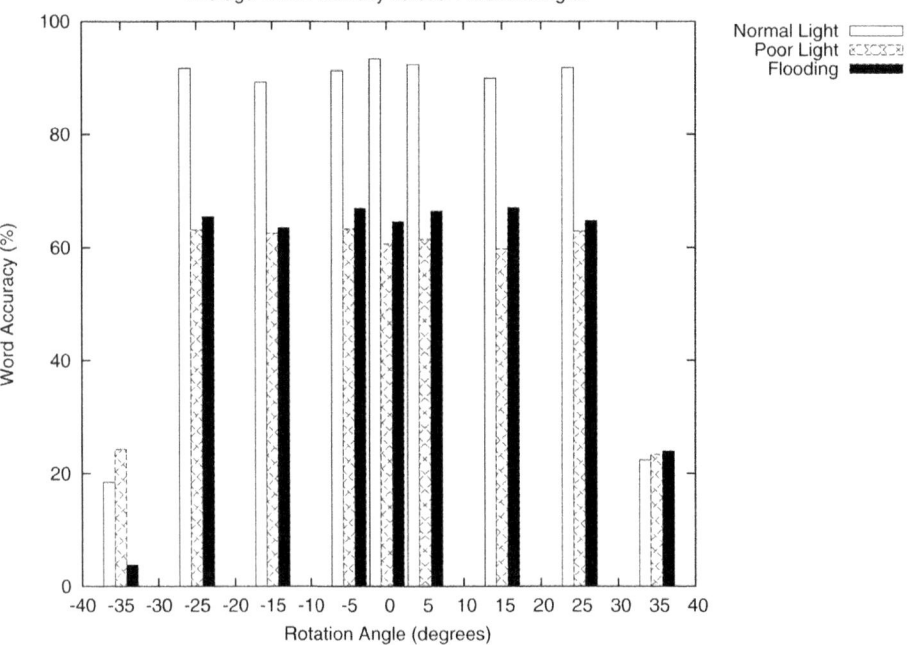

Fig. 11. Average Word Accuracy of OCR output at different rotation angles under three different lighting conditions

other hand, even under normal lighting condition, the performance drops sharply with image rotation at 35°. However, in real-world applications, it is reasonable to believe that general users would not take images at such high degree of rotation.

Misalignment Detection. Since our misalignment detection algorithm is based on the Sauvola's binarization algorithm, the accuracy depends on lighting conditions as well. Figure 12 presents some test cases under both normal and poor lighting conditions. To measure the accuracy, we followed the definitions of three different lighting conditions, and carried out 30 trials under each lighting condition. Finally, the total number of false positives and false negatives are counted and summarized below.

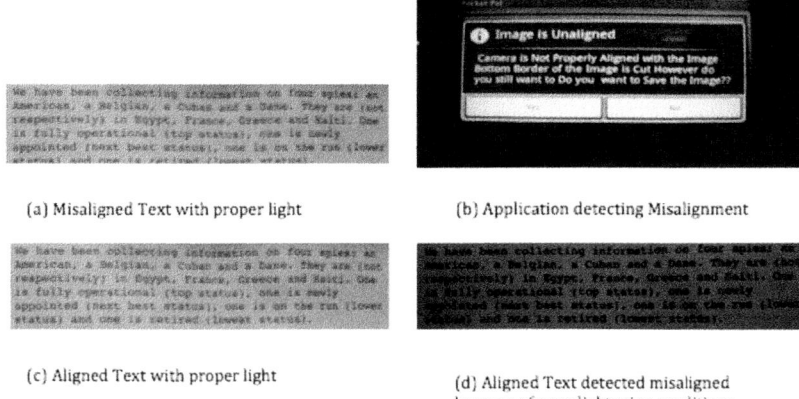

(a) Misaligned Text with proper light

(b) Application detecting Misalignment

(c) Aligned Text with proper light

(d) Aligned Text detected misaligned because of poor lightening conditions

Fig. 12. Screenshots of test cases for misalignment detection

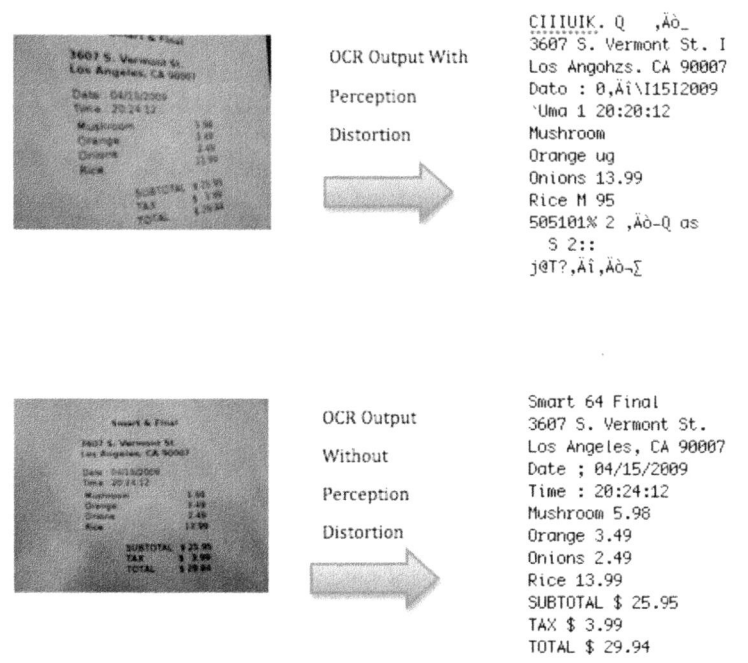

Fig. 13. Comparison of OCR results with and without perception distortion

Perception Distortion. When taking pictures, if the camera is not parallel to the text source being captured, the resulting image suffers from some perspective distortion. However, it is very hard for us to measure the tilting angles directly. Therefore, we tuned the thresholds in the program and then checked the performance. Based on our

Table 3. Experimental results indicating maximum time taken by each of the preprocessing steps

Preprocessing Steps	Max Time Taken(sec)
Misalignment Detection	6
Binarization	3
Skew Detection and Auto-Rotation	2
Character Recognition	3
Total Time Taken	11

experiments, we found the OCR accuracy is highly susceptible to camera orientation. The character accuracy drops sharply if we tilt the camera over 12°. Therefore, we set the threshold for our orientation sensor to [-10°, 10°]. The upper graph in Figure 13 shows a document image captured when the mobile phone is titled 20° and the corresponding OCR text output. The poor OCR accuracy indicates OCR results are strongly correlated to the orientation of the camera. As a contrast, the lower graph presents the image taken with 5° orientation and its OCR result with 100% accuracy. This contrast clearly demonstrates that orientation sensor plays an important role in ensuring good performance.

Timing Performance. We evaluate timing performance in terms of processing time taken by each of the preprocessing steps. Since binarization, skew detection and character recognition are performed at the server, we close all other processes in the server to reduce the measurement error. For misalignment detection, the measurement error is relatively big since we can not shut down some native applications running in the phone. In addition, since we used Wi-Fi as the link between the mobile device and the backend server, the network latency is negligible compared to other processing steps. The result is summarized in Table 3.

It takes maximum eleven seconds to complete the whole process. As expected, due to limited processing power of mobile phone, misalignment detection is the most time-consuming step among the entire process. However, the entire process quite meets the real time processing requirement.

7 Limitations and Ongoing Work

As demonstrated in the previous section, the applications built on our OCRdroid framework can produce accurate results in real time. However, there are still certain areas where we believe our prototype system could be improved. This section discusses several limitations with OCRdroid.

7.1 Text Detection from Complex Background

Our OCRdroid framework only works in the case where the background of the text source is simple and uniform. However, in some cases, the source may contain a very complex background, such as pages of magazines with illustrations, or icons of companies printed on receipts. We are working on applying text detection algorithms to detect regions that most likely contain text and then separate text regions from the complex background.

7.2 Merge Multiple Images Together

In some cases, image documents or receipts are quite long and can not be captured within one single frame due to the limited screen size of mobile phones. We are currently investigating algorithms that can merge the OCR results from images capturing different portions of the same text source and make sure they are merged in a proper sequence without data lost or repetition.

7.3 New Applications on Top of OCRDroid Framework

We are working on a new application on the top of the OCRdroid framework - PocketScan. This application allows users to quickly scan ingredients of any medicines as well as food and check if they contain some particular chemical to which the user is allergic.

8 Conclusions

In this paper, we present the design and implementation of a generic framework called OCRdroid for developing OCR-based applications on mobile phones. We focus on using orientation sensor, embedded high-resolution camera, and digital image processing techniques to solve OCR issues related to camera-captured images. One of the key technical challenges addressed by this work is a mobile solution for real time text misalignment detection. In addition, we have developed two applications called PocketPal and PocketReader based on OCRdroid framework to evaluate its performance. Preliminary experiment results are highly promising, which demonstrates our OCRdroid framework is feasible for building real-world OCR-based mobile applications.

References

1. ABBYY Mobile OCR Engine, http://www.abbyy.com/mobileocr/
2. GOCR - A Free Optical Character Recognition Program,
   http://jocr.sourceforge.net/
3. OCR resources (OCRopus),
   http://sites.google.com/site/ocropus/ocr-resources
4. OCRAD - The GNU OCR, http://www.gnu.org/software/ocrad/
5. OCRdroid, http://www-scf.usc.edu/~ananddjo/ocrdroid/index.php
6. Simple OCR - Optical Character Recognition, http://www.simpleocr.com/
7. Tesseract OCR Engine, http://code.google.com/p/tesseract-ocr/
8. Visual Codes,
   http://www.vs.inf.ethz.ch/res/show.html?what=visualcodes
9. WINTONE Mobile OCR Engine, http://www.wintone.com.cn/en/prod/44/detail270.aspx
10. Bieniecki, W., Grabowski, S., Rozenberg, W.: Image preprocessing for improving ocr accuracy. In: Perspective Technologies and Methods in MEMS Design, MEMSTECH 2007 (2007)
11. Bruns, E., Bimber, O.: Adaptive training of video sets for image recognition on mobile phones (2009)

12. Chen, X., Yang, J., Zhang, J., Waibel, A.: Automatic detection and recognition of signs from natural scenes (2004)
13. Elmore, M., Martonosi, M.: A morphological image preprocessing suite for ocr on natural scene images (2008)
14. Liang, J., Doermann, D., Li, H.P.: Camera-based analysis of text and documents: a survey. International Journal on Document Analysis and Recognition 7(2-3), 84–104 (2005)
15. Luo, X.P., Li, J., Zhen, L.X.: Design and implementation of a card reader based on build-in camera. In: ICPR 2004: Proceedings of the Pattern Recognition, 17th International Conference on (ICPR 2004), vol. 1, pp. I: 417–420. IEEE Computer Society, Los Alamitos (2004)
16. Mistry, P., Maes, P.: Quickies: Intelligent sticky notes. In: International Conference on Intelligent Environments (2008)
17. Niblack, W.: An Introduction to Digital Image Processing. Prentice-Hall, Englewood Cliffs (1986)
18. Ohbuchi, E., Hanaizumi, H., Hock, L.A.: Barcode readers using the camera device in mobile phones. In: CW 2004: Proceedings of the 2004 International Conference on Cyberworlds, pp. 260–265. IEEE Computer Society, Los Alamitos (2004)
19. Otsu, N.: A threshold selection method from gray-level histograms. IEEE Transactions on Systems, Man and Cybernetics 9(1), 62–66 (1979)
20. Rice, S.V., Jenkins, F.R., Nartker, T.A.: OCR accuracy: UNLV's fifth annual test. INFORM, 10, xx–yy (1996)
21. Sauvola, J., Pietikainen, M.: Adaptive document image binarization. Pattern Recognition 33(2), 225–236 (2000)
22. Seeger, M., Dance, C.: Binarising camera images for OCR. In: Sixth International Conference on Document Analysis and Recognition (ICDAR 2001), pp. 54–58 (2001)
23. Sezgin, M., Sankur, B.: Survey over image thresholding techniques and quantitative performance evaluation. Journal of Electronic Imaging 13(1), 146–168 (2004)
24. Shafait, F., Keysers, D., Breuel, T.M.: Efficient implementation of local adaptive thresholding techniques using integral images. In: Document Recognition and Retrieval XV, vol. 6815, 681510 (2008)
25. Ulges, A., Lampert, C.H., Breuel, T.M.: Document image dewarping using robust estimation of curled text lines. In: Eighth International Conference on Document Analysis and Recognition, pp. II: 1001–1005 (2005)
26. Whitesell, K., Kutler, B., Ramanathan, N., Estrin, D.: A system determining indoor air quality from images air sensor captured cell phones (2008)

Gradient Domain Image Blending and Implementation on Mobile Devices

Yingen Xiong and Kari Pulli

Nokia Research Center, Palo Alto, CA 94304, USA
`firstname.lastname@nokia.com`

Abstract. This paper presents an image blending approach which combines optimal seam finding and transition smoothing for merging a set of aligned source images into a composite panoramic image seamlessly. In this approach, graph cut optimization is used for finding optimal seams in overlapping areas of the source images to create a composite image. If the seams in the composite image are still visible, a gradient domain transition smoothing operation is used to reduce color differences between the source images to make them invisible. In the transition smoothing operation, a new gradient vector field is created using the gradients of source images and the seam information. A new composite image can be recovered from the new gradient vector field by solving a Poisson equation with boundary conditions.

Our approach presents several advantages. The use of graph cut optimization over the source images guarantees that optimal seams are found. The gradient domain transition smoothing operation allows smoothing out color differences globally and further improves image quality after merging with graph cut optimization. The final composite image is a global optimal solution. The approach is implemented in two ways called sequential image blending and global image blending. Sequential image blending allows us to use little memory in the whole blending process, which is very important for mobile devices. Global image blending guarantees a globally optimal solution. Experimental and application results in creating mobile image mosaics and mobile panorama are also given.

1 Introduction

As computational capabilities and memory of mobile devices increase and the camera quality of mobile devices improves, mobile image processing and mobile computational photography become increasingly interesting. Many applications which could only work on computer before can now be implemented on mobile devices, including mobile augmented reality [1,2], mobile image matching and recognition [3], and so on. Here, we are interested in creating high quality image mosaics and panoramic images on a mobile device. User can take an image sequence of a wide range of scenes with a mobile phone. Almost immediately after the capture the user can see a panoramic image on the device and share it with friends.

Image blending is the final and a key step in creating high quality image mosaics and panoramic images. A simple copying and pasting of overlapping areas of source images usually produces visible artificial edges in the seam between images, due to differences in camera response (white balance, vignetting, etc.) and scene illumination, or due to geometrical alignment errors. Image blending can hide the seams and reduce color differences between source images.

In our mobile panorama system, two kinds of image blending algorithms are used. A fast and low quality blending algorithm is used to produce a quick result, so that user can preview it as soon as the image sequence is captured. If the user is satisfied with the image sequence, a more refined blending approach is used to produce a high-quality panoramic image. The fine blending approach is typically slower than the simple one, but its blending quality is much higher. Here we focus on the latter one.

In this work, we present a gradient domain image blending approach and its implementation on mobile devices to produce high-quality panoramic images. The approach combines processes of optimal seam finding and transition smoothing, so that it can make full use of their advantages and avoid their disadvantages. We use graph cut optimization to find optimal seams and gradient domain blending for smoothing the transitions. With graph cut optimization, we can create labeling for all pixels and find optimal seams in the overlapping areas of source images. A composite image can be created by copying pixels from corresponding source images with labeling information. As each pixel in the composite image comes only from one source image, the algorithm avoids most blurring and ghosting problems. If all seams in the composite image are invisible, then we use it as our final result. Otherwise, with gradient domain blending, we can reduce the color differences between source images and hide the seams. In the gradient domain image blending, a new gradient vector field is created by copying the gradient values from the constituent images. The gradients across seams are set to zero for smoothing color differences. A new composite image can be recovered from the gradient vector field by solving a Poisson equation with boundary conditions. The composite image is a globally optimal solution, where the color transition is smoothed over the whole image.

1.1 Related Work

Existing methods for image stitching in the literature can be categorized into two main classes[4,5]: optimal seam finding and transition smoothing. Each class has its own advantages and disadvantages.

Optimal seam finding algorithms [4,6,7,8,9] search for a seam in the overlapping area where the differences between source images are minimal. All pixels of the composite image are labeled based on the seams so that for each output pixel there is a unique input pixel. The composite image is produced by copying corresponding pixels from the source images using labeling information. This kind of approach works well when the source images are similar enough. However, when they are too dissimilar for the algorithms to find ideal seams, artifacts may still be created.

Transition smoothing algorithms reduce the color differences between source images for hiding the seam. Alpha blending [10] is one of the simplest and fastest transition smoothing approaches. It uses weighted combination to create the composite image. The main disadvantage of alpha blending is that moving objects will cause ghosting and small registration errors can cause blurring of high frequency details. In order to solve this problem, Burt and Adelson [11] presented a multi-band blending algorithm. They blend low frequencies over a large spatial range, and high frequencies over a short spatial range. Recently, gradient domain image blending approaches [9,5,12,13,14,15,16] have been applied to image stitching and editing. A new gradient vector field is created with the gradients of source images and a new composite image can be recovered from the gradient vector field by solving a Poisson equation. This kinds of algorithms can adjust the color differences due to illumination changes and variations in camera gains for the composite image globally. It can produce high quality composite images. However, the memory and computational cost is also high.

1.2 Organization of the Paper

In Section 2, we introduce the work flow of our approach. The details of the gradient domain image blending approach are described in Section 3. An implementation of the approach on mobile devices is explained in Section 4. Applications and result analysis are discussed in Section 5. A summary of the paper is given in Section 6.

2 Summary of Our Approach

The approach can be divided into two parts, optimal seam finding and transition smoothing processes. Figure 1 shows the work flow of the approach. The input source images are already aligned and warped. We need to stitch them together seamlessly to create a composite image.

The first part of our approach is to find optimal seams in the overlapping areas of the source images. An objective function is created by combining two items: a data property of a pixel and the color differences between corresponding pixels in the overlapping areas. The objective function is minimized to obtain optimal seams by graph cut optimization. Using the results, we create labeling for all pixels. The composite image is created by copying corresponding pixels from source images according to labeling information. If all seams in the composite image are invisible, we use it as our final result. Otherwise, further processing in the next part is needed.

The second part of our approach is to reduce the differences of the source images to make the seams minimally visible. We perform gradient domain blending to smooth the differences. A gradient vector field is created by copying gradient values from the constituent source images with labeling information. We set the gradient values across seams to zero for smoothing color transition. A divergence vector $divG$ is created from the gradient vector field and is used as a guidance

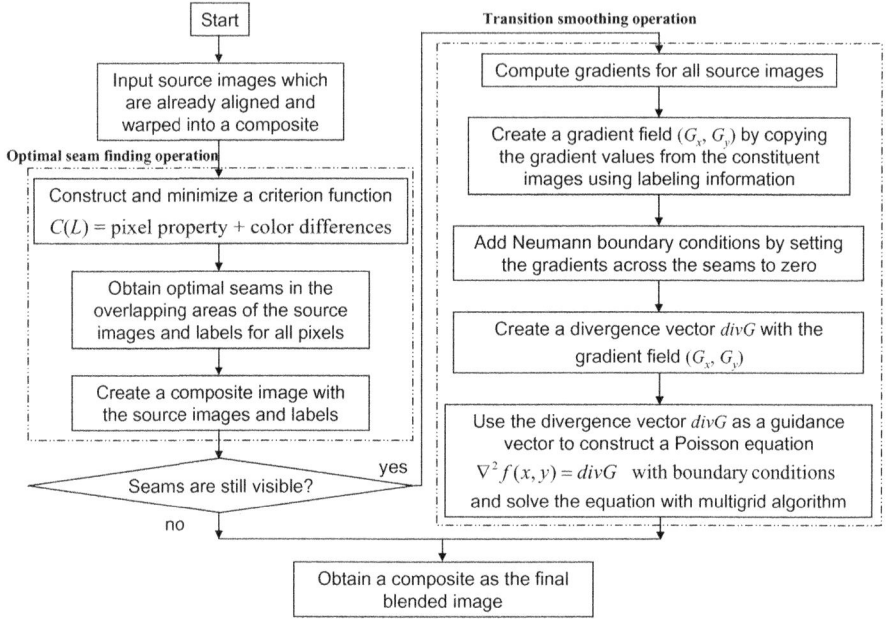

Fig. 1. A work flow of the gradient domain image blending approach

vector to construct a Poisson equation $\nabla^2 f = divG$. A new composite image can be recovered from the gradient vector field by solving the Poisson equation with Neumann boundary conditions. Since the discrete Poisson equation is an over-constrained linear system, we solve for the least-squares optimal results. In order to speed up the performance, we employ a multigrid algorithm for the linear solver.

3 Gradient Domain Image Blending

We use graph cut optimization to find optimal seams and gradient domain image blending to reduce color differences between source images for hiding visible seams.

3.1 Labeling with Graph Cut Optimization

As Figure 2 shows, we assume that the source images $S_0, S_1, \ldots, S_n$ are already registered and warped into a composite image I_b. We apply graph cut optimization to find optimal seams $m_0, m_1, \ldots, m_{n-1}$, using which a mapping or labeling between pixels in image I_b and the source images can be created. With the labeling, we can copy corresponding pixels from the source images to image I_b.

Considering implementation on mobile devices, we create a simple and efficient objective function which includes two items: pixel property P_p and color

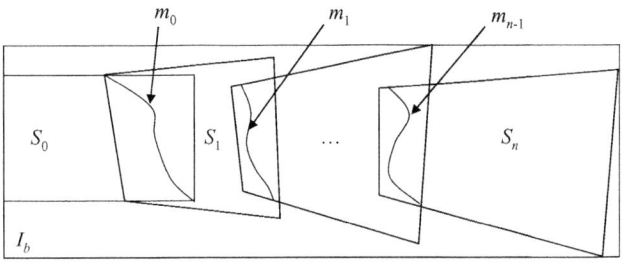

Fig. 2. Optimal seam finding with graph cut optimization

differences D_p between corresponding pixels, both of which depend on pixel labeling L.

$$O(L) = \sum_i P_p(i, L(i)) + \sum_{i,j} D_p(i, j, L(i), L(j)) \qquad (1)$$

where

$P_p(i, L(i))$ depends on the property of pixel i;
$D_p(i, j, L(i), L(j))$ is the color difference.

The pixel property item, $P_p(i, L(i))$ is set it to a very large number when the pixel is invalid, which means that the seam is not allowed to go to invalid areas. Otherwise, we set it to zero, i.e.,

$$P_p(i, L(i)) = \begin{cases} N & \forall i \in \Phi \\ 0 & \text{otherwise} \end{cases} \qquad (2)$$

where

N is a large number;
Φ is an invalid area.

The invalid areas are created in warping and interpolation processes after registration. Figure 3 shows an example. The black regions shown in (a) and (b) are invalid areas. The seam shown in (c) between the two source images is not allowed to go into these areas.

The color differences D_p are defined by the Euclidean distances in RGB space of all pairs of neighboring pixels i, j, i.e.,

$$D_p(i, j, L(i), L(j)) = \|S_{L(i)}(i) - S_{L(j)}(i)\| + \|S_{L(i)}(j) - S_{L(j)}(j)\| \qquad (3)$$

where

S_k is source image k;
$L(i)$ is the label of pixel i.

We could add other items into the objective function to consider other properties when cutting the source images, such as an item to consider edge information,

Fig. 3. An example of graph cut results with source images

and so on, but it would require more computation and memory, and this definition has been sufficient.

The optimal seam finding or labeling problem can be cast as a binary graph cut problem. We apply a max-flow approach [17] to solve it. The approach uses "alpha expansion" to minimize the objective function and obtain a globally optimal solution.

3.2 Transition Smoothing in the Gradient Domain

When source images to be stitched are too dissimilar for the optimal seam finding process to find ideal seams, the seams remain visible in the composite image. Figure 3 (c) shows an example. In this case, transition smoothing is needed to reduce the differences between the source images for hiding the seams.

We apply gradient domain transition smoothing to improve blending quality. In this operation, we create a color gradient vector field (G_x, G_y) by copying the color gradients of corresponding pixels in source images using the same labels which are created by optimal seam finding process. With the color gradient vector field, a divergence vector $div(G)$ can be computed. Suppose $f(x, y)$ is the composite image. We use the divergence vector as a guidance vector to construct a Poisson equation

$$\nabla^2 f(x,y) = div(G) \qquad (4)$$

where

∇^2 is the Laplacian operator such that $\nabla^2 f(x,y) = \frac{\partial^2 f(x,y)}{\partial x^2} + \frac{\partial^2 f(x,y)}{\partial y^2}$;
$div(G)$ is the divergence vector field with $div(G) = \frac{\partial G_x}{\partial x} + \frac{\partial G_y}{\partial y}$.

Equation 4 is a linear partial differential equation. In order to solve this equation, we must first specify boundary conditions. In our case, we use the Neumann boundary conditions,

$$\nabla f \cdot \overrightarrow{n} = 0, \qquad (5)$$

where

∇f is the gradient of $f(x,y)$;
$\overrightarrow{n}$ is the normal on the boundary.

Equation 5 tells us that the gradient in the direction normal to the boundary is zero. The discrete form will be used in actual implementation.

In the color gradient vector field (G_x, G_y), we set gradient values across seams to zero for smoothing color transition between two source images and gradient values on the boundaries to zero for fitting the boundary conditions.

Finally, a new composite image f_b can be recovered from the color gradient vector field (G_x, G_y) by solving the Poisson equation with boundary conditions. We use the result as our final blended image I_b.

4 Implementation

4.1 Discretization of the Poisson Equation

Since the Poisson equation is linear, we use standard finite differences to approximate each item.

For the left hand side,

$$\nabla^2 f(x,y) = \frac{\partial^2 f(x,y)}{\partial x^2} + \frac{\partial^2 f(x,y)}{\partial y^2}, \tag{6}$$

we apply central difference

$$\frac{\partial^2 f(x,y)}{\partial x^2} \approx f(x+1,y) - 2f(x,y) + f(x-1,y) \tag{7}$$

and

$$\frac{\partial^2 f(x,y)}{\partial y^2} \approx f(x,y+1) - 2f(x,y) + f(x,y-1). \tag{8}$$

where we use a unit step size, i.e.,

$$\begin{cases} \triangle x = 1 \\ \triangle y = 1 \end{cases} \tag{9}$$

Now the Laplacian operation for $f(x,y)$ is approximated as

$$\nabla^2 f(x,y) \approx f(x+1,y) + f(x-1,y) + f(x,y+1) + f(x,y-1) - 4f(x,y) \tag{10}$$

For the right hand side, we first use standard finite differences to approximate the gradients of the source images.

$$\begin{cases} G_{sx}(x,y) \approx S(x+1,y) - S(x,y) \\ G_{sy}(x,y) \approx S(x,y+1) - S(x,y) \end{cases} \tag{11}$$

where $S(x,y)$ is the source image.

As described in Section 3.2, we create a gradient vector field $(G_x(x,y), G_y(x,y))$ using the gradients of source images and labels obtained from the optimal seam finding process. With the gradient vector field, we construct a divergence vector field $div(G)$ as follows:

$$div(G) \approx G_x(x,y) - G_x(x-1,y) + G_y(x,y) - G_y(x,y-1) \qquad (12)$$

So the discrete form of the Poisson equation can be described as

$$\begin{aligned} f(x+1,y) + f(x-1,y) + f(x,y+1) + f(x,y-1) - 4f(x,y) \\ = G_x(x,y) - G_x(x-1,y) + G_y(x,y) - G_y(x,y-1) \end{aligned} \qquad (13)$$

For each color channel, we solve the Poisson equation with Neumann boundary conditions and obtain a recovered composite image for this channel. Finally, we combine these single channel results into a final blended image I_b.

We have two implementations for the approach presented in this paper on mobile devices. We call the first implementation sequential image blending and the second one global image blending. In the following sections, we describe these two implementations in detail.

4.2 Implementation 1: Sequential Image Blending

In this implementation, we select a base image S_{base} from the source image sequence $S_0, S_1, \ldots, S_n$ and blend others onto the base image sequentially. The blending order is decided by sorting the offsets of the source images.

Figure 4 shows the process. First, we sort the offsets of the source images to create an order for the blending operation. The size of the final blended image I_b

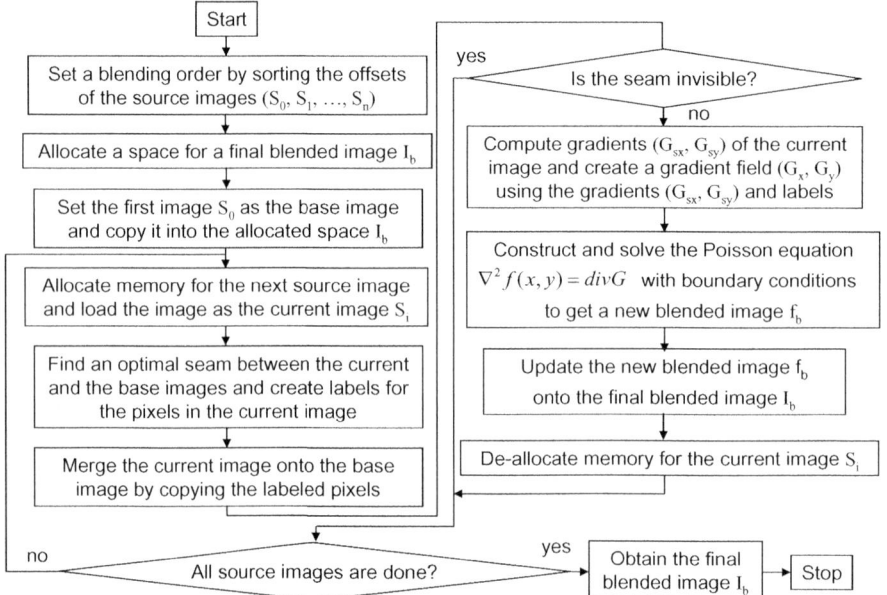

Fig. 4. The work flow of sequential image blending

can be computed by the sizes and offsets of the source images. For convenience, we set the first image S_0 as the base image and put it into the space of the final blended image I_b. Then, the next source image S_i is loaded as the current image. With graph cut optimization, we compute the optimal seam between image I_b and the current image S_i and merge S_i onto image I_b. At the same time, we create labels for all pixels in the overlapping area.

If the seam of the merged image I_b is invisible, the blending process for the current image S_i is done. We can load next image S_{i+1} to replace the current image S_i and repeat the blending process.

Otherwise, if the seam is still visible, then gradient domain smoothing is needed for further processing. In this case, we compute gradients (G_{sx}, G_{sy}) of the current image and create a gradient vector field (G_x, G_y) using labeling information. Then we construct and solve a Poisson equation to obtain a new blended image. We update it onto image I_b. The blending operation for the current image S_i is done, and we continue with the next source image S_{i+1}, until all source images $S_0, S_1, \ldots, S_n$ have been blended into the final image I_b.

One of the main advantages of this implementation is low memory consumption, which is very important for running on mobile devices. There is a disadvantage, though. When more than two images overlap in the same area, the area may be blended several times. Usually, we use this implementation in one dimensional stitching, where only at most two images can overlap at a given pixel.

4.3 Implementation 2: Global Image Blending

In this implementation, we store all source images $S_0, S_1, \ldots, S_n$ into memory, find optimal seams for all overlapping areas of the source images, and merge and blend them together in one time.

Figure 5 shows the process of this implementation. First, we load all source images $S_0, S_1, \ldots, S_n$ into memory and create a space for the final blended image I_b. Then we compute optimal seams for all overlapping areas of the source images and create labels with graph cut optimization. With the labeling information, we merge the source images together to create a composite image I_c. If the seams in the composite image are invisible, we use it as our final blended image I_b.

If the seams are still visible, we apply the gradient domain blending operation to improve the quality of the composite image I_c. In this case, we compute gradients (G_{sx}, G_{sy}) for all source images and create a gradient vector field (G_x, G_y) to construct a Poisson equation. A new composite image can be recovered from the gradient vector field (G_x, G_y) by solving the Poisson equation with boundary conditions, and we use it as our final blended image I_b.

Since the optimal seam finding operation is performed globally, the obtained optimal seams and labeling are global solutions. The whole composite image only needs to be blended once and the transition smoothing operation is performed globally. It can smooth color differences on the whole composite image. The main disadvantage is that all source images need to be loaded into memory at same time, which is problematic on mobile devices with limited amount of RAM.

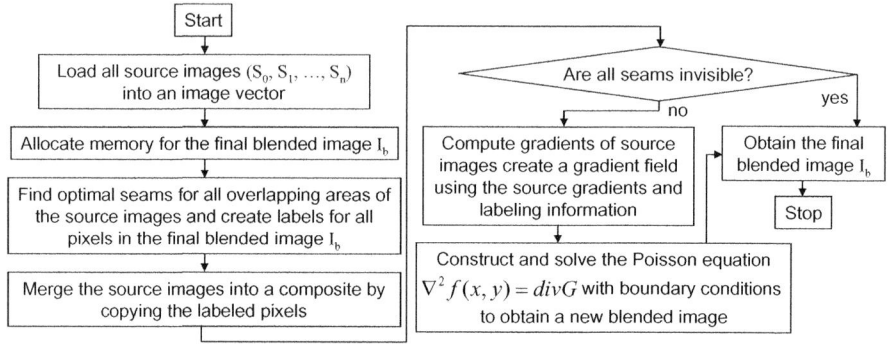

Fig. 5. The work flow of global image blending

4.4 Solving the Poisson Equation

The partial differential equation 4 can be solved after specifying the boundary conditions. As mentioned in Section 3.2, we employ Neumann boundary conditions. In the case of image blending, the boundary conditions are equivalent to dropping any equations involving pixels outside the boundaries of the image.

The approximation of finite differences shown in equation 13 produces a large system of linear equations. One equation corresponds to a pixel in the final blended image. The large system of linear equations is over-constrained. We solve for least-squares optimal vector f using the Full Multigrid Algorithm [18] as our final result.

5 Applications and Result Analysis

We have implemented these algorithms on mobile devices and used them to produce high-quality panoramic images in our mobile panorama system. Here we describe some example applications and results obtained by running them on Nokia N95 8GB mobile phones with an ARM 11 332 MHz processor and 128 MB RAM. It can also be run on other mobile devices. The results are shown by figures. In each figure, the top shows the blending result and the bottom shows the source images. In these applications, the size of source images is 1024×768. Since the seam finding process with graph cut optimization needs more memory and computation, we down-sample the source images with scale factor 0.25 to solve the optimization problem to obtain labeling and scale the results back for the final composite image.

We use different kinds of image sequences captured in different conditions to test the approach and verify its performance.

5.1 Applications to Panorama Stitching for Outdoor Scenes

We apply the approach to create panoramic images for outdoor scenes. Figure 6 shows an example result with 6 source images in the image sequence. First, we

Fig. 6. An outdoor panoramic image created with 6 source images

apply graph cut optimization to find optimal seams and create labeling for the composite image. Since the color and luminance between the source images are very different, the seams are still visible as shown in Figure 6 (a). The graph cut operation takes 323.76 seconds when the source images with the original size are used. However, it only takes 16.4 seconds when the down-sampled source images are used.

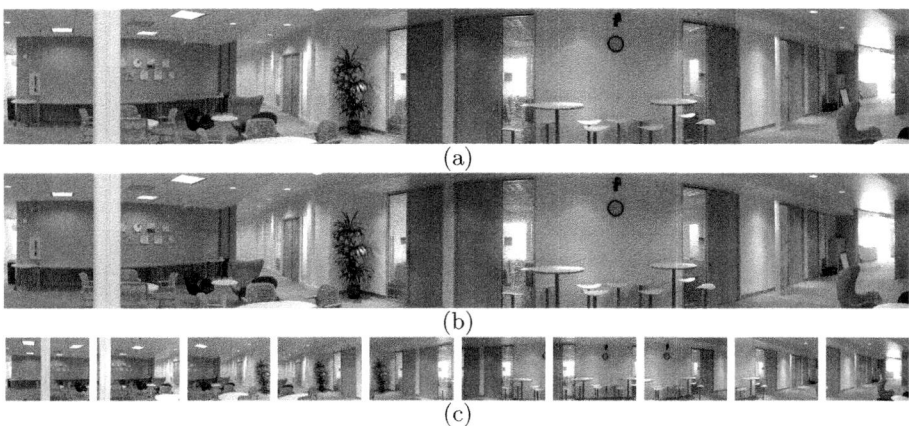

Fig. 7. An indoor panoramic image created with 10 source images

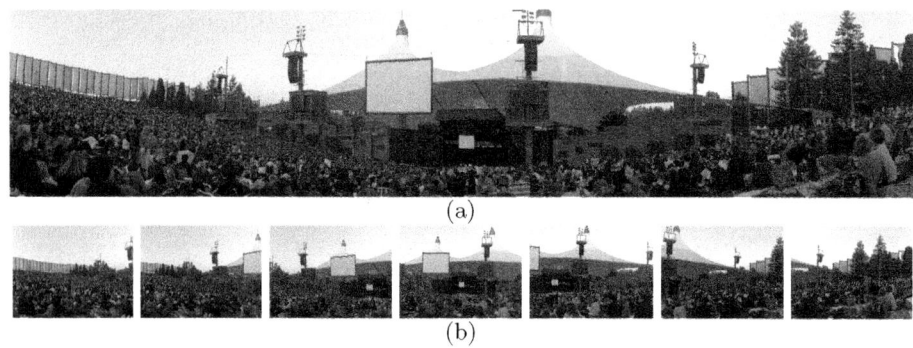

Fig. 8. A panoramic image created with 7 source images captured after sunset

We perform gradient domain image blending to further smooth the color transition between the source images. Figure 6 (b) shows the final result. From the result we can see that all merging artifacts are removed. The color transition between source images is smoothed. All seams are invisible after the gradient domain blending. The gradient domain blending operation takes 54.4 seconds.

5.2 Applications to Panorama Stitching for Indoor Scenes

We applied the approach to panorama stitching for indoor image sequences. Figure 7 shows the source images and results. Figure 7(a) shows the result created by the optimal seam finding process. It takes 15.92 seconds with down-sampled source images. Since the color differences between source images shown in Figure 7(c) are small, some seams are invisible. We apply the gradient domain blending operation to further smooth the color transitions. Figure 7(b) shows the blended panoramic image. From the result we can see that all stitching artifacts are removed. The blending operation takes 80 seconds.

5.3 Applications to Panorama Stitching for Low Light Scenes

We apply the approach to panorama stitching for the scene where the environment light is not sufficient. Figure 8 shows a result created with 7 source images which are captured after sunset. From the panoramic image we can see that the approach still works fine. In panorama stitching, the seam finding process takes 15.92 seconds and the gradient domain operation takes 52.2 seconds.

From the result of this application we can also see another property of the approach. In this case, there are many moving objects in the scene and the approach can find proper seams to avoid ghosting artifacts.

The approach has been tested with many image sequences and applied to different scenes. Figure 9 shows more panoramic images created by the approach. The results are satisfying.

Fig. 9. More example results

6 Conclusions and Discussion

We have presented a gradient domain image blending approach and implemented it on mobile devices. It can be applied to image stitching and image editing for producing high quality mobile panoramic images and composite images.

The approach combines optimal seam finding and transition smoothing processes to provide operations for different source images which are in different conditions. Usually, camera responses and environmental illumination changes when capturing pictures. If the change is small, it may suffice to only use optimal seam finding operation to merge the source images, and the result will be good enough. When the source images are too dissimilar, the seams between the source images in overlapping areas may still be visible. In this case the algorithm also uses the transition smoothing process to further improve the quality of the result image. In this approach, a gradient domain image blending is used for transition smoothing.

We have presented examples in panorama stitching for indoor scenes, outdoor scenes, low light scenes, and others. From the applications we can see that the approach works fine on mobile devices. The results are satisfying.

The main disadvantage of the approach is that memory consumption and computational costs are high. Especially, the optimal seam finding process with graph cut optimization is slow when large source images are used without downsampling.

The future work includes saving memory and speeding up the algorithm. OpenGL ES and parallel processing implementation will be considered.

References

1. Chen, W.C., Xiong, Y., Gao, J., Gelfand, N., Grzeszczuk, R.: Efficient extraction of robust image features on mobile devices. In: ISMAR 2007: Proceedings of the 2007 6th IEEE and ACM International Symposium on Mixed and Augmented Reality, Washington, DC, USA, pp. 1–2. IEEE Computer Society, Los Alamitos (2007)
2. Takacs, G., Chandrasekhar, V., Gelfand, N., Xiong, Y., Chen, W.C., Bismpigiannis, T., Grzeszczuk, R., Pulli, K., Girod, B.: Outdoors augmented reality on mobile phone using loxel-based visual feature organization. In: MIR 2008: Proceeding of the 1st ACM international conference on Multimedia information retrieval, pp. 427–434. ACM, New York (2008)
3. Chen, D., Tsai, S.S., Chandrasekhar, V., Takacs, G., Singh, J., Girod, B.: Robust image retrieval using multiview scalable vocabulary trees. In: Rabbani, M., Stevenson, R.L. (eds.) Visual Communications and Image Processing 2009. SPIE, vol. 7257, 72570V (2009)
4. Gracias, N., Mahoor, M., Negahdaripour, S., Gleason, A.: Fast image blending using watersheds and graph cuts. Image Vision Comput. 27, 597–607 (2009)
5. Levin, A., Zomet, A., Peleg, S., Weiss, Y.: Seamless image stitching in the gradient domain. In: Pajdla, T., Matas, J.G. (eds.) ECCV 2004. LNCS, vol. 3024, pp. 377–389. Springer, Heidelberg (2004)
6. Milgram, D.L.: Computer methods for creating photomosaics. IEEE Trans. Comput. 24, 1113–1119 (1975)
7. Efros, A.A., Freeman, W.T.: Image quilting for texture synthesis and transfer. In: SIGGRAPH 2001: Proceedings of the 28th annual conference on Computer graphics and interactive techniques, pp. 341–346. ACM, New York (2001)
8. Davis, J.: Mosaics of scenes with moving objects. In: CVPR 1998: Proceedings of the IEEE Computer Society Conference on Computer Vision and Pattern Recognition, Washington, DC, USA, p. 354. IEEE Computer Society, Los Alamitos (1998)
9. Agarwala, A., Dontcheva, M., Agrawala, M., Drucker, S., Colburn, A., Curless, B., Salesin, D., Cohen, M.: Interactive digital photomontage. ACM Trans. Graph. 23, 294–302 (2004)
10. Uyttendaele, M., Eden, A., Szeliski, R.: Eliminating ghosting and exposure artifacts in image mosaics. In: Computer Vision and Pattern Recognition (CVPR 2001), pp. 509–516. IEEE Computer Society, Los Alamitos (2001)
11. Burt, P.J., Adelson, E.H.: A multiresolution spline with application to image mosaics. ACM Trans. Graph. 2, 217–236 (1983)
12. Szeliski, R., Uyttendaele, M., Steedly, D.: Fast poisson blending using multi-splines. Technical Report MSR-TR-2008-58, Microsoft Research (2008)
13. Kazhdan, M., Hoppe, H.: Streaming multigrid for gradient-domain operations on large images. ACM Trans. Graph. 27, 1–10 (2008)
14. Fattal, R., Lischinski, D., Werman, M.: Gradient domain high dynamic range compression. In: SIGGRAPH 2002: Proceedings of the 29th annual conference on Computer graphics and interactive techniques, pp. 249–256. ACM, New York (2002)
15. Pérez, P., Gangnet, M., Blake, A.: Poisson image editing. ACM Trans. Graph. 22, 313–318 (2003)
16. Jia, J., Sun, J., Tang, C.K., Shum, H.Y.: Drag-and-drop pasting. In: SIGGRAPH 2006: ACM SIGGRAPH 2006 Papers, pp. 631–637. ACM, New York (2006)
17. Kolmogorov, V., Zabih, R.: What energy functions can be minimized via graph cuts. IEEE Transactions on Pattern Analysis and Machine Intelligence 26, 65–81 (2004)
18. Briggs, W.L., Henson, V.E., McCormick, S.F.: A Multigrid Tutorial. The Society for Industrial and Applied Mathematics, New York (2001)

A Comparative Evaluation of HTML5 as a Pervasive Media Platform

Tom Melamed and Ben Clayton

Hewlett-Packard Labs Europe, Bristol, UK
{tom.melamed,ben.clayton}hp.com

Abstract. This paper introduces the field of pervasive media, the use of pervasive computing to deliver applications and services into peoples' lives, and the problems of creating and distributing pervasive media applications to mobile phones. It determines the requirements of a development platform for pervasive media applications through analysis of a range of pervasive media applications, including games and tourism based applications. It then evaluates a number of platforms against those requirements, including J2ME and native Smartphone development. The paper then introduces HTML5 and through a description of its features evaluates it against the requirements. Finally it concludes that HTML5 is a good solution to the problems of creating and distributing most pervasive media applications and describes its advantages and disadvantages compared to existing solutions.

1 Introduction

Pervasive media is an emerging field that combines new technologies with rich media and experience design to create compelling experiences, based on the user's context. Pervasive media follows on from two main technological fields, firstly computing technology leaving the "beige box" to pervade our lives, from mobile devices, to large screen displays, to smart cards. Secondly, sensing and instrumentation technologies are becoming cheap, ubiquitous and connected. Pervasive media takes advantage of these technologies to create applications and services that offer value to users which would not otherwise have been possible. For example a pervasive media treasure hunt game might run on a mobile phone providing the user with clues and tasks based on their GPS location. However as the examples provided in section 3.2 show there are many different forms of pervasive media application and more will probably emerge as the technologies and tools become more widely available.

Pervasive media points to a world of embedded technology and sensors in the environment capable of enhancing and augmenting everyday experiences based on the user's unique context. However in the short term mobile phones are a very good delivery platform for pervasive media as they are carried by users throughout their normal life and contain an increasingly rich set of sensors and media delivery options. This allows pervasive media applications to interweave with a users normal life offering interactions or content based on the user's context. Context can be physical and locally sensed such as a user's location sensed via GPS or more abstract such as the identities of the user's friends from Facebook.

This paper focuses only on pervasive media applications delivered through a mobile phone. We enumerate the main problems of developing for mobile phones currently, as a motivation for looking at alternative ways of developing such applications. We then look in more detail at the specific requirements of pervasive media applications by studying several of them from different genres and built using different technologies. We analyze the features from this sample set to identify common requirements for the creation of pervasive media applications. We use these features and the main problems of phone development as criteria to comparatively evaluate several different runtime environments available on different phones. Some runtime environments form the actual OS of the phone such as Windows Mobile or Symbian, others run on top and host applications within a virtual machine such as J2ME or Flash Lite but all are candidates to deliver pervasive media applications and so all are considered equally. We then introduce HTML5 which provides the specification for a browser-based runtime environment. We then evaluate its features in relation to the criteria that we had previously identified and in comparison to the other runtime environments. Based on this analysis we highlight some of the current shortcomings of HTML5 for pervasive media application but conclude that for many such applications it provides a good runtime environment.

1.1 HTML5

This paper is focused on the evaluation of one runtime environment in relation to the others, that environment being the web browser. The web browser is rapidly evolving from a renderer of simple html into a runtime environment capable of delivering rich interactive applications across many application domains. Browser development is primarily focused around the emergence of the HTML5 standard [1]. This standard effort grew from the Web Hypertext Application Technology Working Group which formed with a specific goal: "... to address the need for one coherent development environment for Web Applications." [2]. The original HTML5 specification consists of a number of different sections including offline storage of web content [3], the canvas element and others. There is another standard that did not grow from HTML5 called the Geolocation API specification [4], this is also part of the W3C specification effort and is focused specifically on an API for web pages to determine the browser location. For simplicity and brevity this paper will use *HTML5* to refer the full HTML5 specification and use *HTML5+GL* to refer to the combination of HTML5 and the W3C Geolocation specification. At the time of writing the standards are still in the drafting stages, but they are already being adopted by some web browsers. For example the latest versions of Mozilla's Firefox and Apple's Safari browsers already support a large percentage of these specifications, such as media playback and offline storage.

2 Problems with Mobile Development

Pervasive media is a new medium and as such its full potential is yet to be determined but there is already considerable interest within research, creative and commercial sectors. A number of commercial ventures and research projects have undertaken to create pervasive media applications for mobile phones, such as those discussed in section 3.2. This section will list some significant difficulties with current phone development. Most existing solutions contain certain tradeoffs such as the features available and the size of the potential market, or the ease of development and of distribution.

Development for mobile phones often requires different skill sets and familiarity with different runtime environments than those more commonly used on desktop PCs. Phone development requires knowledge of runtime environments such as J2ME, Objective C on iPhone or C++ on Symbian. There are also issues of memory management and constrained system resources that can be cumbersome and unfamiliar to PC developers.

Distribution is often piecemeal and fragmented with many users unfamiliar with the mechanisms available to download applications to their devices. While some Smartphones such as the iPhone and Android platforms are addressing this issue successfully, delivery to many devices is still problematic.

There can be restrictions and costs associated with developing and deploying to certain platforms, such as needing an Apple developer licence and Apple's approval to distribute applications to iPhones.

Device fragmentation continues to be a problem for developers with many different phone operating systems, and programming APIs. Even within a single runtime environment such as J2ME, fragmentation between devices and inconsistencies of implementation can greatly increase the cost and time involved in development.

Access to the device's sensors such as location and accelerometers can be restricted to certain applications or programming languages. Access to the user's context is very important for the execution of pervasive media applications.

An ideal goal would be the ability to write an application once using a well documented and supported language and set of APIs and have this application instantly available to all mobile devices irrespective of device or network. Unfortunately no such solution currently exists; this situation has forced makers of pervasive media applications to find compromises and increased the barriers to entry to this field. For example all of the applications studied in this paper were written for single specific runtime environments, reducing the potential user base.

Emerging web standards are one possible solution to the problems outlined above; the main contribution of this paper is an evaluation of these new technologies with respect to the problems outlined above and the specific requirements of pervasive media.

3 Requirements

In this section we shall enumerate and analyze the requirements for a successful pervasive media runtime environment by first considering general mobile development requirements and then by looking at a number of pervasive media applications and their individual requirements.

3.1 General Requirements

All mobile phone runtime environments require a base level of functionality to be useful for creating popular and or commercially successful applications. While it is possible to build applications on runtime environments without these features the ease of development and chance of commercial success is greatly diminished. For applications built for research purposes these features ease both development and the recruitment of test subjects.

- Large target market of installed devices.
- Good and accessible distribution channels to reach that large installed base.
- Good developer support, including documentation, libraries and community.
- Useable interface options, the user has to be able to find the application within the phone and understand how to use it.
- Efficient execution allowing for complex applications to be run on phones with limited processing power.
- Efficient use of limited resources such as battery life and system memory.

When evaluating different development platforms for pervasive media applications these should also be enumerated and considered.

3.2 Requirements from Existing Applications

In this section we analyse a number of pervasive media applications in order to determine which requirements are common amongst them. We will not try to distinguish which requirements are specific to pervasive media only which are commonly used by such applications. Those are the features that a runtime environment should support in order to support pervasive media applications.

Because all existing pervasive media applications on mobile phones have been constrained by the available technology it is not always easy to determine desired features from existing applications. Constraints within the technology may well have forced the designers of these applications to make certain unknown tradeoffs. However limiting ourselves to applications that have been built does show the features that are immediately desired and useful to pervasive media applications and those that have been used repeatedly by different applications are likely to be useful to other application authors too. Furthermore if pervasive media applications from different subgenres such as games, tourism or theatre and produced by different groups contain the same features then this is an even stronger indication that these features are general case requirements for pervasive media applications.

A number of applications were chosen for study that covered a broad variety of pervasive media genres, created by different groups and focused on different aspects of the technology. As phone technology moves so rapidly it was deemed acceptable to include some applications designed a few years ago for PDAs, while some of the PDAs' features were not then available in mobile phones such as a large touch screen and GPS, all are now available in phones such as the iPhone and Google Android devices. Some of these applications were created using toolkits such as mscape [5] and Mupe [6]. While the features available in these toolkits may have been a restraint on the features used in these applications, those built using toolkits do not appear to have different features or requirements to those built from scratch. For applications to

Table 1. Overview of surveyed applications

	Year	Platform	Toolkit	Loaned device	Genre
Bot Fighters [7,8]	00	Any SMS device	N/A	No	Game
Riot! 1831 [9]	04	Windows Mobile	Mobile Bristol [10]	Yes	Located audio play
Stamp the Mole [11]	07	Windows Mobile	Mscape	No	Game
Uncle Roy All Around You [12,13]	03	Windows Mobile	N/A	Yes	Game
GPS Mission [14]	09	iPhone	N/A	Yes	Game
Insectopia [15,16]	06	J2ME	Mupe	No	Game
'Ere Be Dragons	05	Windows Mobile	Mscape	Yes	Game/ Exercise
Feeding Yoshi [17]	06	Windows Mobile	N/A	Yes	Game
REXplorer [18]	07	Nokia N70	N/A	Yes	Tourist/ Game

be considered relevant they had to use more than just voice telephony and be in some way influenced by the users' context, for this reason applications like Pac-Manhattan (which only used voice) and Day of the Figurines (which was not context-based) were excluded.

The pervasive media applications studied are summarized in Table 1 above. From these applications a number of common features were identified that relate to functionality within any given runtime environment. After the table each relevant feature will be discussed in order of frequency within the sample applications. Some applications where designed to be played in a single session for a limited time period on a loaned device that was then returned at the end of the session.

Local Execution

Any application that actually runs programmed instructions on the phone counts as local execution. While this is common among all phone applications it is not required. Bot Fighters is the only pervasive media application surveyed that did not have local execution; instead all interactions took place via SMS messages sent to a central server.

Interactive User Interface
Again this is a common requirement across most phone applications. Any application that the user controls through either a touch screen or buttons is counted. However because pervasive media applications also use the users context, it is possible to create applications that do not require traditional input methods. Of those surveyed only Riot! 1831 did not use an interactive interface. It should also be noted that 'Ere be dragons did not have any input on the device but did have a dynamic display based on the users context.

Audio
Only applications that play pre-recorded sound files are, for the purpose of this paper, considered to play audio, simple beeps do not count. Many pervasive media applications use sound extensively as it has been found to be a good way to provide information or content to the user without distracting their gaze from their soundings. Sound is used in these pervasive media applications; Riot 1831, Stamp the mole, GPS Mission, 'Ere be dragons and REXplorer.

Location Sensor
This feature is of primary importance to pervasive media as location is a key form of context. This paper does not distinguish between the different location-sensing technologies such as Cell Tower based positioning or the much more accurate GPS. Not all pervasive media applications use a location sensor but the following surveyed applications do: Riot 1831, Stamp the mole, GPS Mission, 'Ere be Dragons and REXplorer.

Server Interaction
Any application that relies on a server interaction for parts of its functionality fits into this category, for example an email program. This is a common requirement for many applications outside pervasive media. Within pervasive media applications servers are used for many reasons such as coordinating a multiplayer game, retrieving context that is not sensed locally or updating a high score table. Within our sample Bot Fighters, Uncle Roy, Insectopia, GPS Mission, and 'Ere be Dragons all used a server.

Non-location Sensors
While location is a key form of context there are many others that can be sensed from some mobile phones, such as acceleration or proximity. Most mobile phone applications do not use such sensors but an increasing number of them do, such as those that use accelerometers to re-arrange the interface for the devices orientation. The sensors used in our sample survey are Bluetooth and WiFi proximity in Insectopia and Feeding Yoshi respectively, heart rate in 'Ere be Dragons and accelerometers in REXplorer.

Local Persistence
Local persistence is characterized as any application that can save state between one instance of being executed and another, for example saving the status of the game or the user history. This is a common requirement among applications from many fields however only three of the sampled pervasive media applications used local persistence, this is probably due to the high number of applications that were created to be loaned out with devices for a single session. Of those that were not designed to run on devices that were loaned out, all but one used local persistence: Stamp the Mole, Insectopia, GPS Mission, and Feeding Yoshi.

Video
Some applications play video as part of their interface or content. For example some games use videos as an introduction to the game. Within the surveyed pervasive media applications video does not appear to be used very often. This could be due to its ability to draw users into the screen. While seated at home this is often advantageous but while moving through a rich environment it can be preferable and safer to not get too absorbed in the screen of the device. Only REXplorer used video in our sample set.

Alert Users
It is often advantageous to alert the user to some opportunity or change within an application. A common example of this is a phone alerting the user to a new email or SMS message. Within pervasive media applications one could imagine the user being alerted when their context matches certain criteria – such as moving to a location where a game can be played.

In our sample only Bot Fighters and Feeding Yoshi alerted users to situations they may want to respond to. This low number of applications may have been due to the number of applications that were designed to be lent out for single sessions where the application is already assumed to be the main focus of the user's attention.

Content Capture
Any application with the facility to record content such as audio, video or photos would fit within this category. Uncle Roy, GPS Mission and REXplorer used content capture, for photo and/or voice recording.

3.3 Requirements Analysis

From our survey of pervasive media applications we can draw a table of feature usage.

While it is not possible to draw strong conclusions from the exact frequency of each feature, overall patterns are worthy of consideration. The most common requirements appear to be local execution and interactive user interfaces. Other common requirements appear to be playing audio, interaction with servers and location sensing. The fact many different and independently produced applications use a similar feature set implies that even if this sample is somewhat small these features would still feature regularly in other applications.

Table 2. Feature usage by surveyed application

	Bot Fighters	Riot! 1831	Stamp the	Uncle Roy	GPS MIssion	Insectopia	Ere Be Dragons	Feeding Yoshi	REXplorer	Total
Local execution		X	X	X	X	X	X	X	X	8
Interactive UI			X	X	X	X		X	X	6
Play Audio		X	X		X		X		X	5
Use Location sensor		X	X		X		X		X	5
Interact with server	X			X	X	X	X			5
Non-location sensors						X	X	X	X	4
Local persistence			X		X	X		X		4
Content capture				X	X				X	3
Alert users	X							X		2
Play Video									X	1

4 Evaluation of Existing Runtime Environments

From the requirements obtained in section 3 above we now have a list of general phone development requirements and required features used in pervasive media. These requirements will now be used to evaluate a number of different mobile phone runtime environments to assess their suitability for building and distributing pervasive media applications.

The ideal runtime environment would address the problems discussed in section 2 above and meet the requirements determined from the analysis in section 3.

There are many existing runtime environments for the creation of pervasive media; all the examples discussed in section 3 were built using one of these platforms. The features of the most common runtime environments are summarized in Table 3 above. Neither mscape nor Mupe are analyzed as runtime environments. This is because they act as abstractions on top of other environments and the small number of devices with mscape or Mupe installed makes them of limited direct interest. Mscape runs within Windows Mobile which is a Smartphone operating system and Mupe runs with J2ME.

It can be difficult to categorize a platform as specifically having or lacking a certain property such as developer support or good distribution. Table 3 below is an attempt at an impartial categorization. Details and justifications will then be described for each runtime environment.

Table 3. Requirements against runtime environments

	Voice /SMS	J2ME	HTML4	Flash Lite	Native Code	HTML5+GL
Large installed base	X	X	X	X	X	Δ
Good distribution	X		X	O	O	X
Consistent implementation	X		O	X		X
Developer support	X	X	X	X	X	X
Launching applications	X	X	X	O	X	X
Efficient execution	X	X	X	X	X	X
Use of limited resources	X	X	X	X	X	X
Local execution		X	O	X	X	X
Interactive user interface	X	X	X	X	X	X
Play audio		O		X	X	X
Use location sensor	O	O			X	X
Interact with server	X	O	X	X	X	X
Use non-location sensors		X		X	X	X
Local persistence		O			X	X
Play video		O			X	X
Alert users		O		O	X	X
Content capture	X	O			X	

X Criteria met.
O Partial complience.
Δ Critera that is expected to rapidly become met.

4.1 SMS and Voice

A user can interact with a service or application by sending and receiving messages or talking through their phone. This is often used for televised competitions or booking cinema tickets. While this is not strictly a runtime environment it does offer a means to deliver a pervasive experience to a phone as Bot Fighters demonstrated. All phones are capable of voice communications and the vast majority can also send and receive SMS messages with most users very familiar with using these features. It is easy for developers to send and receive SMS or voice calls programmatically.

However without any local execution or access to local sensors the scope of applications is limited and latency and the cost of round trips can become prohibitive. While some operators can determine the crude cell location of SMS as used in Bot Fighters this tends to cost the developer on a per usage basis and normally requires a separate agreement with each operator.

4.2 J2ME

Java is the most common runtime for mobile phone development. A version of the J2ME Java profile is installed on millions of phones and there is a large and mature developer community.

However J2ME is a fragmented platform with a many different versions and a number of different combinations of optional APIs for features such as media playback, network access, and access to sensors. This fragmentation makes it very difficult to run the same program on any large subset of that installed base without resorting to the lowest common denominator, which may lack the features required for pervasive media. This means that there is effectively a smaller installed base if one takes these requirements into account.

Wide distribution is also a problem for J2ME as each network operator runs their own distribution and billing channels. These systems can be hard to access and generally require the developer to provide versions of the software that can run on all of their supported phones, making use of optional high-end features problematic.

4.3 HTML4

Most modern phones now come with some form of web browser that can render a subset of HTML, CSS and JavaScript in addition to WAP content. This provides an easy route for developers to deliver content to users, with few issues of signing or operator approval and a wide potential installed base of capable phones. Such pages can show text and images and include simple web forms and being server-based can easily be updated to reflect the current state of the application.

Being confined to the browser means that these pages cannot access much of the device's native functionality such as location-sensing or rich media playback. Also due to the inherent networked nature of this method, without the ability to store data locally, data costs and latency can become issues. As with J2ME incompatibilities between phones can be a problem but as the content is dynamically generated from servers it is possible to match the content to the capabilities of the device's browser without having to issue and distribute new binaries. Another issue is the variance in JavaScript support making some browsers capable of much more interactive interfaces and local execution than others.

4.4 Flash Lite / Silverlight

Flash Lite is a version of Adobe Flash specifically targeted at embedded devices such as phones and set-top boxes. It provides much of the rich interactivity and media capabilities of desktop Flash with compromises for the constrained devices it is targeted at. There is a relatively large development community but the tools used to create Flash Lite content are proprietary and not free. While there are different versions of Flash Lite they are much less fragmented than J2ME. However, due to Apple's App Store terms and conditions restricting virtual machines, Flash will not be available for the iPhone for the foreseeable future.

There is no access to the device's location sensors and earlier versions which make up most of the installed base [19] have limited or no video playback options, depending on the native device. Distribution and execution of flash content varies from device to device, for example some devices can run Flash Lite content from within a browser while others can't.

Silverlight is a technology similar to Flash from Microsoft, though at the time of writing the player is restricted to the desktop, with mobile versions currently in development.

4.5 Native Code

Native applications for Smartphones include software written in C++ for Symbian, Objective C for iPhone, Java for Android, and C++ for Windows Mobile. Programs written for the device's native runtime environment have rich APIs and tend to be able to use all the facilities of the phone, including sensors, media playback, file management, and network access. They generally run faster than software run in managed execution environments such as J2ME or Flash Lite. However porting code between different Smartphone operating systems can be costly due to the different programming languages used and the relatively low level of some of the languages, for example with C++ and Objective C the developer has to handle their own memory management.

There is a wide discrepancy within this group with respect to addressable market and distribution options. For example the iPhone has a relatively small but rapidly growing addressable market (about 45 million devices including the iPod Touch) and simple application distribution once you have obtained a developer license from Apple and bought an Apple Macintosh Computer to develop on. While Symbian has a larger addressable market and fewer restrictions on development it also has a far more fragmented distribution model. Most operators run their own distribution channels for Symbian apps and independent developers also distribute content via their own sites. Nokia, the largest distributor of Symbian devices will soon release an Application Channel similar to Apple's however that software itself will have to be distributed to the existing installed base so its uptake is uncertain.

5 Evaluation of HTML5+GL

As stated in the introduction HTML5 is being specifically designed to provide a single platform for the delivery of web based applications, but it was not specifically designed for pervasive media applications on mobile phones. In this section we evaluate HTML5+GL with respect to the previously determined problems and requirements.

An application based on the HTML5+GL runtime environment consists of a HTML web page created in the HTML5 mark up and some amount of JavaScript as the programming language. This is loaded in to a Browser which constructs a Document Object Model (DOM) from HTML and executes the JavaScript. The APIs for the features described below form part of the HTML5+GL DOM. The program receives inputs via events within this DOM and then processes the data using an arbitrary amount of JavaScript; the responses are then typically changes to the DOM which change the page contents and media being used.

5.1 Large Installed Base

At the time of writing HTML5+GL has a relatively small installed base as only the latest versions of a few desktop and mobile browsers support significant subsets of its functionality.

Increasing competition within the Smartphone sector and consumers' increasing reliance on web-based services such as web-based email has made the mobile web experience a focus of development and competition within the Smartphone sector. The lowering cost of cellular data and the prevalence of web content aimed at mobile phones have also increased the expectation among users that even low-end phones will have web browsers. Generally speaking popular features, applications and services found on Smartphones migrate to non-Smartphones over time; there is no reason that HTML5+GL will not follow the same pattern. An example of this migration is the camera. Current technologies following this trend are GPS and WiFi connectivity both of which are starting to appear in non-Smartphones.

Another significant factor in the uptake of HTML5+GL within mobile web browsers is the prevalence of the Webkit rendering engine across both desktop and mobile web browsers. Webkit is an open source rendering engine for web browsers that is used within Safari and Chrome on the desktop and Mobile Safari, the Android browser, the Nokia S60 browser, the Nokia S40 browser and WebOS from Palm. As the team behind Webkit are involved in drafting HTML5 as well as being at the forefront of adoption of HTML5 it is assumed that these browsers will continue to adopt these emerging standards as Webkit does. The use of Webkit in the Nokia S40 platform is particularly interesting due to the general popularity of this platform and the fact that it is not a Smartphone platform.

The forces driving the adoption of location sensing within mobile phones include the rise of GPS-based navigation and location-based services such as Nokia's Map 2.0 and Google Maps.

Mobile browsers that do not use Webkit are also adopting the draft HTML5 specification or providing very similar specifications. Opera is actively supporting HTML5 and its desktop support of HTML5 is very good. Opera has recently announced that it will soon support a large number of HTML5 features in its mobile browser, however it is doing so via a Google Gears implementation rather than direct HTML5 support. While it will have most of the same features, excluding media playback, the APIs will differ [20]. This difference in the APIs, though not the missing features, can be mitigated by an intermediate JavaScript Layer [21]. It is expected that the APIs for Google gears and HTML5 will converge [22].

As an example of the pace of HTML5+GL deployment within mobile phones Apples recently released iPhone Software 3.0 has updated some significant portion of the 45 million deployed iPhones and iPod Touch devices with a new browser featuring most of the HTML5+GL features discussed in this paper.

5.2 Good Distribution

Most users in the developed world are already familiar with using the internet in some form [23]. On a HTML5+GL equipped mobile phone any user who can access a web page can download a pervasive media application as they are the same thing. The server-side technology for distributing a HTML5+GL pervasive media application should be essentially the same as the technology and requirements of distributing normal web-based content or services as they both consist of the HTTP distribution of HTML, JavaScript, CSS and content files.

5.3 Consistent Implementation

The HTML5+GL specification requires implementations to be consistent across any number of operating systems and devices. Due to the detailed specification and test suites currently being devised by the HTML5 group within W3C [24] all browsers should in time be highly consistent application environments. In the mean time the prevalence of Webkit, which is highly standards compliant, across many mobile browsers will help to establish a de facto standard within the mobile browser space.

5.4 Developer Support

JavaScript used in conjunction with the HTML DOM is already one of the most widely used programming languages in the world with a huge community of support, tools and documentation. For developers with experience in this field, creating pervasive media applications is an incremental step up from the development of a website, and should not require investment in significant new skills.

This also raises the possibility of existing web sites targeted at mobile phones starting to take on pervasive media features such as adapting to the users location. Despite these benefits, a new user learning to develop JavaScript applications may take a significant time due to the number of technologies required (HTML, JavaScript, CSS, and optional server-side components such as PHP).

5.5 Launching Applications

The ease of finding an HTML5+GL application within a phone's menu will be determined by the phone manufacturer or browser developer and not the HTML5+GL specification. However as each application is effectively a web page any interface options available to find and launch bookmarks will also work for HTML5+GL applications.

5.6 Use of Limited Resources

JavaScript is an interpreted language meaning that well-written JavaScript will never run as efficiently as a well-written natively compiled application. However recent developments within desktop browser technology have greatly increased the execution speed of JavaScript with each major browser vendor competing on the basis of JavaScript execution speed [25]. The technology and techniques used in this competition are already starting to reach the mobile browsers [26].

For games and other-graphics heavy applications DOM-manipulation is often used such as dynamically creating or modifying DOM elements. This can result in fairly slow applications as the browser has to keep track of all the page elements to keep the layout up-to-date in real time. Using SVG (Scalable Vector Graphics) is another option but this also can become slow with large data sets. HTML5 includes a new Canvas element that makes drawing and animating graphics considerably faster [27]. Currently there is no support for text, though this is planned.

The offline aspects discussed below also help to reduce the amount of bandwidth required to run applications. However due to its interpreted nature a HTML5+GL application may well never be as efficient as a natively written application. But as phones become more powerful this should become a less significant constraint.

5.7 Play Audio and Video

The HTML5+GL specification contains new elements for the native playback of audio and video. Until now audio and video within web pages have been played through the use of plug-ins such as Adobe Flash. This means that browsers which do not support plug-ins, such as non-Smartphones browsers, but do support HTML5+GL will be able to playback video and audio.

However, at the time of writing attempts to standardize the codecs for the HTML5 audio and video elements have stalled. This means it may still be required to transcode video into different formats on a per-browser and per platform basics in order to make applications truly portable.

The Safari browser on iPhone 3.0 supports both the audio and video HTML5 elements. However, when audio or video is played a full-screen media player is shown, including transport controls. The user then needs to press play manually to access the content. Furthermore, only a single audio file can be played at any one time. These behaviors are contrary to the HTML5 specification, so it is hoped that future versions of the iPhone software will correct them.

5.8 Use Location Sensor

The W3C Geolocation specification provides a mechanism for the browser to query the device's location irrespective of the specific location technology being used. It also allows for the browser to specify conditions based on accuracy and receive information relating to the accuracy of the location determination. Mobile Safari supports this API in iPhone Software 3.0 and above.

5.9 Interactive User Interface

HTML4 is already used to provide rich interactive interfaces to users such as the Google Docs web-based office productivity suite. HTML5 does not remove any of the current APIs for doing this; in fact new APIs like the Canvas API create new opportunities for developers to create compelling application interfaces for users. An interface written in HTML5 will probably always be slower and less responsive than a native interface but there are large categories of application for which an HTML5 interface will be fast enough. This is already being demonstrated by the apps mentioned above and some simple arcade type apps such as Space War [33].

5.10 Server Interaction

Server-side interaction is used by web-based applications using various technologies based around AJAX web development. HTML5 improves on this by adding a property to indicate if the browser is online or not.

Furthermore there are now two technologies within HTML5 that allow web pages from one domain to communicate with servers or elements from another domain. *Access Control* allows an XMLHttpRequest object to make a request to a domain other than the one the page came from. *PostMessage* is a JavaScript method that allows two frames within a page from different origins to communicate in a secure way. Taken together these new features allow a web page to use web services and components from other domains where the servers and web pages within both domains have allowed it.

5.11 Local Persistence

HTML5 contains a number of complementary technologies for local offline storage and the retrieval of content and data [28]. HTML5 specifies client-side SQL database access which allows web pages to store structured data beyond the lifetime of any one page. This is already implemented in the Safari browser on iPhone, while the Android browser supports the same features with a different API via its implementation of Google Gears.

There is also a specification for web pages to be accessible even when the device is not connected to the internet. The *Offline Application Cache* allows the HTML page to specify that it and related content such as images and CSS files are stored locally for retrieval even if the page's original server is not accessible.

5.12 Non-location Sensors, Interruption of Users, and Content Capture

Pervasive media applications need to be able to monitor a user's context in order to deliver valuable services or content to the user. As more phones become available with other sensing technologies like accelerometers, more applications may wish to make use of them. However, non-location sensors are not currently covered within the HTML5+GL specification. From analysis of the sample pervasive media applications this would not appear to be a significant omission as very few of those applications used these features. However we believe that these features may become more significant over time.

Similarly, HTML5 does not directly cover capture of content via cameras/microphones, though this will most likely be a significant contribution to future pervasive applications.

Once the user's context is appropriate for the application to offer them some kind of interaction or service the application may well want to alert the user to this opportunity. This is likely to become more common if more pervasive media applications become available that are designed to run on a users own phone over a long period of time. HTML5 does not have a specific notification API, though audio or pop-up windows may be a reasonable substitution, provided that the originating page is open.

6 Application Trial

In order to test HTML5's suitability for pervasive media applications an existing application was chosen and re-implemented in HTML5+GL. *Stamp the Mole* was chosen as the application because it requires many of the most commonly used features making it a fair representative of a typical pervasive media application.. Additionally, the creative commons licensing of the application and publically-available source code meant it was easier to port and legal to redistribute.

The new application was developed entirely in JavaScript and HTML5+GL using the NetBeans 6.5 IDE [29] to develop the JavaScript. It was tested using Safari 4 on the desktop, and on iPhone running version 3.0 software.

A HTML5+GL based pervasive media application is created in much the same way as a normal web page. A main HTML file is created; this is the scaffolding upon which the rest of the application is constructed. JavaScript libraries are included, by including scripts within the body of the HTML page. Execution starts when the page is loaded and the "onload" event is fired by the browser. From there the JavaScript can register with sensors and start to communicate with servers via AJAX. The JavaScript can process input from user interaction, sensors and server and modify the interface via manipulating the HTML page's DOM or drawing directly to a Canvas object within that DOM. The application continues to execute until it or the user navigates away from the page or closes the window or tab.

To allow maximum code reuse for future work the application was written in three layers. A base layer abstracted out the exact sensor and rendering technology, this meant that for example a flash movie could have been used for the media rendering. And on browsers that did not support the W3C Geolocation API a map, fire eagle [30] call or AJAX call to a location server could be substituted. For this application all media was rendered by direct manipulation of the HTML5 DOM and location was provided by repeated AJAX calls to a server.

The second layer consisted of a library of classes and methods that were expected to be re-used for other pervasive applications using HTML5+GL. They provided utilities such as inclusion testing for spatial regions defined in JSON. The third layer consisted of the Stamp the Mole's application-specific JavaScript that specified the exact behavior of the game.

The application was found to run well and be playable on the desktop and fully functional but less easy to play on the mobile browser. The major difference was the media playback limitations on iPhone already discussed in section 5.7.

While a single application can obviously not be extrapolated to the entire field of pervasive media it does provide an existence proof that such applications with typical features can be developed and deployed using HTML5+GL.

7 Discussion

As has been demonstrated HTML5+GL provides a platform independent and easily deliverable means to create and deploy pervasive media applications that use a limited set of the most common requirements. But there are other features that HTML5+GL does not support and it is not currently very easy to make any strong assertions as to how important they will be in future applications. One distinct possibility is that as phones become more capable more applications will target the user's existing device and not be designed with a single session device loan model. This will mean that features such as persistence, running as a background process to monitor the user's context and alerting the user to get their attention may well become more common requirements. HTML5 already has a persistence model but the other features are not currently supported.

7.1 Background Tasks and Notifications

Currently web browsers do not have a notion of background tasks; however this feature can be very useful for pervasive media applications. A common design pattern for pervasive media applications is to monitor various local or remote sensors to determine the users' context and then to attract the users' attention and deliver the appropriate service, content or interaction based on that context. This can be achieved by multi-tabbed browsers as they do tend to allow JavaScript execution in background tabs. However this requires that the browser stays open and does not provide a facility for the application to alert the user except through audio.

If the user's context is already measured or available off the device on a server then the process which monitors the user's context and decides when to alert the user could reside on that server. The user could then be alerted and the application restarted through for example the sending of an SMS message or WAP push. This does of course assume that the application knows the users phone number which cannot be assumed but could be asked of the user.

7.2 Access to Other Sensors and Content Capture

One of the strengths of the core HTML5 specification is its uniformity and lack of the kind of fragmentation that causes problems for J2ME developers. However just as location is supported by a different and complementary specification other sensors and content capture systems could also be introduced through standardized specifications and then optionally included in browsers. Of course any optional components may lead to fragmentation between devices' browsers, and whilst the time between new features being introduced into the HTML5 spec and their appearance in beta browsers is fast (a few months is common), the total time between an idea and its mainstream adoption is far longer, often years.

Several groups are finding ways to extend the browsers API to provide richer functionality. For example PhoneGap and Titanium Mobile [31] are frameworks that allow applications to be written in HTML and JavaScript on a variety of different Smartphones [32]. These applications are run in a native wrapper that gives access to features usually unavailable to JavaScript such as the accelerometer, the file system, contact databases, and vibration. From these systems it is likely that new standards efforts will result in the development of standard APIs for such services. However in the meantime those who are willing to trade platform independence for richer functionality have a means to do so.

8 Conclusion

Distribution has long been a problem for mobile software; this can observed by the amount of interest, developer support and downloaded applications that the Apple App Store has generated out of all proportion to its installed base. This could be in part due to the publicity and marketing of the iPhone and related factors, but the ease of distribution is a significant factor. With HTML5+GL distribution will be easy as each application will simply be a web page, which most users are already familiar with.

As this paper shows HTML5+GL provides all of the major features apart from background processing required to create pervasive media applications for mobile phones. As detailed in section 4, it is one of the best platforms in terms of coverage of the features that were found to be common requirements from section 3. There are potential issues with access to non-location sensors but this may improve over time through the introduction of new web standards or via proprietary extensions. It is not yet clear how background processes could be integrated with HTML5+GL and this may be a limit on the applications that can be built using this runtime environment but there is still a huge potential for applications that do not require such features.

HTML5+GL is also platform independent, this should provide a real boost for application developers as it reduces the cost of porting and increases the addressable market without compromising the application. If browsers supporting HTML5+GL become as commonplace as current HTML4 browsers, the addressable market will be very large indeed.

This paper has highlighted some of the problems of existing solutions to pervasive media and has shown how with HTML5+GL it should soon be possible to write a pervasive media application once distribute it anywhere and run it everywhere.

References

1. HTML5 specification, http://dev.w3.org/html5/spec/Overview.html
2. Web Hypertext Application Technology Working Group news, http://www.whatwg.org/news/start
3. Web Storage HTML5 specification, http://dev.w3.org/html5/webstorage
4. Geolocation API Specification, http://dev.w3.org/geo/api/spec-source.html
5. Mscape, http://www.mscapers.com
6. Mupe, http://www.mupe.net
7. Ready, aim, text, The Guardian, http://bit.ly/XFABk
8. Sotamaa, O.: All The World's A Botfighter Stage: Notes on Location-based Multi-User Gaming. In: Proceedings of Computer Games and Digital Cultures Conference 2002, pp. 35–44. Tampere University Press (2002)
9. Reid, J., Hull, R., Cater, K., Clayton, B.: Riot! 1831: The design of a location based audio drama. In: Proceedings of UK–UbiNet (2004)
10. Mobile Bristol, http://www.mobilebristol.com
11. Stamp The Mole, http://www.mscapers.com/msin/ABA0000044
12. Uncle Roy is All Around You, http://www.blasttheory.co.uk/bt/work_uncleroy.html
13. Benford, S., Flintham, M., Drozd, A., Anastasi, R., Rowland, D., Tandavanitj, N., Adams, M., Row Farr, J., Oldroyd, A., Sutton, J.: Uncle Roy All Around You: Implicating the City in a Location-Based Performance. In: Proc. Advanced Computer Entertainment, ACE (2004)
14. GPS Mission, http://gpsmission.com

15. Peitz, J., Saarenpaa, H., Björk, S.: Insectopia - Exploring Pervasive Games through Technology already Pervasively Available. In: Advancements in Computer Entertainment, Salzburg, Austria (2007)
16. Insectopia design document, http://bit.ly/dUXfu
17. Bell, M., et al.: Interweaving mobile games with everyday life. In: CHI 2006: Proceedings of the SIGCHI conference on Human Factors in computing systems, pp. 417–426. ACM Press, New York (2006)
18. Ballagas, R., Kuntze, A., Walz, S.: Gaming Tourism: Lessons from Evaluating REXplorer, a Pervasive Game for Tourists. In: Proceedings of the 6th International Conference on Pervasive Computing. LNCS. Springer, Berlin (2008)
19. Flash Lite Devices Worldwide, http://bit.ly/CcvoK
20. Gears-enabled Opera Mobile 9.5 technology preview, http://bit.ly/4bEQP
21. HTML5 offline format implemented using Google Gears, http://bit.ly/O6HIH
22. Google Gears as a bleeding-edge HTML5 implementation, http://bit.ly/y7B6T
23. Internet Users Per 100 Inhabitants (1997-2007), http://bit.ly/ewYuW
24. HTML5 Editor Ian Hickson on features, pain points, adoption rate, and more, http://bit.ly/eK37C
25. SquirrelFish Extreme promises to speed JavaScript in Safari 4.0, http://bit.ly/OIiN7
26. JavaScript to get 3x speed boost in iPhone OS 3.0 – Ars Technica, http://bit.ly/9MpgR
27. Canvas vs SVG Performance, http://bit.ly/tHf7s
28. Offline Web Applications, http://bit.ly/WwZXb
29. NetBeans IDE, http://www.netbeans.org
30. Fire Eagle Location Service, http://fireeagle.yahoo.net
31. Appcelerator Titanium, http://titaniumapp.com
32. PhoneGap - Cross platform mobile framework, http://phonegap.com
33. Space War, http://virtualgs.larwe.com/iPhone/space/

Study of Usability of Security and Privacy in Context Aware Mobile Applications

Neha Pattan and Deepthi Madamanchi

Carnegie Mellon University, USA
Neha.Pattan@sv.cmu.edu, Deepthi.Madamanchi@sv.cmu.edu

Abstract. Mobile devices, such as smart phones, are becoming increasingly powerful with more memory, processing capacity and interface to several hard and soft sensors, making it possible to easily develop and use context aware applications on them. Context aware applications make use of context from several sources including location, light, sound, as well social context and users' behavioral patterns to make more informed decisions for the user. Since these applications are indeed aware of all the user's personal information and context, it is important that they be designed with security and privacy in mind. In this paper, we discuss a study conducted to understand user perception of security and privacy features implemented in context aware mobile applications, improvements that can be made and design guidelines that will help improve the usability of these features.

Keywords: Context-aware, mobile applications, security, privacy, usable security, usability.

1 Introduction

Context aware mobile applications are a class of applications that examine and react to an individual's changing context [1]. Although security and privacy are significant factors that need consideration in these applications, one area of research that has not received adequate attention is the usability of these aspects.

We conducted a usability study to understand how users perceive security and privacy features provided by their cell phones and whether or not these features are indeed usable. We used Nokia Friend View [2], a location-based social networking application, as a case study for context aware mobile applications. We also evaluated a paper prototype version that we created, in which security is made explicit and the application gives a clear indication of security and privacy features.

2 Related Work

Chen and Kotz [3] identified two key problems with security and privacy of context-aware systems: 1) ensuring the accuracy of location information and identities, and 2) establishing secret communications. They observed that these problems are not satisfactorily addressed by existing context aware mobile applications. We built on their recommendations and made sure that we give the user the ability to share location information with specific friends/ groups of friends in lieu of broadcasting the user's location to the whole buddies' list. Iachello et al.[4] discussed eight design guidelines for enhancing the privacy of social location disclosure applications and services. These papers give useful guidelines on how to design software for context aware mobile applications. They, however, do not perform a quantitative study of the usability of applications designed using these guidelines.

3 Methodology

The study was conducted over a period of seven weeks with 15 participants in the following three phases.

3.1 Pilot Study

The study was conducted using the mental models technique. Participants were asked to use the Nokia Friend View application and discuss how they felt while using the application to share their location and status messages. Each interview lasted for about 20 minutes.

3.2 Paper Prototyping

Based on information gathered from the pilot study, we first designed an initial paper prototype to address some of the issues that came up from the pilot study and then improved it iteratively. To arrive at the final version of the improved prototype, we conducted six iterations with four participants. At the end, we had a prototype with more explicit cues for secure and insecure connections, and for privacy of user's location and status information. These cues are discussed more in section 5, Design Implications.

We used three scenarios for testing: i) The user logs into the application, ii) the user updates his location and iii) the user updates his status message.

3.3 Usability Testing - Comparative Study between Application and Improved Prototype

This phase consisted of performing a comparative study of usability of the original Friend View application and the paper prototype we designed. In this phase, each participant was given both the Friend View application and the improved prototype. They were given about 5-10 minutes to get acquainted with these. Each participant was then handed out a survey to fill in his/her feedback on security and privacy features in both – the original application and the improved prototype. The participants were allowed to use the prototype or original application at any time while answering the survey.

4 Results and Analysis

The survey questions consisted of statements about the user feeling very secure while using the application to perform a certain task, or feeling that the application respected their privacy. Users' responses were recorded on a five-point Likert scale.

4.1 Perception of Security

In general, we observed that participants responded more positively to the prototype when asked whether the application made them feel very secure and respected their privacy.

Security While Logging in. About 62% of the users either disagreed or strongly disagreed that the original Friend View application made them feel secure while submitting their credentials. The rest neither agreed nor disagreed. On the other hand, about 87% users either agreed or strongly agreed that the prototype made them feel secure while logging in, while only 8% remained neutral.

Security While Sharing Location. While testing the user perception for sharing location, we found that all participants either agreed or strongly agreed that the prototype made them feel secure. On the other hand, only one out of 16 participants agreed that Friend View application made them feel secure while sharing their location.

Security While Posting a Status Message. Similar results were observed while testing user perception for security while posting a status message. About 81% participants responded positively to the prototype, while only 6% responded positively for Friend View application.

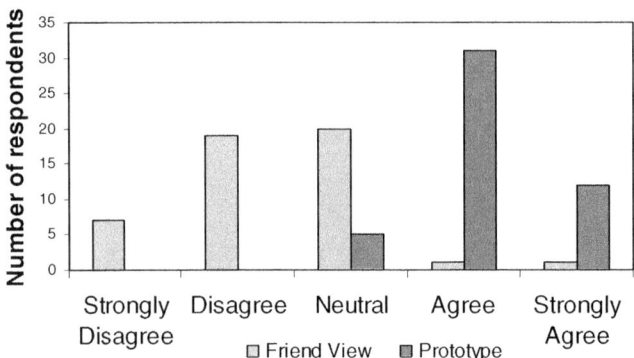

Fig. 1. Average response of participants to security scenarios: (i) while logging in, (ii) while sharing location and (iii) while posting status message to friends in the application

4.2 Perception of Privacy

We observed that the majority of users responded that they either disagreed or were neutral when asked if the original application respected their privacy. For the prototype, over 93% participants either agreed or strongly agreed.

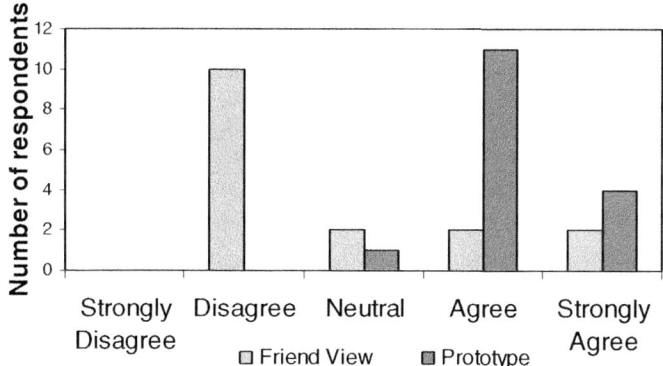

Fig. 2. Response of participants to application respecting user's privacy

5 Design Implications

Through the initial phase of mental model interviews and second phase of prototype design, we were able to understand several important factors that can contribute towards better usability of security and privacy features in context aware mobile applications. We have outlined these factors as design guidelines in this section. This list of guidelines is in no way exhaustive or complete. It is a start in the direction of improving usability of context aware applications and will be worked on for improvement in the future.

5.1 Security Guidelines

1. Security indicator should be part of native interface
2. Security indicator should be more dynamic
3. Indicating secure connection is just as important as indicating insecure connection

5.2 Privacy Guidelines

1. User should be able to see his/her most recent action

If a user is given the option of sharing his location with everyone, a group of contacts or a selected set of contacts, it is important that the application indicate what the user's most recent action was. Users tend to find this information useful to understand the causes for any present actions.

6 Conclusion

In conclusion, we believe that usability of security and privacy aspects in context aware mobile applications is important and needs more attention while designing these applications. In all our tests, we observed that users preferred the improved prototype over the original Friend View application. Indeed, we tested only the Friend View application, which is a relatively safe application to use since it only involves friends and does not consider many dimensions of users' contexts. We therefore assume that these privacy and security aspects will definitely have more gravity in applications which involve sharing more personal and important information with strangers. A best example of it is a ridesharing application.

Acknowledgments. We would like to extend our gratitude towards Dr Cynthia Kuo for her guidance and help in designing and conducting this user study.

References

1. Schilit, B., Adams, N., Want, R.: Context Aware Computing Applications. In: Proceedings of Workshop on Mobile Computing Systems and Applications, December 8-9, pp. 85–90 (1994)
2. Nokia Friend View Application, Nokia Beta Labs,
   `http://betalabs.nokia.com/betas/view/nokia-friend-view`
3. Chen, G., Kotz, D.: A Survey of Context Aware Mobile Computing Research. Dartmouth Computer Science Technical Report (2000)
4. Iachello, G., Smith, I., Consolvo, S., Chen, M., Abowd, G.: Developing Privacy Guidelines for Social Location Disclosure Applications and Services. In: Proceedings of the 11th international symposium on Modeling, analysis and simulation of wireless and mobile systems, pp. 229–238 (2008)
5. Barkhuus, L., Anind, D.: Is Context-Aware Computing Taking Control Away from the User? Three Levels of Iteractivity Examined. In: Dey, A.K., Schmidt, A., McCarthy, J.F. (eds.) UbiComp 2003. LNCS, vol. 2864, pp. 150–156. Springer, Heidelberg (2003)

Friendlee: A Mobile Application for Your Social Life

Anupriya Ankolekar, Gabor Szabo, Yarun Luon, and Bernardo A. Huberman

Social Computing Lab
Hewlett-Packard Laboratories
1501 Page Mill Road
Palo Alto, CA, 93404, USA
gabors@hp.com

Abstract. We have designed and implemented Friendlee, a mobile social networking application for close relationships. Friendlee analyzes the user's call and messaging activity to form an intimate network of the user's closest social contacts while providing ambient awareness of the user' social network in a compelling, yet non-intrusive manner.

1 Introduction

The viral growth of online social networking applications in the past few years has facilitated, like never before, forming new friends online and keeping in touch with old friends and past colleagues. These sites typically make it easy to declare friends, or add like-minded people as friends, and then follow their activities or posts online. While these declared networks appear large and thriving, it has been recently shown [2] [4] that much of the activity in these networks is driven by a more intimate group of users. Twitter networks of friends and followers, for example, are sustained [2] by an underlying sparse network of friends who interact frequently and reciprocate each other's attention. Even in the social networks formed through mobile phone calls and text messages--these overlap substantially with a user's 'real' social network [3]--an analysis of phone communication logs [4] reveals that people interact with only a small fraction of the people actually present in their phonebook.

These kinds of intimate social networks with the closest, most meaningful ties, such as between close friends, family, relatives and even close colleagues, are characterized by high frequency of interaction, but also by a great need to feel connected, to be in touch [6], and a need for sharing detailed activity and context information [5]. However, scarcely any of the online social networking applications[1] support users adequately in staying connected with this core group of people [7]. To address this problem, we have developed Friendlee, an application that analyzes the user's call and messaging activity to identify the user's closest social contacts. Friendlee enhances the mobile phone, providing the user with an ambient awareness of her intimate network.

---

[1] Even beyond social networking applications, there are no applications that support intimate networks, with a few notable exceptions, e.g.7.

Since Friendlee also keeps track of the businesses the user has called frequently, we are able to automatically identify the user's preferred services, which can then be used as recommendations to their social network. Close friends and colleagues remain among our most influential sources of practical advice and recommendations about services, such as health insurance and restaurants, as well as about people, in both social and professional settings[9]. Several studies [1] have shown that people find recommendations generated by their social networks in taste-related domains to be as useful and interesting as the ones generated through traditional collaborative filtering approaches.

In addition, like most social networking applications, Friendlee allows users to browse the connections (and preferred services) of people in their intimate network. People in close relationships are often already peripherally aware of each other's contacts: the high degree of clustering (forming cliques) in social networks makes it very likely that friends of friends are friends or at least partially know about each other [8].

 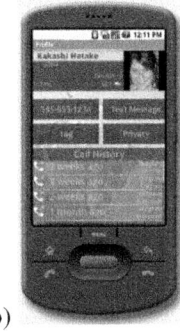

(a) (b)

Fig. 1. Friendlee's user interface, showing (a) the screen with the list of contacts and (b) a profile screen

2 Design

2.1 Behavior-Based Intimate Social Network

Friendlee analyzes the user's call and messaging history to identify the people she is closest to, based on phone conversation frequency, recency and duration. Using these variables, the connections of a user are assigned a relative weight that determines the 'closeness' of that contact to the user with respect to other contacts. The strongest connections (the ones with the largest weights) are displayed prominently, allowing the user to have instant access to them without wading through a large phonebook. By using phone conversations as an indicator of close social interaction, Friendlee trims the user's large casual social network into a core intimate one.

The contact list screen (shown in Fig. 1(a)) is the primary screen of Friendlee and displays the user's intimate social network in reverse order of relationship strength. In design, the screen is similar to a mobile phonebook or instant messaging contact list.

2.2 Ambient Awareness of Intimate Social Network

The user can easily share key aspects of her context, namely her location at different granularities (country, city or GPS-based street address), her status message and her phone status (on/off/available/ringer/silent/vibrate), local time and weather as well as who her other family, friends and colleagues are. Such ambient awareness of people's closest connections helps them feel emotionally close and also facilitates communication (e.g., knowing whether this is a good time to call). To protect privacy, people have access to this context-sharing functionality only for connections made by mutual consent.

2.3 Browse Connections of Close Contacts

A key differentiator of this application from existing ones is the ability to browse the connections of close friends. This allows users to reach out and be aware of their social network beyond their immediate relatives and friends. A significant proportion of these connections will already be known to the user; however, the user may not have any means to contact them herself. To safeguard privacy, users can always restrict visibility of chosen contacts to specific categories of people.

2.4 Search and Get Recommendations for Businesses from Social Network

Favored businesses also constitute part of a user's true daily social network, from the local take-out to the user's cable company. While browsing a friend's connections, people also see their preferred businesses, getting implicit recommendations about e.g. the dentist their friend likes to go to. In addition, Friendlee allows users to search their social network for people and businesses. Search results are ranked by social distance from the user.

2.5 Category-Based Privacy Model for Sharing Context Information with Contacts

Friendlee allows users to classify their contacts into categories, such as 'Colleague' or 'Family', which helps the user both navigate quickly to a desired contact, as well as define privacy settings for various categories, e.g. sharing location information only with family. Categories are also used to define privacy settings for which of the user's contacts are visible, e.g. my colleagues may view my other colleagues, but not my family.

3 Implementation

Friendlee consists of three components: (1) a phone-based client that represents Friendlee's user interface and gathers user information, such as personal status, call and messaging history, (2) a Web-based interface where users can access and change the same information as on the client, and (3) a backend server that stores a centralized copy of all user information within a large database. The client synchronizes several times a minute with the server, providing it with up-to-date information about the user's call history and context, such as location, phone status, etc. The server propagates context information of users (including current local time and weather conditions) through the user's social network taking into account her privacy policies. The server is also responsible for calculating the strength of relationships in the social network based on communication history and thus the 'social distance' between any two people.

We have developed a prototype of Friendlee for the Android and Windows Mobile operating systems. In addition, we have developed a Web-based interface that users can access on a desktop. We plan to develop a simplified mobile Web browser so that restricted functionality is available on phones that support Web browsing. The server is implemented in Perl and uses a MySQL database for storage. The client-server connections are currently 'stateful' (TCP/IP), but we could also support stateless connections using HTTP and SMS. In next steps, these prototypes will be used as part of a field study to assess the usability and usefulness of Friendlee's user interface and recommendation algorithms.

References

1. Groh, G., Ehmig, C.: Recommendations in taste related domains: collaborative filtering vs. social filtering. In: Proceedings of the International ACM Conference on Supporting Group Work (GROUP 2007), pp. 127–136 (2007)
2. Huberman, B.A., Romero, D.M., Wu, F.: Social networks that matter: Twitter under the microscope. First Monday 14(1) (January 2009)
3. Kuitto, E.: The friendship practices and the conceptions of friendship among the high school girls and boys. Master's thesis in sociology, University of Helsinki (2001)
4. Lugano, G.: Mobile Social Networking in Theory and Practice. First Monday 13(11) (November 2008)
5. Neustaedter, C., Elliot, K., Greenberg, C.: Interpersonal Awareness in the Domestic Realm. In: Proceedings of OZCHI 2006, pp. 15–22 (2006)
6. Smith, E., Mackie, D.: Social Psychology, 2nd edn. Psychology Press, New York (2000)
7. Vetere, F., Gibbs, M.R., Kjeldskov, J., Howard, S., Mueller, F., Pedell, S., Mecoles, K., Bunyan, M.: Mediating Intimacy: Designing Technologies to Support Strong-Tie Relationships. In: Proceedings of CHI 2005, pp. 471–480 (2005)
8. Watts, D.J., Strogatz, S.H.: Collective Dynamics of 'Small-World' Networks. Nature 383, 440–442 (1998)
9. Wellman, B., Gulia, M.: The Network Basis of Social Support: A Network is more than the Sum of its Ties. In: Wellman, B. (ed.) Networks in the Global Village: Life in Contemporary Communities, ch. 2. Westview Press (1998)

A Prototype for Resource Optimized Context Determination in Pervasive Care Environments

Nirmalya Roy[1], Christine Julien[1], Archan Misra[2], and Sajal K. Das[3]

[1] Dept. of Electrical and Computer Engineering, The University of Texas at Austin
{nirmalya.roy,c.julien}@mail.utexas.edu
[2] Applied Research, Telcordia Technologies, New Jersey
archan@research.telcordia.com
[3] Dept. of Computer Science and Engineering, The University of Texas at Arlington
das@uta.edu

Abstract. In this demonstration we demonstrate an early prototype that shows the importance of context determination in the presence of mobile entities in pervasive care environments. We use our novel context model to build a framework for resource-constrained sensor networks. We then use this context model to use a user's mobility to infer his *activity*, which we refer to as his *context state*. Because the context state is inferred from actual sensed context, we use the prototype to demonstrate the tradeoff between context inferencing accuracy and communication overhead. The ability to sense the environment and accurately infer context can help monitor the user in a pervasive care environment.

1 Introduction

Because mobile applications operate in unpredictable and changing environments, the ability to sense and act on the changing context is essential to any application. An emerging scenario for mobile computing entails an assisted living environment, in which a mobile user and his surroundings are sensed by a variety of devices in the environment and carried by the user. These applications enable *aging in place*, promoting healthy independence. Energy-efficient determination of an individual's context (both physiological and activity) is an important technical challenge for assisted living environments. Given the expected availability of multiple sensors, context determination may be viewed as an estimation problem over multiple sensor data streams. We have developed a formal and practically applicable model to capture the tradeoff between the accuracy of context estimation and the overheads of sensing [2]. In particular, we use *tolerance ranges* to reduce an individual sensor's reporting frequency while ensuring acceptable accuracy of the derived context. In our vision, applications specify their minimally acceptable value for a Quality-of Inference (QoINF) metric. We introduce an optimization technique allowing our framework to compute both the best set of sensors to use to infer the particular context metric *and* their associated tolerance values, that satisfy the QoINF target at minimum communication cost. In this demonstration we show a prototype of this tolerance range based sensor data reporting system implemented on a SunSPOT sensor testbed.

2 Implementation

We use the SunSPOT [1] (Sun Small Programmable Object Technology), a small, wireless, battery-powered experimental platform. We employ two types of devices: *free-range* devices that are embedded in the environment and perform sensing and communication tasks, and a *base-station* device that simply communicates with the free-range devices and is connected to a larger format device (e.g., laptop) for supporting a user interface. Each free-range SunSPOT contains a processor, radio, sensor board, and battery, and the base-station SunSPOT contains the processor and radio only. The SunSPOT uses a 32-bit ARM9 microprocessor running the Squawk VM and programmed in Java, supporting IEEE 802.15.4 radio. In our demonstration we use the accelerometer sensor available with SunSPOT sensor board.

Our demonstration includes two scenarios. In Case 1, the free-range devices reports each data sample to the base-station regardless of the user's actual requirements. This provides the highest quality information but also incurs the highest cost. Using this sensor data, our application infers the user's context state, such as sitting, walking, or running. In Case 2, we use an application-specified QoINF to limit the sensors' data reporting frequencies to reduce the communication overhead while still meeting the target accuracy for estimating the context state. In this case, the sensor reports only if the change in the sensed value is outside of the sensor's specified tolerance range. This is a simple example with intuitive behavior and expected performance outcomes. The goal of the demonstration is to exercise our QoINF programming framework and demonstrate how application developers can have a significant impact on the lifetime of their network by trading accuracy for overhead.

The sensor application on the free-range device contains two java class files, one to maintain the radio communication connection and another to read and send the accelerometer data. The host application on the base-station consists of three main Java classes, one to handle the radio connection to the remote SPOT sensor, one to display the data collected, and the third to manage the GUI.

2.1 Case 1: No Tolerance Range

In this first case, we do not ask the application to specify any tolerance, and the motion sensor constantly reports the data samples used to infer the user's context state. We use the measured acceleration to determine the context state. If the person is sitting, the acceleration due to the earth's gravity will be $1g$ along the positive Z-axis, and $0g$ along the X and Y axes. By monitoring the deviation of acceleration from the $1g$ of gravity, we can infer whether the user is in motion. As seen in Figure 1, we can easily confirm the different context states from this measurement (sitting ($1g$), walking ($2.5g$), or running ($4.5g$)). We have shown how this information can be used to support a variety of context-aware tasks in an assisted living environment [2].

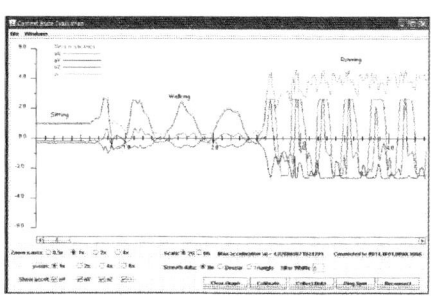 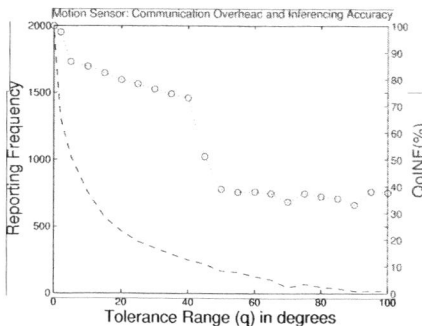

Fig. 1. Context Sensing with Accelerometer

Fig. 2. Overhead and Accuracy vs. Tolerance

2.2 Case 2: Using Tolerance Range

In the second case we consider a tolerance range derived from a QoINF function provided by the application, which causes the sensor to report only if changes in the sensed values exceed the given tolerance range. This reduces the communication between the free-range sensors and the base-station, ultimately increasing the life of the energy constrained devices. We use the following simple logic to reduce the sensor's communication overhead.

```
SendSensorData (input range) {
  last_reported_valuex=acc.getRawX();
  last_reported_valuey=acc.getRawY();
  last_reported_valuez=acc.getRawZ();
  //value of x, y and z-axis acceleration fluctuation in voltage
  if (ABS(last_reported_valuex-acc.getRawX()) >= range OR
      ABS(last_reported_valuey-acc.getRawY()) >= range OR
      ABS(last_reported_valuez-acc.getRawZ()) >= range){
        Datagram.writeByte(acc.getRawX());
        Datagram.writeByte(acc.getRawY());
        Datagram.writeByte(acc.getRawZ());
        //sending the packets or reporting the data samples
  }
}
```

2.3 Measuring QoINF Accuracy and Communication Overheads

To study the potential impact of varying the tolerance ranges, we collected traces for the SunSPOT motion sensor for a single user engaged in a mix of three activities (*sitting*, *walking*, and *running*) for a total of $\approx$ 6 minutes (2000 samples at $5.5Hz$). We then used an emulator to mimic the samples that a sensor would have reported for different tolerance ranges and compared the context inferred in both cases to the ground truth. Figure 2 shows the total number of samples reported (an indicator of the reporting overhead) and the corresponding QoINF achieved (defined as $1 - error\ rate$) [2] for different values of the tolerance range.

A QoINF accuracy of $\approx 80\%$ is achieved for a modestly large tolerance as shown in Figure 2. Moreover, using this tolerance range reduces the reporting overhead dramatically. This suggests that it is indeed possible to *achieve significant savings in bandwidth, if one is willing to tolerate marginal degradation in the accuracy of the sensed context.*

3 Demonstration Details

In the demonstration we will show two SunSPOTs reporting data samples for the user in the same state for both of our cases. In Case 1 we show the remote sensor reporting every data sample regardless of any tolerance range or user context state. This data gathering process is demonstrated by a continuous graph at the base-station as shown in Figure 1 and with a steady blue LED at the free-range sensor.

In Case 2 the sensor reports its data sample only if there is a tangible change. The data gathering process is demonstrated by a graph at the base-station and with a flashing blue LED at the free-range sensor. We do this for three different SunSPOTs sensor each using a different tolerance range (0, 20, and 100). For tolerance range 0, the blue light will flash constantly, similar to Case 1. In this case we achieve a near 100% inferred context state accuracy. For tolerance range 20, the blue light flashes only with a change in the context state and for 100 it flashes rarely, even with a frequent change in the context state. This confirms that, with a medium tolerance range, we can achieve a moderate context accuracy with much reduction in overall sensor sample reporting frequency, whereas a high tolerance range can nullify the reporting overhead but is very error prone.

Acknowledgments

This work was funded in part by Air Force Office of Sponsored Research (AFOSR), Grants #FA9550-09-1-0155 and #FA-9550-07-1-0157. The conclusions herein are those of the authors and do not necessarily reflect the views of the sponsoring agencies.

References

1. Project SunSPOT. Sun Small Programmable Object Technology,
http://www.sunspotworld.com
2. Roy, N., Misra, A., Das, S.K., Julien, C.: Quality-of- Inference (QoINF)-Aware Context Determination in Assisted Living Environments. In: Proc. of WiMD 2009 in conjunction with MobiHoc 2009 (May 2009)

RFID-based Distributed Shared Memory for Pervasive Games

Michel Simatic[1] and Annie Gentès[2]

[1] Institut Télécom, Télécom & Management SudParis, 9 rue Charles Fourier,
91011 Evry Cedex, France
`Michel.Simatic@it-sudparis.eu`
[2] Institut Télécom, Télécom ParisTech, 46 rue Barrault, 75013 Paris, France
`gentes@telecom-paristech.fr`

Abstract. The goal of our work is to give a user equipped with an RFID-enabled mobile handset (mobile phone, PDA, laptop...) the ability to access contents of distant elements of the system (tags or handsets), without physically moving to them and without using a Wireless Area Network. Our solution consists in an RFID-based Distributed Shared Memory (RDSM). After describing RDSM, we present its demonstration: a pervasive game which relies on RDSM to analyze data stored in distant elements (without any Wireless Area Network). We conclude by presenting the analysis of the players' feed back.

Keywords: Distributed memory, RFID, NFC, Vector clocks, Pervasive game.

1 Introduction

RFID tags are interesting components for pervasive games. They are inexpensive, easy to deploy and robust. Moreover a player equipped with an RFID-enabled handset can easily interact with the tags. Finally we can take advantage of Near Field Communication (NFC) technology, a subclass of RFID where a reader and a tag can communicate only if they are a few centimeters away from each other. Using NFC guarantees that the player will not be bothered by other tags while interacting with a given tag.

Our goal is to provide an architecture where a mobile handset can get the value of data stored in distant RFID/NFC tag. Classical solutions are based on the use of a server [7]. But a Wireless Area Network is required to communicate with a server. This induces installation and/or operational costs. Our goal is to do with neither a server, nor Wireless Area Network.

There are pervasive games which use RFID/NFC. But they do not meet all of our constraints. For instance, in *Save the Princess!* game, each time a player enters a room, *TinyLIME* middleware informs their terminal about the data stored in the tag located in the room [3]. The terminal displays the virtual contents associated to this room without communicating with a server. But it cannot give any information about data stored in the tag of a nearby room. On the contrary, in *PAC-LAN* game, players are able to know what happened on a distant NFC tag [5]. Each time a tag is scanned by a player's mobile phone, the mobile sends the tag identifier to a central server. The server sends back game-related information to the mobile. And this information may concern distant tags. But, to achieve this, the game uses a server and a Wireless Area Network.

This paper presents a demonstration of our solution, the RFID-based Distributed Shared Memory (RDSM) [7]. Data are stored in a memory which is distributed among the different tags and mobile handsets (mobile phone, PDA, laptop...). Moreover, this memory is replicated on each tag and each handset of the system. Thus a handset can get the value of the data stored on a distant tag or handset. It just has to query its own replica. Whenever a mobile handset meets a tag (or another handset), their respective replica are made consistent by comparing the vector clock values coupled to the replicas.

Section 2 details RDSM. Then section 3 presents the proof of concept. Finally section 4 we present some conclusions.

2 Description of RDSM

The system we consider is made of two types of elements: RFID tags and mobile handsets (See Figure 1). Each element holds data which can only be modified by this element[1]. We note DM the distributed memory made of all of these data. Each element e holds a replica DM_e of the distributed memory. As in [4], the main usage of this replica is to offer a local view of DM to e: e can make queries on the contents of any elements of the system at any time. We note $DM_e[e']$ the view element e has of the contents of DM hold by element e'. In particular, $DM_e[e]$ contains the part of DM hold by e. Each element e holds also a vector clock VC_e which is used to propagate operations done on DM.

To do so, whenever a mobile handset comes in contact with a tag (or another handset), these two elements compare vector clock values and update their own replica of DM (see [7] for algorithms). Thus, each element takes advantage of the knowledge of the other one to get more recent information concerning DM evolutions.

Usually, a vector clock element is a logical clock, incremented upon each update of its associated data [6]. In our demonstration, to save space on each tag, $VC_e[e]$ holds the timestamp of the last update or query done on $DM_e[e]$. Meanwhile, $VC_{e',e'\neq e}[e]$ holds the timestamp of the last operation done on $DM_e[e]$ which element e' is aware of.

RDSM faces staleness and scalability issues. They are discussed in [7].

3 Demonstration

The demonstration of RDSM is based on *Plug: Secrets of the museum* (PSM) pervasive game [8], developed in the context of the *PLUG* project [1]. 48 virtual

---

[1] In the case of data hold by a passive tag, these data can only be modified by a handset which is in contact —via RFID/NFC protocol— with the tag.

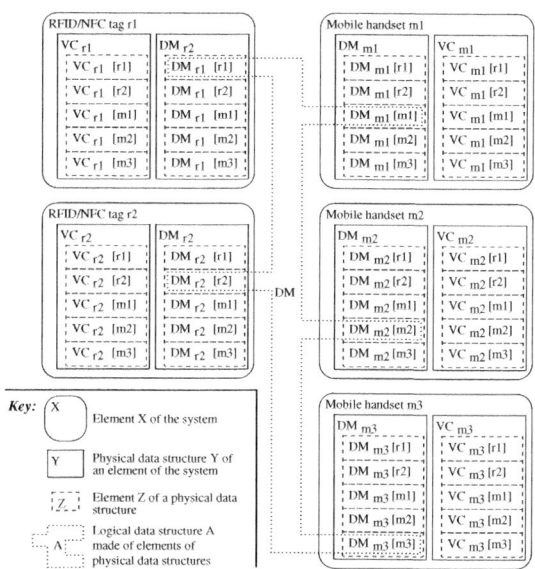

Fig. 1. Data present in a system made of 2 RFID/NFC tags and 3 mobile handsets

playing cards represent objects of a French museum called *Musée des arts et métiers*. These cards are dealt between 16 NFC tags (1 card per *Mifare* tag, each of them being equipped with 1 Kbyte of RAM) and 8 mobile phones (4 cards per Nokia 6131 NFC mobile). The players' goal is to collect cards of the same family on their mobile. Players use their mobile to swap a card with a tag or another mobile. They can also look at the hint function that indicates which tags and/or mobiles contain a card interesting for the player.

PSM uses data structure presented in section 2 as follows. DM contains the 48 cards. If e and e' are any element of the system, $DM_e[e']$ is 1 byte representing the value of the card stored. $VC_e[e']$ is 1 short (2 bytes) representing the real-time clock —formated as the number of seconds since the beginning of the game session— of the last update or query on $VC_{e'}[e']$ as seen by e. Thus DM_e and VC_e occupy 144 bytes in the memory of e.

The demonstration consists in having several users/players. Each one is given one of the eight mobiles to do exchanges with tags and other mobiles. This updates DM_{mobile} and VC_{mobile}. Their contents can be displayed by entering a cheat code. Thus, users can check the results of the hint function. They can also check if there are stale data in DM_{mobile}.

4 Conclusion

In this paper, we have presented an RFID-based Distributed Shared Memory (RDSM). As illustrated by the demonstration, RDSM is a good candidate for pervasive games which rely on RFID/NFC tags, must provide players hints on data hold by distant elements of the system, with no (or limited) use of a Wireless Area Network. Such game is based on the mobility of contents and players to update the overall vision of the state of the game. Mobility and the potential for reading and writing tags create the game momentum.

The analysis of the players' feed back showed that three main features characterize mobility when it is connected to pervasiveness [2]. First, mobility appears as a way to read and collect information. Second, it is a tool to virtually mark the environment and the artifacts. Moving can be akin to "writing" a new scenario. Third, people become part of the network propagating and refreshing information not only on their mobiles but also on the RFID displays.

References

1. PLUG: PLay Ubiquitous Games and play more (August 2009), http://cedric.cnam.fr/PLUG/
2. Gentès, A., Jutant, C., Guyot, A., Simatic, M.: Designing mobility: pervasiveness as the enchanting tool of mobility. In: Proceedings of the 1st international ICST Workshop on Innovative Mobile User Interactivity (IMUI 2009), San Diego, USA, ICST (October 2009)
3. Mottola, L., Murphy, A.L., Picco, G.P.: Pervasive games in a mote-enabled virtual world using tuple space middleware. In: NetGames 2006: Proceedings of 5th ACM SIGCOMM workshop on Network and system support for games, p. 29. ACM Press, New York (2006)
4. Murphy, A.L., Picco, G.: Using LIME to Support Replication for Availability in Mobile Ad Hoc Networks. In: Ciancarini, P., Wiklicky, H. (eds.) COORDINATION 2006. LNCS, vol. 4038, pp. 194–211. Springer, Heidelberg (2006)
5. Rashid, O., Bamford, W., Coulton, P., Edwards, R., Scheible, J.: PAC-LAN: mixed- reality gaming with RFID-enabled mobile phones. Computers in Entertainment 4(4), 4–20 (2006)
6. Saito, Y., Shapiro, M.: Optimistic replication. ACM Comput. Surv. 37(1), 42–81 (2005)
7. Simatic, M.: RFID-based replicated distributed memory for mobile applications. In: Proceedings of the 1st International Conference on Mobile Computing, Applications, and Services (Mobicase 2009), San Diego, USA, ICST (October 2009)
8. Simatic, M., Astic, I., Aunis, C., Gentes, A., Guyot-Mbodji, A., Jutant, C., Zaza, E.: "Plug: Secrets of the Museum": A pervasive game taking place in a museum. In: Natkin, S., Dupire, J. (eds.) Entertainment Computing – ICEC 2009. LNCS, vol. 5709, pp. 302–303. Springer, Heidelberg (2009)

Coarse In-Building Localization with Smartphones[*]

Avinash Parnandi[2], Ken Le[1], Pradeep Vaghela[1], Aalaya Kolli[2], Karthik Dantu[1], Sameera Poduri[1], and Gaurav S. Sukhatme[1,2]

[1] Computer Science Department
[2] Ming Hsieh Department of Electrical Engineering
University of Southern California, Los Angeles, CA 90089, USA
{parnandi,hienle,pvaghela,kolli,dantu,sameera,gaurav}@usc.edu

Abstract. Geographic location of a person is important contextual information that can be used in a variety of scenarios like disaster relief, directional assistance, context-based advertisements, *etc.* GPS provides accurate localization outdoors but is not useful inside buildings. We propose an coarse indoor localization approach that exploits the ubiquity of smart phones with embedded sensors. GPS is used to find the building in which the user is present. The Accelerometers are used to recognize the user's dynamic activities (going up or down stairs or an elevator) to determine his/her location within the building. We demonstrate the ability to estimate the floor-level of a user. We compare two techniques for activity classification, one is naive Bayes classifier and the other is based on dynamic time warping. The design and implementation of a localization application on the HTC G1 platform running Google Android is also presented.

1 Introduction

Indoor localization is a challenging problem facing the ubiquitous computing research community. Accurate location information is easily obtained outdoors using GPS. However, when we enter a building the reception of GPS signals become weak or is lost causing the inability to determine additional location information within the building. Existing solutions for indoor localization include techniques that use WiFi [4], RFID, Bluetooth [2], ultrasound [16], infrared [19] and GSM [17] [13] etc. Many of these solutions rely on external infrastructure or a network of nodes to perform localization. Note that improving the localization accuracy or increasing the availability of the services provided by these systems require the scaling of infrastructure which can be costly and a challenge on its own.

Our goal is to develop a localization system that is independent of external infrastructure. To this end, we built an indoor localization application that relies primarily on the sensors embedded in a smartphone to coarsely locate a person within a building. Our approach is based on user activity modeling. We use the GPS receiver to determine which building the user entered, and we use the accelerometer to determine what activities the user is performing within the building. We localize the user by coupling

---

[*] This work was supported in part by NSF grant CCR-0120778 (CENS: Center for Embedded Networked Sensing), and by a gift from the Okawa Foundation. It was initiated as a project for the graduate course CS 546: Intelligent Embedded Systems taught at USC in Spring 2009.

successive activities like climbing the stairs with some background knowledge. For example, using timestamps we can determine how long the user traversed the stairs or the elevator allowing us to infer which floor he/she is on. The accuracy and capabilities of our localization thus depends on how well we can identify and describe the user's contexts and the knowledge we have about the building.

The rest of the paper is as follows. Section 2 discusses the related work. Section 3 describes the implementation and data collection. Section 4 describes our feature set and classification algorithm, Dynamic Time Warping (DTW) – an alternate technique we explore for better classification at a greater computational cost, and our methodology to detect elevator usage. Section 5 discusses our experimental results in detail. Section 6 lists our conclusions and some directions to extend our work.

2 Related Work

As mentioned in Sec. 1, indoor localization has been the subject of much research. We sample some representative pieces of work from this literature.

Pedestrian indoor localization system developed in [20], does localization using particle filter estimation based on inertial measurement units' data. This work, though requires a detailed map of the building, is close to ours because both use on-board inertial sensors while most other systems depend heavily on external infrastructure. An active badge location system was developed in [18]. In this system, the badges emit a unique id via infrared sensors and the building is instrumented with infrared receivers. All these received ids are relayed to a central server that then tracks every active badge. The RADAR system [4] was a building wide tracking system using wireless LAN. The work in [13] relies on the GSM cell towers id and signal strength for indoor localization. The E911 emergency tracking system locates people who call 911 by doing sophisticated radio signaling and computing the time of arrival, and time difference of arrival from cellphone to perform accurate triangulation. The major difference between all these system and ours, is the fact that all of these require extensive external infrastructure and information, like cell tower id, signal strength measurement capability, access point locations which are not readily available to be used on a mobile phone. Our work does not depend on external infrastructure and uses embedded sensors to solve the localization problem. While in [20] they use the IMU data and estimate the current location using a particle filter on an offline system, we use an accelerometer for activity recognition with time stamps and utilize this information to do online indoor localization. Another important contribution of our work is the elevator ride detection, which to the best of our knowledge has not been greatly explored before.

Activity recognition using inertial sensors has been a topic of great interest to the research community for the last two decades [7]. It promises to have great impact on a variety of domains like elder care [6], fitness monitoring [9, 14], and intelligent context-aware applications [5]. Recently several researchers have used the accelerometers embedded in smartphones to detect activities such as walking, standing, running and sitting with the goal of developing context-aware applications [8, 10]. Our goal in doing activity detection is localization. We focus on activities that may cause a person's location (floor) to change within a building.

3 Implementation

3.1 Hardware

The sensing device we use is the HTC G1 smartphone equipped with GPS receiver and tri-axial accelerometer. The software was implemented on the Android platform, Google's operating system for mobile devices which includes an extensive API for application development. Notably, the API includes ways for accessing and using the phone's sensors [1].

3.2 Software

In this section we'll describe the localization application that was implemented in the smartphone. The application is made up of multiple services that run ubiquitously in the background. Each service provides a specific functionality and each turns on or off automatically depending on which state the application is in. The *Main* service provides the state machine functionality of the application, this is displayed in Figure 1. Its primary responsibility is to turn on and off other services based on the current state of the application and the actions that occur. Lets briefly walk through each of the states.

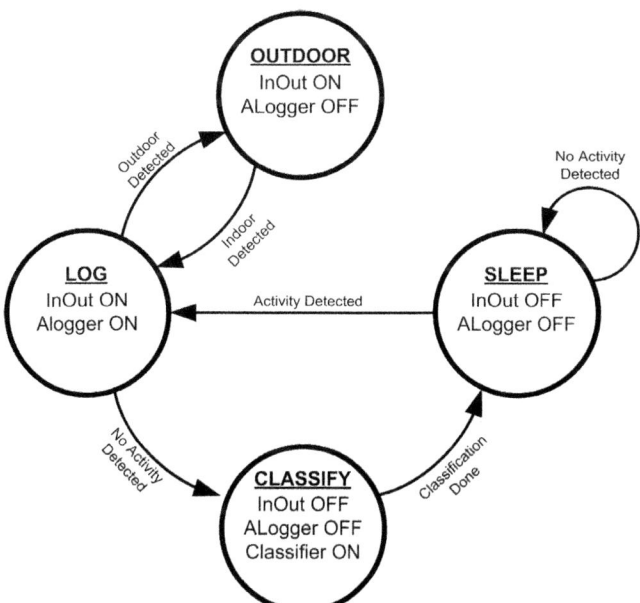

Fig. 1. State diagram of context sensing application

When the user is outdoors, the application is in the OUTDOOR state. In this state the *Indoor/Outdoor Transition Detection* service, abbreviated as InOut in Figure 1, is on to determine when an outdoor to indoor activity occurs using the phone's GPS receiver. When the user enters a building the *Indoor/Outdoor Detection* service logs the location (latitude and longitude) of where the outdoor to indoor activity occurred and transmits a signal to the Main service.

When an outdoor to indoor signal is detected the application goes into the LOG state. The *Unlabeled Activity Logger* service (ALogger) turns on and begins logging "Unlabeled" accelerometer data into a file.

Fig. 2. Outdoor to indoor activity indicating which entrance of the building was used

When the application is in the LOG state and the user becomes still for a specific duration, the application transitions into the CLASSIFY state. Stillness is detected by the *Simple Activity Detection Service*, which is a lightweight module that determines whether the user is performing an activity or not. This service remains on throughout the life of the application for the purpose of triggering various state transitions.

In the CLASSIFY state the *Classification+Filter* service (Classifier) is turned on. This service retrieves the log file created by the *Unlabeled Activity Logger*, analyzes the data sequence, and classifies/labels each instance in the sequence with the most probable activity (e.g. *stairs_up*, *walk*). Occasionally, the classification will be wrong. For this reason, we implement an in-line filtration algorithm that removes misclassified activities. The result is a sequence of identified activities tagged with timestamps indicating when the activities occurred as shown in Figure 2.

Once classification is done, the application transitions to the SLEEP state where everything is turned off except for the *Simple Activity Detection* service. This is the most efficient state, and occurs when the user is staying still.

When the *Simple Activity Detection* service detects an activity again, the application transitions back to the LOG state where the processes described above repeat.

3.3 Data Collection and Analysis

Table 1 shows the list of activities that we focus on perceiving for our initial implementation.

We are interested in these activities since they all cause location change within a building except for the *still* activity. We began by collecting raw accelerometer data

Table 1. List of activities used to estimate floor-level

Activity Label	Description
Still	The user is still
Walk	The user is walking
Stairs_up	The user is going up a staircase
Stairs_down	The user is going down a staircase
Elevator_up	The user is going up in an elevator
Elevator_down	The user is going down in an elevator

using a simple logging application. To keep the data collection trials consistent we strapped the phone to the user's ankle as shown in Figure 3; this kept the y-axis of the phone aligned to the lower leg at all times. Figure 4 shows acceleration data in the y-axis while performing various activities. As you can see from the activity labels, we found that there were distinguishable characteristics in the accelerometer data between the different activities. Note that the still activity is represented by the flat sections in the graph. Our next task was to find a way to quantize the data into various features that would allow us to distinguish the different activities in numerical terms.

The section on Classification will go into further detail on the other features that were explored.

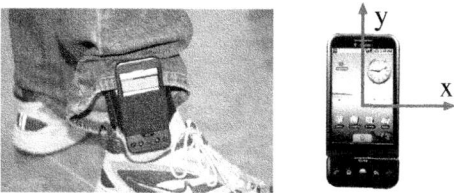

Fig. 3. Position of the phone for data collection with the axis

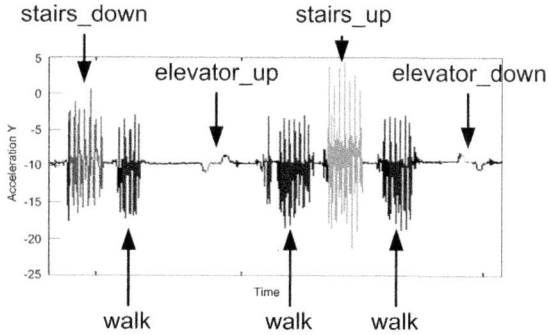

Fig. 4. Acceleration data along the Y axis

3.4 Feature Extraction and Classification

Features are extracted from the accelerometer data every X amount of samples, X was chosen to be 50 within our implementation. There were four sampling rates provided by the Android API.

We chose to operate at the fastest rate which from our observations sampled at approximately 75 samples/sec max and 40 samples/sec on average.

We used the Weka machine learning library in Java for activity classification on the phone. The classifier is used within the Classification+Filter service to classify/label the unlabeled log file, which is stored in *Attribute Relation File Format*. This file is created during feature extraction in the Unlabeled Activity Logger service.

4 Classification

In this section we describe the classification algorithms used to differentiate between the activities listed in Table 1. Two different approaches are presented and compared. The first approach, using a naive Bayes classifier, computes the posterior class probabilities based on a set of carefully chosen features. While this method gives good results for detecting *stairs_up*, *stairs_down*, and *walking* activities (average accuracy of 97% for a user specific trained classifier and 84% for a generally trained classifier), it does not perform well for elevator activity detection. The second approach, Dynamic Time Warping (DTW), classifies activities based on matching raw data with a reference *template* for each activity. This approach gives competitive results but is computationally expensive and not suited for implementation on a phone. We designed a light-weight DTW inspired approach that uses templates and a convolution operation to detect *elevator_up* and *elevator_down* activities.

4.1 Naive Bayes Classification

Naive Bayes classification [11] is a simple and well-known method for classification. Given a feature vector f, the class variable C is given by the *maximum aposteriori* (MAP) decision rule as

$$C(f) = argmax_C \left\{ p(C) \prod_{i=1}^{k} p(f_i|C) \right\} \quad (1)$$

The accuracy of the classifier depends on the choice of features. After testing a wide range of features [3] [15], we selected the following four features based on classification accuracy and computational cost. Details are shown in the table in Figure 5. The raw acceleration data is divided into windows of fixed size samples. Each window gets feature extracted and classified independently. As expected, the acceleration data along the y-axis (vertical direction) contained most information about the activity.

We use the following four features for the classification:

1. mean (along y).
2. variance (along y).

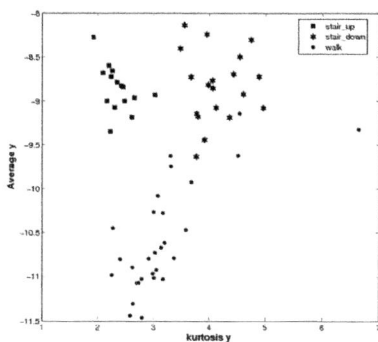

Feature	stair_up vs stair_down	stair_up vs walk	stair_down vs walk
kurtosis y	3.92	31.11	11.32
mean y	44.92	9.55	16.28
energy y	43.00	14.66	19.31
skew y	15.53	49.77	20.12
eccentricity y	45.00	23.11	32.05
variance y	47.00	36.44	21.83
correlation yx	31.30	34.00	49.23
correlation yz	31.30	27.55	52.47
kurtosis x	21.84	48.89	18.58
average x	43.46	50.00	36.79
energy x	33.53	28.89	16.58
skew x	35.38	48.00	26.50
eccentricity x	42.69	52.88	47.13
variance y	45.30	38.66	46.92
correlation xy	31.30	34.00	49.23
correlation xz	41.15	38.66	31.45

Fig. 5. Cluster plot and Misclassification rate of features

3. skewness (along x) which is the degree of asymmetry in the distribution of the acceleration data.
4. kurtosis (along y) is a measure of how much a distribution is peaked at the center of the distribution.

Figure 5 shows a cluster plot of a set of training data along a pair of selected features (average y and kurtosis y). It can be seen that the feature pair confidently separates the three activities. We estimate the likelihood distributions to be used in equation 1 as Gaussian distributions that minimize the least squares error of the training data.

4.2 Dynamic Time Warping

Dynamic time warping (DTW) is an algorithm for measuring similarity between two signals which may vary in time or speed [12]. It has been used successfully to classify activities that have a cyclic pattern. A reference signal called a *template*, which is one cycle of the activity, is generated for each activity based on training data and new data is compared with the templates to find the best match on the basis of their Euclidean distance.

We extracted templates manually from the training data. Figure 6 shows an example of the stairs_up template. Classification of data is performed by using a sliding window of y acceleration data and finding its (Euclidean) distance to each of the activity templates. The window is labeled with the activity that gives the least distance. The DTW based approach had an average accuracy of 87.5%.

4.3 Elevator Activity Detection

As mentioned earlier, when using naive Bayes classifier, it was difficult to find features that allow classification of *elevator_up* and *elevator_down* activities. Preferring to use naive Bayes over DTW for efficiency, we needed to develop an auxiliary method for detecting elevator rides. The method we developed specifically for elevator detection

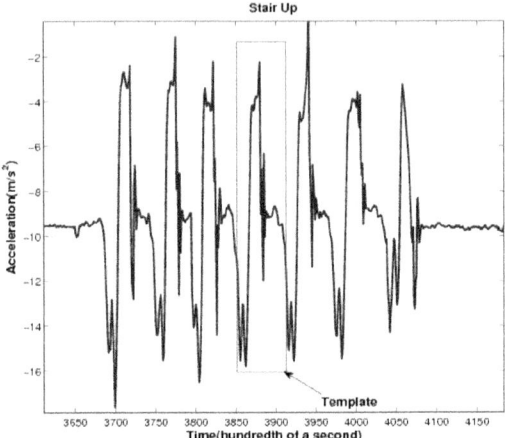

Fig. 6. *stairs_up* activity template

performs template matching using a convolution operator. The first step in this method is to filter out all data that has a standard deviation above a certain threshold. Observing Figure 4, we note that elevator activity has significantly less variance than other activities. With a correctly chosen threshold, the filtered data leaves us only with elevator peaks as shown in Figure 7.

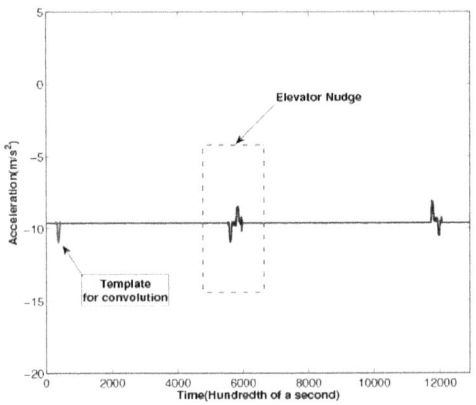

Fig. 7. Acceleration data post filtering

The filtered data is then convolved with the elevator activity template. A typical output of the convolution step is shown in Fig. 8. When the incoming elevator data is convolved with the standard elevator template, the points in the output curve corresponding to the elevator minima get amplified and are shown as the maxima in Fig. 8 (the minima gets amplified because our template has downward facing tip) while the elevator maxima gets registered as a minima in the convolution output.

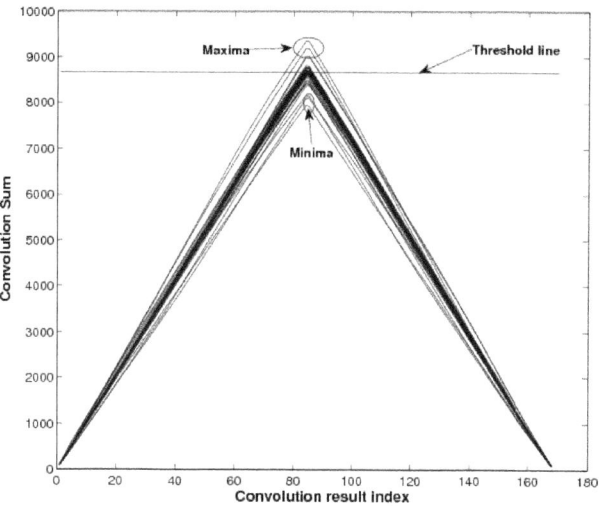

Fig. 8. Convolutional elevator detector: the minima and maxima with respect to the threshold

5 Experiments and Results

This section goes over the various experiments we carried out and their results.

5.1 Indoor Localization

We performed our indoor localization experiments in a five story building named Ronald Tutor Hall (RTH). The table in Figure 9 shows an example of pre-observed times taken to transition between various floors within the building using the stairs.

Floor Transition	Duration
First to Third	51 sec
Third to First	42 sec
First to Second	24 sec
Second to First	21 sec

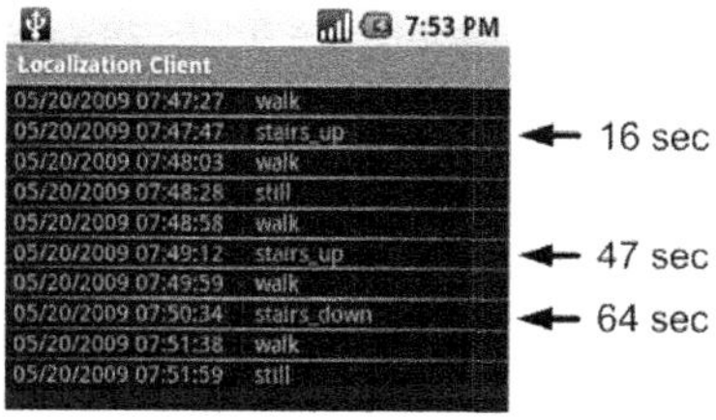

Fig. 9. Pre-observed times and Output of user's context history

The snapshot in Figure 9 shows the output of our localization application after a user performs a trial walk throughout the building. Coupling our knowledge of the approximate times taken to transition between floor levels with the context history outputted from the application we can infer what floor the user is on and at what time. For example, assuming the user is initially on floor 1 at time 7:47:27, we can infer that he/she is on floor 1 at time 7:48:03, on floor 3 at 7:49.59, and on floor 1 again at time 7:51:38. Here we simply divide the duration of *stairs_up* or *stairs_down* activity with the average time to transition up or down one floor level to determine the number of floor transitions. The inferences made match with the actual floors the user was on during the trial run. The results here show the potential possibilities of performing localization using inference with dynamically retrieved contextual knowledge about the user and static background knowledge about the building.

5.2 Activity Classification

To test activity classification we asked 10 users with different physique and walking styles to perform composite activities. We tested two cases. The first case, *User Specific*, is where the classifier is trained and tested uniquely for each user. In this case, the trial begins by placing the application in training mode and asking the user to perform each activity several times. After training the classifier specifically for the user, the application is switched to operational mode and the user performs the same set of activities to see how well they can be classified. In the second case, *General*, the classifier is generically trained with a couple people and tested on all. In this case, we begin by selecting only a couple of users to train the phone. We then switch to operational mode, and ask every user to exercise the set of activities to see how well the generically trained classifier performs.

Results from our tests show that the User Specific case classified with 97% accuracy and the General case classified with 84% accuracy. In either case, after running the classified activities through the filter in the Classifier+Filter service all misclassified activities were removed.

5.3 Elevator

Using the elevator detection method we developed, we were able to detect elevator rides pretty well. The key part about an elevator ride (whether going up or down) is that convolution maxima and minima always occur in pair, at the start and end point. This is evident from Figure 7. By keeping track of these max-min pairs and their corresponding timestamps we can find out how much time the user spent in the elevator. To test elevator detection we began by extracting an elevator ride template from a trial ride of one user. We then convolved this template with the filtered acceleration data of test runs by 10 other users. Our algorithm detected the elevator rides in each of these runs with no errors.

5.4 Analysis

In Figure 10 we compare the performance of the generally trained classifier, user specific trained classifier, and DTW for each of the 10 users. For the classifiers we use the four features mentioned earlier. Naive Bayes classifier and DTW results are compared with ground truth. The number of false positives divided by total number gives the error percentage. Note that convolution based elevator detection is used in the classifier cases. DTW with generic templates performed better than the generally trained classifier for most users. The user specific trained classifier performed the best on all users. Computationally, DTW is much more expensive and time consuming than the naive Bayes classifier. Since a mobile device must use its resources efficiently and since there was not a clear advantage for using DTW, naive Bayes classification is our method of choice.

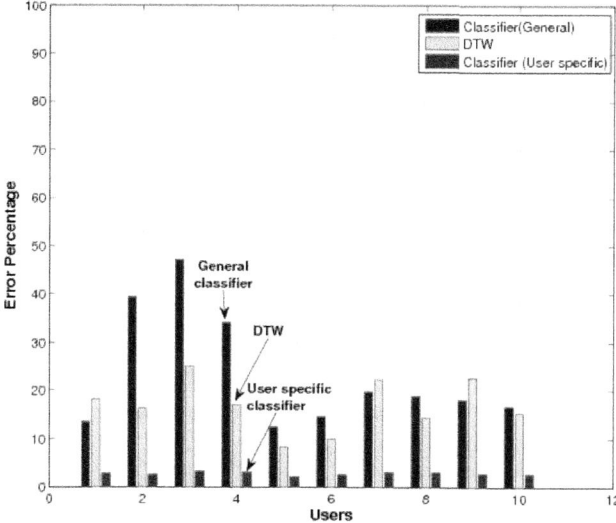

Fig. 10. Percentage of misclassification between Classifier and DTW

6 Conclusion and Future Work

This paper presents an indoor localization system that demonstrates the ability to perform floor level detection using mobile phones. Specifically, we use the accelerometers of the phone to detect activities like *stair_up*, *stair_down*, *elevator_up* and *elevator_down* and further investigate these activities to detect movement between floors. We propose two algorithms (DTW and naive Bayes classifier) to perform this activity detection. These algorithms offer a trade off between detection accuracy and computational requirement (energy efficiency).

Our initial results for floor-level detection are encouraging. We have manually computed the average times needed to traverse up/down a floor for one building but this may vary significantly across different users and different buildings. Our next step is to implement a learning algorithm to adapt this (with user feedback) for the buildings that the particular user visits. We would also like to implement a coordinate frame transformation on the phone so that the application becomes independent of where the phone is on the body.

References

1. Google Android, http://www.android.com/
2. Aalto, L., Göthlin, N., Korhonen, J., Ojala, T.: Bluetooth and wap push based location-aware mobile advertising system. In: MobiSys 2004: Proceedings of the 2nd international conference on Mobile systems, applications, and services, pp. 49–58. ACM, New York (2004)
3. Baek, J., Lee, G., Park, W., Yun, B.-J.: Accelerometer signal processing for user activity detection, Berlin, Germany, vol. 3, pp. 610–617 (2004)
4. Bahl, P., Padmanabhan, V.N.: RADAR: An in-building RF-based user location and tracking system. In: International Conference on Computer Communications (INFOCOM), pp. 775–784 (2000)

5. Choudhury, T., Borriello, G., Consolvo, S., Haehnel, D., Harrison, B., Hemingway, B., Hightower, J., Klasnja, P., Koscher, K., Lamarca, A., Landay, J.A., Legrand, L., Lester, J., Rahimi, A., Rea, A., Wyatt, D.: The mobile sensing platform: An embedded activity recognition system. IEEE Pervasive Computing 7(2), 32–41 (2008)
6. Jeon, A., Kim, J., Kim, I., Jung, J., Ye, S., Ro, J., Yoon, S., Son, J., Kim, B., Shin, B., Jeon, G.: Implementation of the personal emergency response system using a 3-axial accelerometer. In: 6th International Special Topic Conference on Information Technology Applications in Biomedicine, ITAB 2007, vol. X, pp. 223–226 (November 2007)
7. Jeon, A., Kim, J., Kim, I., Jung, J., Ye, S., Ro, J., Yoon, S., Son, J., Kim, B., Shin, B., Jeon, G.: Implementation of the personal emergency response system using a 3-axial accelerometer, Tokyo, Japan, pp. 223–226 (2008)
8. Krause, A., Ihmig, M., Rankin, E., Leong, D., Gupta, S., Siewiorek, D., Smailagic, A., Deisher, M., Sengupta, U.: Trading off prediction accuracy and power consumption for context-aware wearable computing. In: ISWC 2005: Proceedings of the Ninth IEEE International Symposium on Wearable Computers, Washington, DC, USA, pp. 20–26. IEEE Computer Society, Los Alamitos (2005)
9. Mathie, M., Coster, A., Lovell, N., Celler, B.: Accelerometry: providing an integrated, practical method for long-term, ambulatory monitoring of human movement. Physiological Measurement 25(2), 1–20 (2004)
10. Miluzzo, E., Lane, N.D., Fodor, K., Peterson, R., Lu, H., Musolesi, M., Eisenman, S.B., Zheng, X., Campbell, A.T.: Sensing meets mobile social networks: the design, implementation and evaluation of the cenceme application. In: SenSys 2008: Proceedings of the 6th ACM conference on Embedded network sensor systems, pp. 337–350. ACM, New York (2008)
11. Mitchell, T.M.: Machine Learning. McGraw-Hill, New York (1997)
12. Muscillo, R., Conforto, S., Schmid, M., Caselli, P., D'Alessio, T.: Classification of motor activities through derivative dynamic time warping applied on accelerometer data, August 2007, pp. 4930–4933 (2007)
13. Otsason, V., Varshavsky, A., LaMarca, A., de Lara, E.: Accurate gsm indoor localization, Berlin, Germany, pp. 141–58 (2005)
14. Preece, S., Goulermas, J., Kenney, L., Howard, D., Meijer, K., Crompton, R.: Activity identification using body-mounted sensors-a review of classification techniques. Physiological Measurement 30(4), R1–R33 (2009)
15. Ravi, N., Dandekar, N., Mysore, P., Littman, M.L.: Activity recognition from accelerometer data, Pittsburgh, PA, United states, vol. 3, pp. 1541–1546 (2005)
16. Savvides, A., Han, C.-C., Srivastava, M.B.: Dynamic fine-grained localization in ad-hoc networks of sensors. In: International Conference on Mobile Computing and Networking (MOBICOM), pp. 166–179 (2001)
17. Varshavsky, A., de Lara, E., Hightower, J., LaMarca, A., Otsason, V.: GSM indoor localization. Pervasive and Mobile Computing 3(6), 698–720 (2007)
18. Want, R., Hopper, A., Falcao, V., Gibbons, J.: The active badge location system. ACM Transactions on Information Systems 10(1), 91–102 (1992)
19. Ward, A., Jones, A., Hopper, A.: A new location technique for the active office. IEEE Personal Communications 4(5), 42–47 (1997)
20. Woodman, O., Harle, R.: Pedestrian localisation for indoor environments. In: UbiComp 2008: Proceedings of the 10th international conference on Ubiquitous computing, pp. 114–123. ACM, New York (2008)

Delay Analysis of Large-Scale Wireless Sensor Networks

Jun Yin[1], Yun Wang[2], and Xiaodong Wang[3]

[1] Dominican University, River Forest, IL, USA
[2] Southern Illinois University Edwardsville, Edwardsville, IL, USA
[3] Qualcomm Inc. San Diego, CA, USA

Abstract. For wireless sensor network applications, the latency from the sensing of the event to the reporting through the network is critical. In this paper, we try to characterize the delay incurred by sensed packets traversing across the network. We derive the source and destination distance, hop count distribution under typical sensor traffic patterns. Then how various network parameters, such as the node density and the transmission range, impact on delay and delay distribution is investigated. The results of our research can provide insights in designing both flat and cluster-based sensor networks to meet the specified delay requirement.

Keywords: wireless sensor networks, clustering, delay, diameter, hop count, Voronoi diagram.

1 Introduction

Recent technical advances have enabled the large-scale deployment and applications of wireless sensor nodes. These small in size, low cost, low power sensor nodes is capable of forming a network without underlying infrastructure support. Such a wireless network is called a wireless sensor network (WSN). It consists of spatially distributed autonomous sensor nodes to cooperatively monitor physical or environmental conditions, such as temperature, sound, vibration, pressure, motion or pollutants, at different locations[1,2,3]. Sensor nodes consist of sensing, data processing and communication components. Their functions are to sense data from the environment, process the data, and finally transmit the data to the destination. Because of sensor nodes' reliability, accuracy, cost effectiveness, and easy deployment, WSN is emerging as a key tool for various applications including home automation, traffic control, search and rescue, and disaster relief.

In general wireless ad hoc networks, we are concern with the throughput and spatial re-use. Sensor networks make less demand on the throughput performance; rather delay is an important quality of service (QoS) parameter for sensor networks [4]. In lots of sensor network applications, sensed data needs to be transmitted and forwarded to the sink node in a timely manner. A sink node acts like a gateway between the sensor network and the exterior network.

Due to the large network size, packets need to traverse large number of intermediate nodes to reach destination, or the sink node. Furthermore, most sensor networks adopt power management schemes[5,6,7], where nodes are powered on and off periodically. These schemes save energy at the cost of extra delay for each hop, and the end-to-end delay incurred by sensed packets becomes much larger consequently. It is important to characterize the delay for the delivery of sensed packets and to determine how the parameters, such as network size, transmission range and node density, affect the delay performance.

Different from existing works focusing on a new MAC or routing protocol design to reduce delay and energy consumption, we study the delay properties of large scale sensor networks in an analytical framework. Without loss of generality, we assume an uncoordinated sensor MAC protocol in our work. By obtaining distance distribution between source and destination nodes in sensor networks under typical traffic patterns, we characterize the delay incurred by sensed packets. Furthermore, we extend our analysis to a two-tier architecture where there are two types of sensor nodes. The more powerful nodes function as clusterheads [8], and traffic is aggregated within the cluster to the clusterhead. We determine how the introduction of the clustering architecture impacts on the delay performance. Our work has application in choosing network parameters, such as node density, transmission range, and the number of clusterheads to meet the delay requirements.

The paper is organized as follows. In the next section, related works are discussed. In section III, we characterize the distance distribution between source and destination pairs for typical traffic patterns in the sensor networks: (1) random source to random destination and (2) multiple source to a single central sink node traffic. Delay properties are then derived based on the distance distribution. We also extend the analysis to the two-tier structure, where the more powerful nodes are deployed as clusterheads. The last part is the conclusion.

2 Related Work

Several sensor MAC protocols, most notably STEM [5] and SMAC [6], have been proposed for use in sensor network. These schemes trade the delay of each hop with the energy saving. Specifically, the tradeoff between the energy saving and the delay has been identified by Schurgers et al. [5]. Yu et al. [9] analyzed the tradeoff between the latency and the energy consumption from the perspective of data aggregation. Lu et al. [10] tries to improve the end-to-end delay by assigning the time slots to sensor nodes optimally. Dousse et al. [11] study the delay using blinking Poisson boolean model. Gamal et al. [12] shows the tradeoff between the delay and the energy efficiency in the physical layer. The most relevant work was done by Zorzi et al. [13], and they study the multi-hop performance by analyzing the hop count in a geographic random forwarding routing scheme.

To the best of our knowledge, delay analysis for typical traffic patterns in the sensor networks has not been treated in the literature. In this paper, we base our analysis on a general assumption of an uncoordinated sensor MAC protocol, where synchronization is not pre-assumed. It should be noted that for this MAC protocol, the delay of each hop is a random variable. Our analysis can also be extended for special cases where each hop delay is fixed.

3 Delay Analysis

In the large-scale sensor networks, the hop count the packet needs to traverse in the network to reach the destination and the delay of each hop in the MAC layer determine the overall end-to-end delay. Compared with the delay introduced by energy management scheme in typical sensor MAC protocols [5][6], the queueing delay and the packet processing delay can be neglected.

In this section, we analyze the S/D distance distribution under typical traffic pattern in the sensor network, and try to characterize the delay based on the results in the last section.

3.1 Delay of Each Hop

General sensor networks are mostly based on uncoordinated MAC protocols since it is expensive to achieve synchronization [14]. Under completely uncoordinated MAC such as STEM[5], the delay of each hop d is a random variable between 0 and T_s, where T_s is the sleeping/wakeup interval. $\mathsf{E}(d)$ denotes the average delay of each hop and $\mathcal{V}(d)$ is the variance of the delay of each hop. Thus,

$$\mathsf{E}(d) = \frac{T_s}{2}, \tag{1}$$

and,

$$\mathcal{V}(d) = \int_0^{T_s} [s - \mathsf{E}(d)]^2 \frac{1}{T_s} ds = \frac{T_s^2}{12}. \tag{2}$$

The delay of each hop can be seen as an independent and identical random variables with mean $\frac{T_s}{2}$ if the MAC is completely uncoordinated. The end-to-end delay S_h between two nodes being h hops away, can be expressed as sum of independent and identically distributed random variables. If h is large, the probabilistic distribution of S_h can be approximated by central limit theorem [15]:

$$\mathsf{P}[S_h] = \frac{1}{\sqrt{2\pi \mathcal{V}(d)}} e^{-\frac{[S_h - \mathsf{E}(d)]^2}{2\mathcal{V}(d)^2}},$$
$$\mathsf{E}[S_h] = h\mathsf{E}(d), \text{ and}$$
$$\mathcal{V}[S_n] = h\mathcal{V}(d). \tag{3}$$

3.2 Delay between Random Source and Destination

In general sensor network applications, sensor nodes perform sensing and the sensed packets are forwarded to sink node in a multi-hop manner. On the other hand, in complex sensor networks performing in-network query or aggregation, there might exist traffic between random pairs of traffic source and destinations. This type of traffic is treated in this section and the former case will be discussed in the next section.

The distribution of the distance between random source and destination pairs was first investigated in [16]. It is further shown in [17] that the PDF can be closely approximated via Gaussian distribution with mean $\frac{L}{2}$ and standard derivation $\frac{L}{3.5}$, when nodes are randomly deployed in a square area of $L \times L$. Here we adopt the PDF introduced in [18]. For a rectangle area of size $a \times b$, the PDF of the distance between random S/D, denoted by $\mathsf{P}_{S/D}[\gamma]$, is as follows[18]:

$$\mathsf{P}_{S/D}[\gamma] = \frac{4\gamma}{a^2 b^2}\left(\frac{\pi}{2}ab - a\gamma - b\gamma + \frac{1}{2}\gamma^2\right). \tag{4}$$

For our deployed square area, $a = b = L$. For distance γ, the required hop count can be approximated via $\lceil \frac{\gamma}{F(\theta, r_0, \lambda)} \rceil$, where function F is F_{topo} for energy efficient routing with power control case or F_{2D} for shortest path routing with power control [19].

The hop count between random S/D pairs is a random variable, denoted by H.

$$\begin{aligned}\mathsf{P}[H = h] &= \mathsf{P}_{S/D}\left[\frac{\gamma}{F(\theta, r_0, \lambda)} = h\right] \\ &= F(\theta, r_0, \lambda)\mathsf{P}_{S/D}\left[hF(\theta, r_0, \lambda)\right]. \end{aligned} \tag{5}$$

Assuming that packets incur the same fixed delay at each hop, we can then drive the PDF of the delay incurred for all the random S/D traffic in the network from Equ. (5). For uncoordinated MAC with random delay of each hop, we can obtain the average delay.

The average delay incurred by all the random S/D pairs, denoted by $\mathsf{E}\,(Delay)$ can be expressed as:

$$\begin{aligned}\mathsf{E}\,(Delay) &= \mathsf{E}\left[\mathsf{E}(d\,|hopcount = h)\right] = \mathsf{E}\left(h\frac{T_s}{2}\right) \\ &= \frac{T_s}{2F(\theta, r_0, \lambda)} \int_0^{\sqrt{2}L} \mathsf{P}[\gamma]\gamma d\gamma \\ &= \frac{2\sqrt{2}L(\frac{2}{3}\pi - \frac{8}{5})}{F(\theta, r_0, \lambda)}. \end{aligned} \tag{6}$$

In some circumstances, we do care about the maximum delay. The diameter is defined as the maximum hop count in all the shortest paths between all the node pairs in a network [20]. If the diameter of the networks is D, the maximum delay a sensed packet might suffer can be expressed as:

$$MaxDelay = DT_s. \tag{7}$$

For the random deployment in an $L \times L$ square area, the diameter can be approximated by:

$$D = \frac{\sqrt{2}L}{F(\theta, r_0, \lambda)}. \tag{8}$$

We perform simulation in a graph simulator implemented in C++. 500 nodes randomly deployed in a area of 1000×1000. Floyd algorithm [21] is used to compute the shortest path between any pair of nodes in the network. Fig. 1 shows the average hop count between random source and random destination over 50 randomly generated topologies.

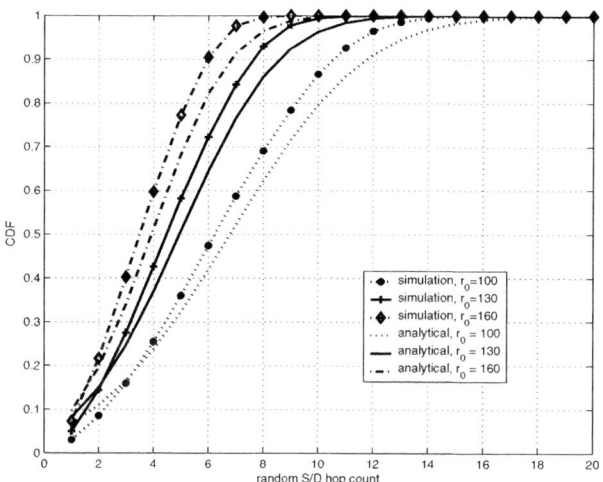

Fig. 1. Hop count distribution between random pairs

3.3 Delay from Multiple Sources to One Sink

In this part, we look into the case where a sink node is located at the center of the deployed area and sensor nodes send sensed packets to the sink node via multi-hop forwarding. We first consider a flat sensor network organization, and then perform the analysis for a two-tier architecture.

Flat architecture. To facilitate our analysis, we suppose that the nodes are deployed in a circle area with radius R, and the node density is λ. The distance from a node to the central sink node is a random variable, denoted by v_1.

$$\mathsf{P}\left[v_1 \leq \xi\right] = \frac{\xi^2}{R^2}. \tag{9}$$

And,

$$\mathsf{P}\left[v_1 = \xi\right] = \frac{2\xi}{R^2}. \tag{10}$$

The PDF of hop count in the flat architecture can be derived using Equ. (5), where $P_{S/D}$ should be replaced by Equ. (10).

The average distance to the sink node $E(v_1)$ can be expressed as:

$$E(v_1) = \int_0^R \frac{2\xi^2}{R^2} d\xi = \frac{2R}{3}. \tag{11}$$

The average delay can be derived as:

$$\begin{aligned} E(Delay) &= E[E(d \mid hopcount = h)] \\ &= \frac{T_s}{2} \int_0^R P[v_1] \frac{v_1}{F(\theta, r_0, \lambda)} dv_1 \\ &= \frac{T_s R}{3F(\theta, r_0, \lambda)}. \end{aligned} \tag{12}$$

Two-tier architecture with two types of sensor nodes. In the flat architecture, sensed packets, need to be forwarded to the sink node via intermediate nodes. For large sensor networks, the delay and the energy consumption in flat architecture is unbearable. To improve the scalability of sensor networks, hierarchical architecture has to be introduced [22,23]. Suppose there are two kinds of nodes, $K1$ and $K2$. Nodes $K1$ are more powerful than nodes $K2$. We consider a two tier sensor structure, where $K1$ and $K2$ type sensors act as clusterheads(CHs) and normal sensor nodes respectively. Sensor nodes $K2$ always choose the nearest $K1$ nodes as their clusterhead, as shown in Fig. 2. We assume that the $K1$ nodes have enough power conserve and are connected to each other through high bandwidth links and the delay incurred by data transmissions between CHs can be negligible. The density of nodes $K1$ and $K2$ are λ_1 and λ_2 respectively.

Regular clusterhead placement
We first consider a regular $K1$ CH placement where the node deployment area is partitioned into hexagonal areas and CH is located at the center of the hexagon. $K2$ nodes are randomly deployed. Suppose the radius of a cell is Z. The hexagonal area covered by one cell is $\frac{3\sqrt{3}Z^2}{2}$ (shown as Fig. 3). Thus, for a deployment area $L \times L$, the number of cells N can be approximated as:

$$N \approx \frac{L \times L}{\frac{3\sqrt{3}Z^2}{2}}. \tag{13}$$

We define the random variable of the distance from a node to its CH in the two tier architecture as v_2. The average distance for a single cell has been derived as Equ. (11):

$$E(v_2) = \frac{2Z}{3} \approx \sqrt{\frac{8L^2}{27\sqrt{3}N}} = \sqrt{\frac{8}{27\sqrt{3}\lambda_1}}, \tag{14}$$

where λ_1 is the node density of $K1$ nodes.

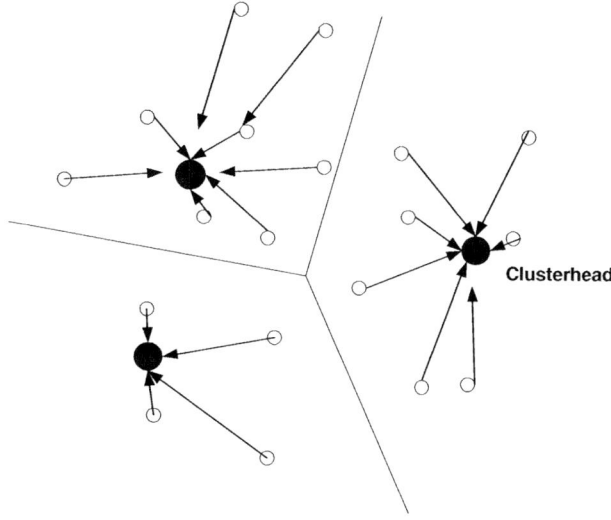

Fig. 2. Two tier clustering based on Voronoi diagram

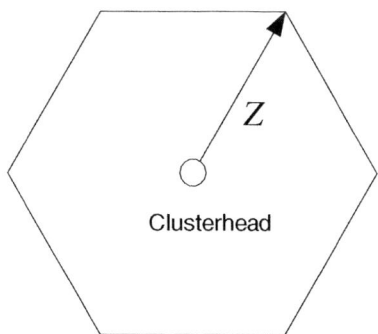

Fig. 3. Hexagon cell in regular deployment

Probabilistic analysis
When both $K1$ and $K2$ nodes are randomly deployed, Voronoi diagram [24,3] can be used to cluster the $K2$ sensor nodes into different Voronoi cells, where a $K2$ sensor node chooses its nearest $K1$ sensor node as its CH. Sensed data from $K2$ nodes are transmitted to their CH via multi-hop forwarding. Denote the distance from a $K2$ node to its CH by v_2. The probability of $\mathsf{P}\left[v_2 \leq \xi\right]$ can be expressed as:

$$\begin{aligned}\mathsf{P}\left[v_2 \leq \xi\right] &= 1 - \mathsf{P}\left[v_2 > \xi\right] \\ &= 1 - e^{-\lambda_1 \pi \xi^2}.\end{aligned} \quad (15)$$

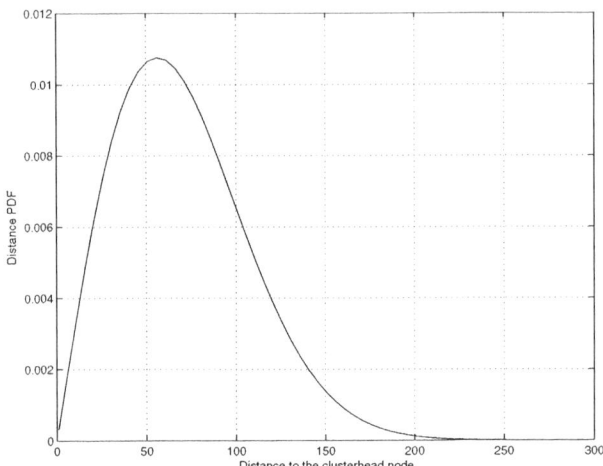

Fig. 4. Distance distribution of different distance from the clusterhead

Thus,

$$P(v_2 = \xi) = 2\pi\lambda_1 \xi e^{-\lambda_1 \pi \xi^2}. \tag{16}$$

Fig. 4 shows the probability density function of distance in the two tier sensor networks, where the number of $K1$ nodes is set to 50, and the number of $K2$ nodes is set to 300. The transmission range of $K2$ nodes, denoted by r_0, is set to 80.

The average of v_2, denoted by $E(v_2)$, can be expressed as:

$$E(v_2) = \int_0^R 2\pi\lambda_1 \xi^2 e^{-\lambda_1 \pi \xi^2} d\xi. \tag{17}$$

When $R \to \infty$, $E(v_2)$ can be derived as Equ. (18) [25].

$$E(v_2) = \frac{2}{\sqrt{\lambda_1}}. \tag{18}$$

From Equ. (18), we can see that for a large two tier sensor network, the average distance decrease on the order of $\lambda_1^{-\frac{1}{2}}$. This result is consistent with Equ. (14) derived for regular $K1$ nodes deployment.

$$\begin{aligned} E(Delay) &= E\big[E(d\,|hopcount = h)\big] \\ &= \frac{T_s}{2} \int_0^{2\sqrt{L}} P[v_2] \left[\frac{v_2}{F(\theta, r_0, \lambda)}\right] dv_2 \\ &= \frac{T_s}{F(\theta, r_0, \lambda)\sqrt{\lambda_1}}. \end{aligned} \tag{19}$$

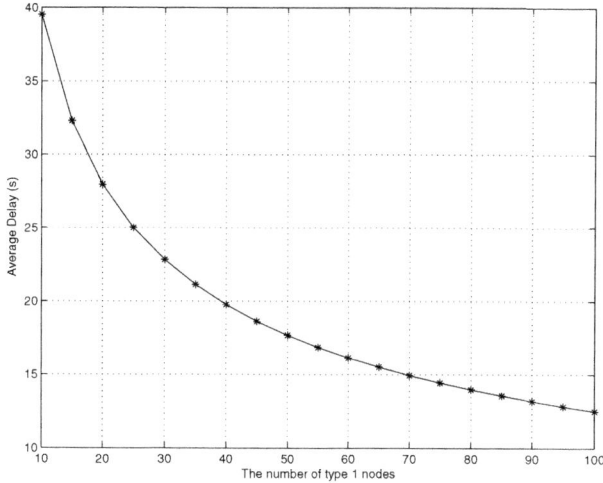

Fig. 5. Average delay variation with the deployment of more $K1$ nodes

Fig. 5 shows the numeric results on the average delay in a network where the number of type $K2$ nodes is 300, and the number of CH nodes ($K1$) increases from 10 to 100. r_0 is 80 and the sleeping interval T_s is 10 seconds.

4 Conclusion

In this paper, we have presented the analysis for the delay properties in the large-scale sensor networks. We quantified both the relationship between the transmission range and delay of each hop, and the delay properties. The S/D distance distributions under typical sensor traffic patterns are derived. Furthermore, the delay is analyzed for the two-tier architecture; how the introduction of more powerful nodes as CHs can improve the delay performance is investigated. Our work can provide guidelines in choosing network parameters to meet the delay requirements.

In our work, we have assumed a deterministic link model and the dynamics of wireless link [26] is not considered. And our delay analysis does not consider the circumstances where an incrementally constructed data aggregation tree [20] is used to deliver the sensed packet. The investigation into this topic will be our future work.

References

1. Estrin, D., Govindan, R., Heidemann, J., Kumar, S.: Next century challenges: scalable coordination in sensor networks. In: MobiCom 1999: Proceedings of the 5th annual ACM/IEEE international conference on Mobile computing and networking, pp. 263–270. ACM Press, New York (1999)

2. Agrawal, D.P., Zeng, Q.-A.: Introduction to Wireless and Mobile Systems. Brooks/Cole Publishing (2003)
3. Ghosh, A., Das, S.K.: Review: Coverage and connectivity issues in wireless sensor networks: A survey. Pervasive Mob. Comput. 4(3), 303–334 (2008)
4. Dousse, O., Tavoularis, C., Thiran, P.: Delay of intrusion detection in wireless sensor networks. In: MobiHoc 2006: Proceedings of the 7th ACM international symposium on Mobile ad hoc networking and computing, pp. 155–165. ACM, New York (2006)
5. Schurgers, C., Tsiatsis, V., Ganeriwal, S., Srivastava, M.: Topology management for sensor networks: exploiting latency and density. In: Proceedings of the 3rd ACM international symposium on Mobile ad hoc networking & computing, pp. 135–145. ACM Press, New York (2002)
6. Ye, W., Heidemann, J., Estrin, D.: An energy-efficient MAC protocol for wireless sensor networks. In: Proceedings of IEEE Infocom 2002, pp. 1567–1576. IEEE Press, Los Alamitos (2002)
7. Ahdi, F., Srinivasan, V., Chua, K.C.: Topology control for delay sensitive applications in wireless sensor networks. Mob. Netw. Appl. 12(5), 406–421 (2007)
8. Heinzelman, W.R., Chandrakasan, A., Balakrishnan, H.: An application-specific protocol architecture for wireless microsensor networks. IEEE Transactions on Wireless Communications 1(4), 660–670 (2002)
9. Yu, Y., Krishnamachari, B., Prasanna, V.K.: Energy-latency tradeoffs for data gathering in wireless sensor networks. In: Proceedings of the Twenty-third Annual Joint Conference of the IEEE Computer and Communications Societies (INFOCOM 2004), pp. 244–255 (2004)
10. Lu, G., Sadagopan, N., Krishnamachari, B.: Delay efficient sleep scheduling in wireless sensor networks. In: Proceedings of the Twenty-forth Annual Joint Conference of the IEEE Computer and Communications Societies, INFOCOM 2005 (2005)
11. Dousse, O., Mannersalo, P., Thiran, P.: Latency of wireless sensor networks with uncoordinated power saving mechanisms. In: Proceedings of the 5th ACM international symposium on Mobile ad hoc networking and computing (Mobihoc 2004), pp. 109–120. ACM Press, New York (2004)
12. Gamal, A.E., Nair, C., Prabhakar, B., Uysal-Biyikoglu, E., Zahedi, S.: Energy-efficient scheduling of packet transmissions over wireless networks. In: Proceedings of IEEE Infocom 2002, pp. 1773–1782. IEEE Press, Los Alamitos (2002)
13. Zorzi, M., Rao, R.: Geographic random forwarding (GeRaF) for ad hoc and sensor networks: multihop performance. IEEE Trans. on Mobile Computing 2(4), 337–348 (2003)
14. Elson, J., Estrin, D.: Time synchronization for wireless sensor networks. In: IPDPS 2001: Proceedings of the 15th International Parallel & Distributed Processing Symposium, Washington, DC, USA, p. 186. IEEE Computer Society, Los Alamitos (2001)
15. Leon-Garcia, A.: Probability and Random Processes for Electrical Engineering, 2nd edn. Addison-Wesley, Reading (1993)
16. Miller, L.E.: Distribution of link distances in a wireless network. Journal of Research of the National Institute of Standards and Technology 106(2), 401–412 (2001)
17. Comaniciu, C., Poor, H.: On the capacity of mobile ad hoc networks with delay constraints. In: Proceedings of the IEEE CAS Workshop on Wireless Communications and Networking (2002)

18. Bettstetter, C., Eberspacher, J.: Hop distances in homogeneous ad hoc networks. In: Proceedings of Vehicular Technology Conference, VTC 2003-Spring, vol. 4, pp. 2286–2290 (2003)
19. Wang, X.: QoS issues and QoS constrained design of wireless sensor networks. In: Department of Electrical and Computer Engineering and Computer Science of the College of Engineering, University of Cincinnati Ohio, USA (2006)
20. Krishnamachari, B., Estrin, D., Wicker, S.B.: The impact of data aggregation in wireless sensor networks. In: ICDCSW 2002: Proceedings of the 22nd International Conference on Distributed Computing Systems, Washington, DC, USA, pp. 575–578. IEEE Computer Society, Los Alamitos (2002)
21. Cormen, T.H., Stein, C., Rivest, R.L., Leiserson, C.E.: Introduction to Algorithms. McGraw-Hill Higher Education, New York (2001)
22. Mhatre, V.P., Rosenberg, C., Kofman, D., Mazumdar, R., Shroff, N.: A minimum cost heterogeneous sensor network with a lifetime constraint. IEEE Transactions on Mobile Computing 4(1), 4–15 (2005)
23. Lazos, L., Poovendran, R., Ritcey, J.A.: Analytic evaluation of target detection in heterogeneous wireless sensor networks. ACM Trans. Sen. Netw. 5(2), 1–38 (2009)
24. Meguerdichian, S., Koushanfar, F., Potkonjak, M., Srivastava, M.: Coverage problems in wireless ad-hoc sensor networks. In: Proceedings of IEEE INFOCOM, vol. 3, pp. 1380–1387 (2001)
25. Bettstetter, C.: On the minimum node degree and connectivity of a wireless multihop network. In: MobiHoc 2002: Proceedings of the 3rd ACM international symposium on Mobile ad hoc networking & computing, pp. 80–91. ACM Press, New York (2002)
26. Zuniga, M., Krishnamachari, B.: Analyzing the transitional region in low power wireless links. In: Proceedings of the First IEEE International Conference on Sensor and Ad hoc Communications and Networks (SECON). IEEE, Los Alamitos (2004)

ProVer: A Secure System for the Provision of Verified Location Information

Michelle Graham and David Gray

School of Computing, Dublin City University, Dublin, Republic of Ireland
{mgraham,dgray}@computing.dcu.ie

Abstract. With location becoming one of the few key pieces of context regarding a user and services springing up to take advantage of this, location verification is an increasingly important system aspect, allowing these services to trust in a device's supplied location. We propose to remove the burden of providing this feature from Location Based Services (LBS) through the introduction of ProVer, a system to provide verified location information regarding any mobile user employing a wireless device. ProVer does not rely on a pre-existing infrastructure to verify a device's location but instead employs evidence from neighbouring devices to confirm a specific device's presence, allowing almost unlimited reach in the system's coverage. ProVer also ensures a device's personal information, such as identity and location, remain private and secure, allowing devices to participate in the system with confidence.

1 Introduction

In recent years, there has been a definite shift towards mobile computing, with more and more services moving to a mobile setting. For this reason, the issue of context has become a key factor, with location being a crucial part of the context of any situation. Many services have since been introduced into the public domain capitalising on a user's location [1], and these Location Based Services (LBS) continue to grow in popularity, with the market seeing no sign of reaching saturation point. One area of LBS development involves the employment of more reliable location information than simple GPS coordinates or user provided data. For example, those services which employ location as a form of access control require their users to prove themselves to be within a specific area in order to access the content requested [2]. This form of "verified" location information is costly, requiring those LBS employing it to somehow verify the current location of a user, usually through the use of some sort of infrastructure. The need to possess such a capability reduces the number of different possible LBS through discouraging the development of those services employing verified location information.

To address this issue, we have developed ProVer, a system for the Provision of Verified location information. ProVer can provide secure, verified location information, which can then be employed by any LBS. This approach removes the need for LBS to possess the location verification capability themselves, abstracting this process away from their system and allowing them to focus on employing the location information within their services. ProVer does not rely on a pre-existing infrastructure, but instead employs other devices in the vicinity of the requesting device, thus allowing the system's coverage to extend far beyond any infrastructure-bound approach.

In this paper, we introduce ProVer as a method of abstracting the task of location verification away from LBS. This work outlines the complete ProVer system, including both a protocol for the protection of evidence gathering and a trust-based approach to verification once evidence has been supplied by the claiming device. In Section 2, we discuss other work on location verification. In Section 3, we outline the functionality of the system and its approach to location verification. In Section 4, we list a small selection of possible applications for ProVer and the benefit of employing an intermediary approach to location verification. Finally, in Section 5 we conclude the paper.

2 Related Work

With the ongoing increase in popularity of mobile computing and networking, a great deal of focus has been placed on the development of positioning and localization systems to capitalise on a device's location as user context information. These systems have traditionally been based upon an existing infrastructure [3,4,5] of trusted receiver devices, allowing location to be calculated relative to these based upon a variety of techniques, such as the measurement of radio frequency (RF) signal strength [6] and time difference of arrival [7]. These techniques have since been adapted to the area of location verification, with a heavy bias towards infrastructure dependence. In [8], Waters and Felten presented the Proximity Proving protocol, a protocol to securely verify a device's location based upon RF time of arrival (ToA). In this protocol, a device proved its location to a trusted entity in its vicinity, whose verdict could then be passed to a central Verifier. However, the timed segment of the Proximity Proving protocol is not tied to the identity of the device making a location claim, leaving it vulnerable to a collusion attack. Sastry et al [9] proposed a similar approach in the Echo protocol, employing both Ultrasound (US) and RF in its ToA calculations. However, due to the employment of US on the response leg of the exchange, this approach is vulnerable to wormhole attacks. In [10], Vora and Nesterenko proposed the use of sensors not only as location verifiers within a particular acceptance zone, but also as rejectors. This approach is heavily reliant upon infrastructure, requiring sensors within the area to function as verifiers in addition to those forming a perimeter around the acceptance zone. Capkun and Hubaux have presented several approaches to verification [11,12], though their focus is primarily on infrastructure or trusted-entity based systems. However, Capkun and Hubaux have also proposed an ad hoc adaptation of their verifiable multilateration technique [13], allowing for its employment in non-infrastructure dependent systems.

3 System Outline and Security

The ProVer system is comprised of three distinct phases; the initialisation stage, where a device requests to have its location verified, the evidence gathering stage and the verification stage, where a final verdict is reached. In the initialisation stage, a device wishing to have a location verified (Claimant) sends a request to the central entity (Verifier). The Verifier processes this request and supplies the Claimant with a list of devices in its vicinity (Proof Providers). The Claimant then begins the evidence gathering stage (section 3.2), communicating with each of the supplied Proof Providers to gather its evidence. When this process has been completed, the Claimant forwards the resulting evidence to the Verifier for use in the verification stage (section 3.3). For clarity's sake, ProVer's exchange sequence is depicted in 1. In order to participate within ProVer, either as a Claimant or a Proof Provider, a device is assumed to possess wireless ad-hoc communication capabilities, a set of asymmetric encryption keys as well as sufficient power to compute encryptions, digital signatures and decryptions in a timely manner. It is also assumed that participants possess a number of pseudonyms for use as identities, although a single core identity may also be used.

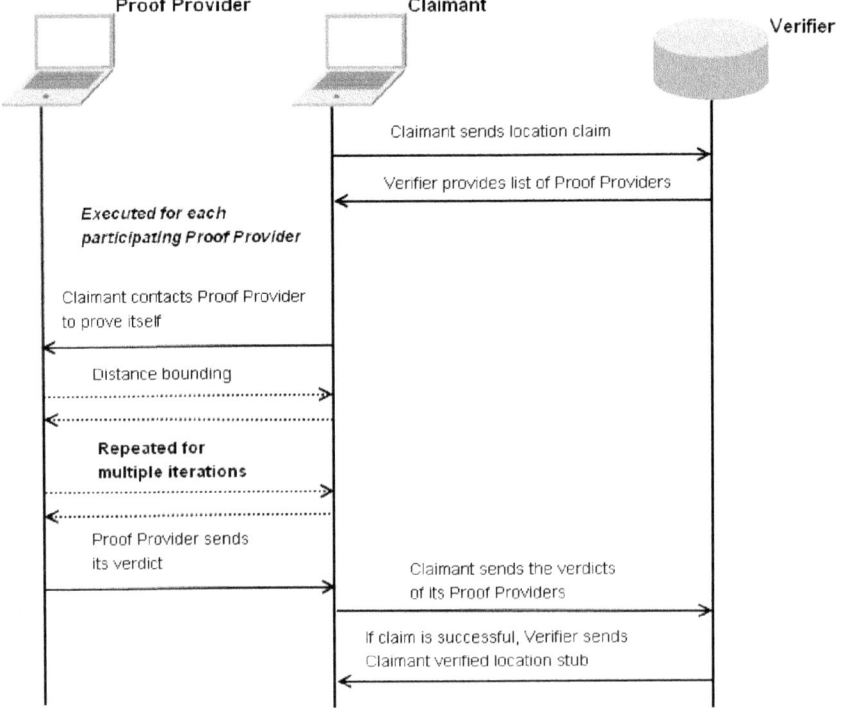

Fig. 1. ProVer exchange overview

3.1 Initiating a ProVer Session

When a device wishes to provide a verified location to a Location Based Service, it contacts the ProVer system, requesting to have its location verified. Upon receiving an initialisation request from a Claimant, the Verifier sends a message to the area seeking devices in the Claimant's vicinity to volunteer as Proof Providers. This is achieved through the employment of geographic routing [14], where a message is routed to an area rather than a specific address. Through transmitting the request in this manner, the Claimant is removed from the volunteer gathering process, protecting it from manipulation. Geographic routing also ensures that those devices which respond are in the correct area, i.e. that area mentioned in the Claimant's location claim. Those devices which respond to this volunteer request are then placed in a volunteer pool for this claim, from which the Verifier selects the final Proof Providers for use within the evidence gathering process. This selection can be done based on a number of criteria, such as those with the highest trustworthiness value or most suitable location, or it may simply be random. With the selection of Proof Providers from the volunteer pool completed, the Verifier then composes a message containing the identities of each and forwards this to the Claimant. The Claimant can then proceed to the evidence gathering stage of the verification process.

3.2 Protecting the Evidence Gathering Process

Once the Claimant receives its list of Proof Providers from the Verifier, it contacts each device and attempts to prove its presence in the area. This is achieved through the use of distance bounding [15], where a series of challenge-response exchanges are timed in order to confirm that the device responding to the challenges is within a reasonable distance. If the Claimant can respond to the Proof Provider's challenges with the correct, digitally signed response within an acceptable time limit, it is deemed to be within communication range of the Proof Provider and therefore in the vicinity, thus proving its presence. We employ a binary metric [16] in this process to confirm the absence of a proxy attack, in order to ensure that the Proof Provider be deceived into believing that the Claimant is in the area when it is not. Each Proof Provider, upon completion of the distance bounding portion of the process, compiles an evidence message and transmits this to the Claimant for forwarding to the Verifier. This message is digitally signed by the Proof Provider and composed of the Claimant's identity, a timestamp, the Proof Provider's current location and finally, a verdict regarding the Claimant's presence in the area. The Proof Provider's verdict is a binary number, with "1" indicating a positive decision and "0" a negative.

The ProVer system employs an extended version of the Secure Location Verification Proof Gathering Protocol (SLVPGP) [17], thus offering a choice between three levels of security due to its tiered design. This protocol has been designed to protect the evidence gathering process from external attackers, as well as preventing its manipulation by malicious entities, both internal and external. Through the employment of digital signatures on the messages preceeding and following distance bounding, the integrity of information is preserved, while encryption prevents the leakage of sensitive personal information regarding participants. ProVer extends the SLVPGP through the inclusion of a final step in which the Verifier provides a verified location stub to the Claimant. This location stub is digitally signed by the Verifier and therefore cannot be undetectably tampered with. It contains the identity of the Claimant, in addition to a timestamp and the verified location, thus tying the particular verification to a specific time. In versions two and three of the SLVPGP, this signed location stub is also encrypted with the Claimant's public key, protecting its contents from eavesdroppers.

3.3 Computing the Veracity of a Claim

Upon completion of the evidence gathering stage, the Verifier receives the total collected evidence from the Claimant, for use in the calculation of its verdict. Before calculating the value of each piece of evidence, the Verifier first confirms that the evidence provided is legitimate. With each proof cleared for inclusion, the Verifier then assigns the evidence contained within the message a weight. As in the calculation of the maximum possible trust value for the claim, this weight is calculated based on the trustworthiness of the Proof Provider from which the evidence was obtained. Within the ProVer system, trustworthiness is calculated using a probability expectation formula, with a device's behaviour history within the system used as inputs for the function according to the formula

$$E(P) = \frac{\alpha}{\alpha+\beta}$$

with α representing the positive events in the behaviour history and β representing the negative events. This approach is drawn from Josang and Ismail's beta reputation system [18]. More recent events within a device's behaviour history are given more weight, through the employment of a fading factor, drawn from Buchegger and Le Boudec's reputation engine [19]. We modify their fading factor to permanently decrease the importance of older events based on the passing of time, through their removal from the behaviour history record. This modification prevents a device from retaining its trustworthiness value despite the passing of time, simply due to lack of activity. While Buchegger and Le Boudec also provide an approach to time-based fading in addition to activity-based, they enact time-based fading over the entire history of the device, which we feel too harsh an approach. The parameters employed within this calculation, α and β, are drawn from the device's behaviour history, in which that device's positive and negative behaviour is recorded for a particular role. If a device is deemed to have behaved positively in an exchange, i.e. it is honest and its verdict is in agreement with the overall verdict for that claim, it receives a positive entry in the behaviour history. If it is deemed to have behaved dishonestly, i.e. its verdict disagrees with the overall outcome of the claim, then it receives a negative entry. However, if the claim receives an "unsure" verdict, the behaviour histories of the devices involved are not updated as no judgement can be made of their behaviour.

With the trustworthiness of each Proof Provider known, its value is multiplied by the value of the evidence it provided (either a "1" or "0") to produce the weighted evidence value. These values are then summed together, along with the trustworthiness value of the Claimant, to provide a total trust value for the claim. This value is placed on the possibility scale and the claim's final verdict is extracted. The possibility scale is composed of the maximum possible trust value for the claim (calculated by multiplying the trustworthiness values of all Proof Providers supplied to the Claimant by a positive verdict value and adding this to the Claimant's weighted trustworthiness value) and is broken up based on two thresholds, 40% and 70%. The employment of thresholds in this situation and not fixed values allows the verification process to be flexible, customising itself to the unique parameters of each claim. The first segment of the scale runs from 0% - 40%, with claims falling between these thresholds receiving a "not possible" verdict. The second segment runs from 40% to 70%, with claims falling between these thresholds receiving an "unsure" verdict. The final segment runs from 70% to 100%, with claims falling between these thresholds receiving a "possible" verdict and a signed verified location stub for use with other LBS. Once the overall verdict has been reached and the Claimant informed, the Verifier updates the behaviour histories of the devices involved.

3.4 Security of the ProVer System

As mentioned previously, the ProVer system is an extension of the SLVPGP, a protocol designed to protect the gathering of proof for location verification. In [17], the authors present a number of protocol versions, each more secure than the last. For each extension of the SLVPGP, a final verification provision stub has been designed, in keeping with the level of security provided by that extension. This approach reduces the increase in overhead costs due to data transmission and computation to a level similar to that already incurred by that extension.

The SLVPGP has been fully model checked using Casper [20] and FDR [21], confirming its security for the environment described. The ProVer extension does not impact this level of security, due to its conformity to the design standards present for each version of the protocol. Therefore, due to this fact and that the system design (the Verifier's powers and structure) does not change in adapting itself to perform as ProVer, ProVer is a secure method of verifying a location, and based upon the version employed, does not leak secure information regarding any of the participants. However, due to the method of communication and evidence gathering employed (RF transmissions received by all devices within range), information regarding the location of a device is leaked to others within its vicinity. An examination of this information leakage and its impact is available in [22].

4 Applications

As discussed briefly in section 1, we envision many possible applications for ProVer in the mobile user market. With the ever-increasing market dominance of smartphones, mobile networking capabilities are now commonly used, with new Location Based Services released regularly to capitalise on this growth.

The introduction of ProVer to the market not only allows the development of more highly functioning LBS, relying on verified information without the need for individual positioning facilities, but also allows currently available LBS to expand their services to a new level.

The linking of location and time information within the ProVer proof stub has potential for use within promotional offers, such as in-store promotions offered to customers on the premises during a specific period. A second advantage of tying location and time in this manner is the provision of access control conditional upon being at a specific location at a specific time. If the location stub was not tied to a time, it would have no expiration and thus could be used infinitely many times.

The generic nature of the location information provided by ProVer allows for its inclusion in a multitude of systems and services. A location stub is comprised of a digitally signed message containing a timestamp and a device's identity and location (in GPS format). As the location information provided is in a standardised form and not a proprietary format, it can be employed within any LBS. Rather than relying upon device based localization techniques (such as cell ids), the generic location information supplied by ProVer would allow for a device's location information to be built up without reliance on infrastructure or costly network queries.

5 Conclusion and Future Work

In this paper, we present ProVer, a system for the Provision of Verified location information. We envision ProVer as a third party location verifier, whose verified proofs can be supplied to Location Based Services (LBS) rather than those services needing to verify or compute locations themselves. This effectively abstracts the need for location verification functionality away from LBS themselves, reducing the complexity of introducing new LBS to the marketplace. ProVer does not rely on a pre-existing infrastructure for location verification, but instead employs neighbouring regular devices as evidence sources, infinitely extending the system's reach while drastically reducing its cost.

In future work, we intend to conduct a complete simulation of the ProVer system, including distance bounding, mobility in nodes and verification of claims. In addition to this, we intend to investigate the impact of heavy traffic on the effectiveness of distance bounding. Currently, our assessment of distance bounding and the binary metric has been based on results generated within a relatively empty network. We wish to confirm that these results hold true for a network with increased levels of network traffic.

References

1. LBSZone (2009), http://www.lbszone.com
2. Ardagna, C.A., Cremonini, M., Damiani, E., di Vimercati, S.D.C., Samarati, P.: Supporting location-based conditions in access control policies. In: ASIACCS 2006: Proceedings of the 2006 ACM Symposium on Information, computer and communications security, pp. 212–222 (2006)

3. Want, R., Hopper, A., Falcao, V., Gibbons, J.: The active badge location system. ACM Trans. Inf. Syst. 10(1), 91–102 (1992)
4. Ward, A., Jones, A.: A new location technique for the active office. Personal Communications of the IEEE 4(5), 42–47 (1997)
5. Correal, N.S., Kyperountas, S., Shi, Q., Welborn, M.: An uwb relative location system. In: IEEE Conference on Ultra Wideband Systems and Technologies, pp. 394–397 (2003)
6. Bahl, P., Padmanabhan, V.N.: Radar: An in-building rf-based user location and tracking system. In: INFOCOM 2000. Nineteenth Annual Joint Conference of the IEEE Computer and Communications Societies, pp. 775–784 (2000)
7. Priyantha, N.B., Chakraborty, A., Balakrishnan, H.: The cricket location-support system. In: MobiCom 2000: Proceedings of the 6th annual international conference on Mobile computing and networking, pp. 32–43. ACM, New York (2000)
8. Waters, B., Felten, E.: Secure, private proofs of location. Technical report, Princeton University (January 2003)
9. Sastry, N., Wagner, D.: Secure verification of location claims. In: WiSe 2003: Proceedings of the 2nd ACM workshop on Wireless security, pp. 1–10 (2003)
10. Vora, A., Nesterenko, M.: Secure location verification using radio broadcast. IEEE Transactions on Dependable and Secure Computing 3, 377–385 (2006)
11. Capkun, S., Cagalj, M., Srivastava, M.: Secure localization with hidden and mobile base stations. In: Proceedings of the 25th IEEE Conference on Computer Communications (INFOCOM 2006), pp. 1–10 (2006)
12. Čapkun, S., Rasmussen, K., Čagalj, M., Srivastava, M.: Secure location verification with hidden and mobile base stations, IEEE Educational Activities Department, vol. 7, pp. 470–483 (2008)
13. Capkun, S., Hubaux, J.P.: Securing position and distance verification in wireless networks. Technical report, EFPL (2004)
14. Mauve, M., Widmer, A., Hartenstein, H.: A survey on position-based routing in mobile ad hoc networks. IEEE Network 15(6), 30–39 (2001)
15. Brands, S., Chaum, D.: Distance-bounding protocols (extended abstract). In: Helleseth, T. (ed.) EUROCRYPT 1993. LNCS, vol. 765, pp. 344–359. Springer, Heidelberg (1994)
16. Graham, M., Gray, D.: Can you see me? the use of a binary visibility metric in distance bounding. In: Liu, B., Bestavros, A., Du, D.-Z., Wang, J. (eds.) Wireless Algorithms, Systems, and Applications. LNCS, vol. 5682, pp. 378–387. Springer, Heidelberg (2009)
17. Graham, M., Gray, D.: Protecting privacy and securing the gathering of location proofs - the secure location verification proof gathering protocol. In: Proceedings of the 1st International ICST Conference on Security and Privacy in Mobile Information and Communication Systems (MobiSec 2009) (June 2009)
18. Josang, A., Ismail, R.: The beta reputation system. In: e-Reality: Constructing the Economy (June 2002)
19. Buchegger, S., Boudec, J.Y.L.: A robust reputation system for mobile ad-hoc networks. Technical report (2003)
20. Lowe, G.: Casper: A compiler for the analysis of security protocols. In: IEEE Computer Security Foundations Workshop, p. 18 (1997)
21. Roscoe, A.W.: Modelling and verifying key-exchange protocols using csp and fdr. In: Computer Security Foundations Workshop, p. 98 (1995)
22. Rasmussen, K.B., Čapkun, S.: Location privacy of distance bounding protocols. In: CCS 2008: Proceedings of the 15th ACM conference on Computer and communications security, pp. 149–160. ACM Press, New York (2008)

Designing Mobility: Pervasiveness as the Enchanting Tool of Mobility

Annie Gentes, Camille Jutant, Aude Guyot, and Michel Simatic

Institut Télécom – Telecom ParisTech
46 rue Barrault Paris F75013
Institut Télécom – Telecom SudParis
9 rue Charles Fourier Evry F79011
{gentes,camille.jutant,aude.guyot,
michel.simatic}@telecom-paristech.fr

Abstract. PLUG – Play Ubiquitous Games -, is a research project that deployed a fully distributed RFID architecture in the Museum of Arts and Crafts in Paris. A pervasive game was designed: "Plug: the Secrets of the Museum" (PSM) where players had to find virtual representations of the Museum artifacts, scatter them around or tidy them in the right spots, or swap them with other players. The analysis of the players' feed back showed that three main features characterize mobility when it is connected to pervasiveness. First, mobility appears as a way to read and collect information. Second, it is a tool to virtually mark the environment and the artifacts. Moving can be akin to "writing" a new scenario. Third, people become part of the network propagating and refreshing information not only on their mobiles but also on the RFID displays.

Keywords: Mobile entertainment, mobile multimedia, mobile and context-aware games, novel user experience and interfaces.

1 Introduction

Mobile and pervasive applications depend on different parameters pertaining to concrete and specific situations. These applications are not a purely virtual world as in videogames but a mix of reality and fiction [15], that rely not only on technological rationale but also on physical and cultural contexts on the one hand and types of mobility on the other hand. Designing pervasive applications means taking the mobility in context or risk contradict what people expect from the situation, their feel for it, what they deem an appropriate behavior. This paper is an attempt at rendering the different planes of experience of a mobile pervasive game that we developed in the context of a museum, so that we might infer some guidelines for the conception of mobile pervasive applications "on the threshold between tangible and immaterial space" [18].

In the project PLUG, an interdisciplinary team including curators of the Museum of Arts and Crafts in Paris (Musée des Arts et Métiers) where the game was deployed, game designers (Tetraedge), media and design researchers, and researchers in computer science (CNAM, Institut Telecom) organized a game based on RFID tags that was to entertain the visitor, create a special bond with the museum, and contribute to a better understanding of the artefacts. As we tested the game, we soon realized how much the mobility introduced by our pervasive game challenged traditional mobility in the Museum.

Based on Michel de Certeau's distinction between « place » and « space », we shall argue that how one circulates in a « place » (i.e. an « instantaneous configuration of relations ») generates a « practiced space » (i.e. « informed by use ») [5]. Mobility says something about one's relation to a specific « place ». Moves are not only a means to an end but a meaningful activity. This led us to analyze more specifically the questions of meaningful forms of mobility and how pervasive applications bring their own logic of mobility that can be confusing for the player/visitor.

After a description of the game, we present and analyze the results of our user tests based on observations, questionnaires and in-depth qualitative interviews that explore reactions, comparisons, recollections that the experience triggered[1]. The analysis of this feedback gives us insight into the dynamic between user and environment as an essential feature to mobile systems. Four types of mobility appeared that are enhanced by a pervasive system: first there is traditional gaming mobility such as running over obstacles or racing, where physical artifacts become anchors for the trajectory; second, the pervasive system turns mobility as a tool to "double read" the environment; third players can leave their mark and write in context; four pervasiveness provides a tool for sociability based on the co-presence of the actors.

2 PLUG, the Secrets of the Museum

2.1 Other Mobile Experiments in Museums

Pervasive technologies present an interesting potential for museums because they improve two characteristics of the visit [8]: the visitor's autonomy – as in "Visit +" in the Cité des Sciences et de l'Industrie in Paris where the user chooses to record data stored on a personal account and can retrieve after the visit [17] - and the precision of the information conveyed – as in "Soundspot", an entertainment system experimented in The Museum of Nature and Human Activities in Hyogo which is "a location/user-dependent audio guide system. It can track the positions of visitors and provide audio information in the limited spots where they stand" [6]. In particular, RFID technology has been especially used by science museums to improve and augment visitor experiences – the Exploratorium in San Francisco with the eXspot system (visitors carry keepsake RFID cards) and the Electronic Guidebook Project [12]; or the Tech Museum in San Jose with the "TechTag" (children wear RFID tag wristbands that trigger exhibits and collect information) – but also to manage their collections (with location tracking systems) - the Industry and Science Museum in Chicago. A few games have also been implemented by museums. The closest project to our experiment is Via Mineralia in the Terra Mineralia Museum in Freiberg that asks players to find a specific exhibit, answer some questions and earn points, with a PDA coupled to an RFID reader [11].

---

[1] These tests were conducted during two days (22 and 23 November 2008). Twelve game sessions were held, with a total of 96 teams and 150 players.

But all these uses of RFID presuppose the same type of mobility for all visitors: a linear progression from one artefact to the other. At each stop, information is delivered in "audioguide fashion", with top down communication. Visitors are not allowed to answer or to leave their mark within the Museum.

2.2 Plug, the Secrets of the Museum

The game "Plug: Secrets of the Museum" looks like the British « Happy Family » card game (or « Jeu des Sept Familles », in French) coupled with a quest based on RFID tags. The purpose of PSM is to discover the Museum by retrieving and collecting cards that represent 16 real objects in the Museum. The "card deck" is composed of four thematic families (for instance, the *Ghost Busters* family puts together famous scientists of the Museum).

Eight players can play together for one hour. They are equipped with a handset able to read/write RFID tags (in our project: Nokia 6131 NFC mobile phone) and with a map of the Museum (so that they can locate themselves and the real objects more easily). 16 passive RFID tags are spread throughout the Museum. Each tag is located on a display beside one of the real objects. Each tag contains 1 virtual card. To gain points, players must prove: their **collector**'s **ability**, by gathering four cards of the same family on their handset; their **public-spiritedness**, by storing a card to its reference RFID display; their **generosity**, by swapping cards with other players; their **curiosity**, by answering quiz on the Museum artefacts.

At the start of the game, each player discovers her "hand" of virtual cards in the handset. She can "zoom" to get detailed information on the object. Then she sets out to find the tags and discover their contents. Players press their phones on the tag. The handset displays the virtual card stored in the tag. The player can, thus, exchange this virtual card with one of the four cards located in her handset. A player can also exchange cards with another player through the handset. The exchange takes place in peer-to-peer NFC mode [9].

The device enables the « reading » and the « writing » of virtual cards, gathered and exchanged among mobile phones and tags on display. The game difficulty stems from this double mobility: content mobility related to user mobility.

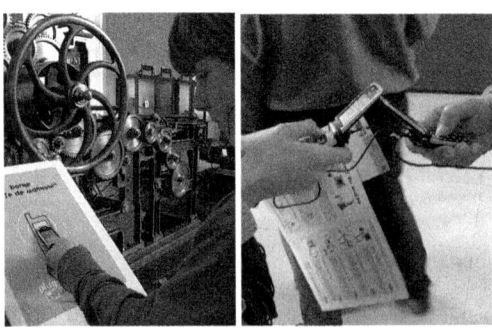

Fig. 1. Experiments in the Museum

3 Pervasiveness as Chaos

Today, questions surrounding mobility and the nature of the museum visits are an integral part of museography. To understand the changes brought about by mobile pervasive applications, one must keep in mind that museums already foster certain types of mobility. The question is how do users juggle with traditional mobility and mobility induced by the mobile pervasive game at the same time.

The first answer is that players consider that their behaviour transgress traditional museum visits: they touch the displays, talk together, swap objects, and moreover rush to gather as many cards as possible.

"I was glancing at security agents, I was worried that they would call me back to order. After all, we run like mad men! It is rather against the Museum's (rules). Where mobiles are forbidden, and we cannot run nor shout".

Speed is considered as an inappropriate answer to the cultural situation and inadequate to get an in-depth knowledge of the artefacts. But this shortcoming of our game is rather independent of the impact of pervasiveness. Other reactions point towards the specific challenges raised by pervasive games.

3.1 The Museum Conventions Disrupted by Scrambled Visits: Pervasiveness Challenging Orientation

Even though it would be an oversimplification to imply that museum exhibitions format visits, it is important to accept the true measure of the norms and conventions involved, more or less acknowledged by visitors [13]. In addition to behavioral conventions, the visit relies on signs that organize the relation to and between the artifacts. The visitor recognizes formal aspects, and shares with the curator a common culture: an arrow is understood as an indication to "move further on in this direction"; a cartel signifies "stop here to find out more". The signs contribute to create meaning so as to help the visitor move in space (following, if the case may be, a story-line), and to understand what kind of relation to establish with objects on display (contemplation, comparison, manipulation...)" [10]. A whole set of signs strive to remove any ambiguity on what to look at and how to get at it.

In PSM, our scenography was also pretty explicit, as the 16 objects and RFID displays were located on a map given to the players before the start of the game. A majority of players explored all the destination points, and therefore visited almost all of the rooms in the museum, discovering some of them: *"It allowed me to go to places I'd never have visited otherwise."*

But the information is not "fixed" as players can change the contents of the display. In other words they can find a locomotive under the RFID tag of a loom and the loom card can be on somebody else's mobile phone. Because virtual cards are stored in both the RFID displays and handsets, players become mobile nodes of the network and change their trajectories to catch either the right player or the right display. In other words, mobility is not just a by-product of the game but rather its driving force with an unpredictability of moves that is related to the mobility of contents and players.

As a consequence, testers considered that their trajectory was less "logical" than during a more conventional visit. *"Generally, in a museum, there's a "direction" and, well, here, we forgot about that entirely. You don't notice the path taken, it's completely arbitrary when compared to the traditional visit. It's very free."* Standard visits were felt as rather disciplined, whereas PSM circuits were considered much more disorganized, "scrambled" as was shown by the drawings that the players made of their moves in the Museum. The two drawings below illustrate how the players represent a normal visit to the museum and a visit with PSM[2].

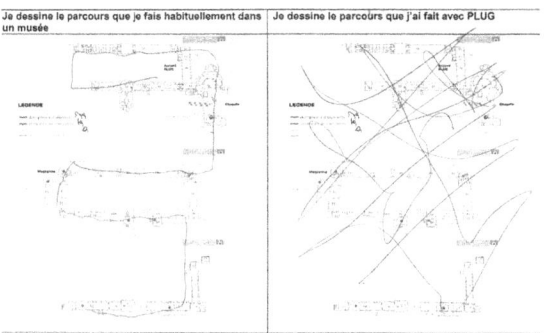

Fig. 2. Representations of PSM trajectories

Indeed, observing the deployment of the game, we realized how much we disrupted the expected and institutionalized mobility within the museum. The move from one RFID tag to another opens up a whole range of options during the visit. If pervasiveness is not used to replicate the mobility in place in the Museum, it introduces a mobility with its own rules that can change the way people orientate themselves or decide where they want to go. The difficulty for players is to partly disregard their usual orientation process.

3.2 From Univocal Text to Double Narrative

As the orientation process is challenged, the narrative of the Museum is also disrupted. In a Museum, objects function as words, articulated so as to produce meaning [2]. Davallon underlines that "understanding is rooted in the way objects are presented" [3]. Creating an exhibition has been compared to directing a play or writing a scenario [7]. To this extent, the exhibition must provide the visitor with "a consistent story-line that engages the viewer to pursue her visit, so as to discover how it ends" [4]. This linear de-ambulation, — with its characteristic ambiance, tempos, breathing spaces, — is punctuated by panels and captions that " help us wend our way through the narrative structure of the show. A panel must be situated at each important stage of the intrigue, thereby signaling the beginning of a new episode" [16]. Ideally, one unit of meaning segues into another, without any back-tracking, and is followed through until the end of the tour.

With PSM, we did not want to contradict the Museum discourse. On the contrary, we were bent on reinforcing the understanding of the Museum and did so by introducing

---

[2] Players were asked to draw first their normal trajectory, then the trajectory induced by the game.

questions and information about the artifacts. Nonetheless we offered another "grammar" that overlapped the existing one. Each artifact no longer belonged to a group of similar technical objects (as they are organized in the rooms of the Museum) but was related to a human quest: embellishing nature, mastering time... Though in no way contradictory to the "first" narrative of the Museum, pervasiveness introduces not so much an ambiguity as is argued by Bjork [1], but a double reading based on a new layer of text. In our experiment, it was felt as unsettling though not necessarily in a negative way. It shows that there is no definite narrative and it highlights the fact that the Museum can present its artifacts in many different ways. This is no surprise to museum professionals but for the users it undermines the assumption that there is one and one only possible discourse. It opens the way to questioning the choices and to participating in their making.

4 "Writing" in the Museum Space: A Way of Creating Meaning

If players testified of the increasing complexity of the experience they also considered that the pervasive system gave them a new tool to handle their mobility as it allowed them not only to read but to write and to create their own narrative.

Testers explained that, unlike audio-guides (the only other mobile device they had encountered) "PSM" pervasive structure allowed them to carry and share contents as they could remove the virtual cards and deposit them at another display. Testers asserted that the NFC mobile enabled them to *write*. *"Here I transpose something onto the display, whereas usually, it's not like that, it's the other way around. Usually you're told, "Go here, listen to the commentary!' You're not the one to carry the information, unlike here."*

Some compare the phone to a magic wand. *"While there it's me, I can give something at any point in the trajectory. That's unique."*

By marking the place and its artifacts, the visitor becomes a collector and the author of her visit. *"I bring my contribution to the system. I complete it, right, it's fun but it's also as if there were a gap, something missing and 'bingo!' I find the missing bit, I finish the Lego and 'wham!' I manage to make it whole, like a construction. There's participation, there's choice, you have to decide, choose the card, find it, identify it. Then there's the swapping in the Museum. In a way we've appropriated ourselves of the Museum space, we know that we've left a small trace of ourselves..."*

Mobility was felt as a way to reorganize and rewrite their access to culture through an active personal but also collective participation.

5 Shared Mobility: Collective Writing

The chaos brought about by the moves of virtual cards and players is not only reduced by the act of collecting and rewriting the overall access to cards. Mobility was finally perceived by the players as a global question involving everybody and soon they tried to organize a global view of the situation and they tended to stabilize the positioning of cards.

The game encouraged social interactions to gain card but also to score in "generosity". People met, looked at each other's screens and, if they wanted to exchange a card, brought their phones together." *It allows you to interact with other people. If you have an audio-guide...there's the aspect, well here I am and now I'll put on my earphones...two hours can go by and nobody speaks to you, you're isolated from each other. Whereas here, there's the interactive side, you can fool around together..."*

Such trust is made possible by the specific environment. The context guarantees the civility of the exchanges. First, people share the Museum that is considered a safe place. It is a closed space, with security and guards. Contrary to other experiments that we did in railway stations [7], nobody expressed the fear of being mugged or harassed. It is also safe because being in a Museum signifies sharing a common interest. People who choose to go to the Museum belong to a real, if transient or ephemeral community. There is the social assumption that museum goers are "of the same world". The game itself, as it is issued by the Museum, benefits from a qualified aura. It cannot be a superficial entertainment. Playing the game is therefore playing under the authority of the Museum that allows and even encourages relationships between visitors. Moreover, the device (phone plus RFID) and the gestures that are associated to it set players apart. They signal a "shared categorical identity" [14]. Because of this shared situation and common identity, players agree to communicate with strangers.

We also observed two types of behaviors that turned the game from an individual quest to a collective rewriting. First, some players call the others to let them know what card they came across. This gives a more comprehensive view of the situation. Second, because civil behavior is promoted within the game (by tidying the cards in the Museum), the individual activity is not only rewarded, it introduces a possibility of anticipation that leads to a collective interest. The players know that if everybody plays "civic", it will be easier to know where the cards are and to reduce the chaos induced by the everchanging position of cards. In our tests, they banked on this collective behavior.

6 Managing a Double Sociability

Mobility is therefore perceived as a condition to meeting people not only as potential enemies/allies in the game but also as people engaged in the same meaningful mobility and therefore potentially in relation to each other. This mobility also sets apart from other visitors of the Musem. Players involved in pervasive systems have to decide how to ignore or involve other people. What kind of relationships is established between gamers and non gamers? Is it a way to influence others to get into the play (out of sheer curiosity)? Beyond games how can mimicry be a social tool for mobile services?

Pervasive games create situations where players must manage a double sociability: the community of players and the community of non players sharing the same social space. They cannot ignore the rules of mobile behavior in the Museum.

7 Conclusion: Design Lessons in Pervasive Applications

Designing pervasive applications can be a lot of fun but it includes a lot of damage control. If the pervasive application does not replicate exactly the audioguide (and in that case why should we even bother to think of developing pervasive games), the users have to show a capacity in running two levels of constraint on a certain number of issues. How to read a double narrative - one already in place in the Museum and one that belongs to the game? How to superimpose or twin together the Museum information, our own and that of the other players in a collective writing? How to deal with at least two communities, one "in" the other "out of" the game? Each decision involves a certain type of mobility but mobility itself then has a double meaning. First, mobility is meaningful because it involves some kind of relationship to the context that can be defined as a tangible reality (building, walls, windows, openings, obstacles) but also a social one. Second, mobility is meaningful because there is a "pervasive agenda". Our game tried to rely on the institution as a cultural entity where people adopt what they consider the proper behavior but also the correct situation of communication. Museums are places of civil behavior that are disrupted by the proposal of a game and both institution and players won't push the situation too far from the requirements of the situation. This engages a political model of mobility and pervasiveness.

7.1 The Museum Conventions Disrupted by Scrambled Visits: Pervasiveness Challenging Orientation

If mobility is perceived as a way to create a meaningful experiment by linking some places, objects together in a more or less loose association, a form of narrative, it is important that designers should first consider space as a "reservoir of resources" to use Heidegger's words. The space around us contains, organizes and feeds our activities. These resources have to be defined as they are not necessarily useful for an activity but meaningful in a narrative. In any event, contents must be organized as items that can be reorganized through the moves of the users. Designing mobility means designing bits of information that can be "walked" together in a creative way, leaving the user to build her phrase, her narrative as in an Exquisite Cadaver or Queneau's A Thousand Billion of Poems. Each fragment can be re-configured. Each time the user is confronted to a new piece of information, image or text, she considers it not only in its own right but also as it is going to be part of a composition that makes sense both of the item and its context.

7.2 Creating Double Layers of Text

Designing pervasiveness means studying the museum as a second page. Players did not want to annoy the other visitors.

"There were lots of people in front of the display. I had to push them a little to be able to plug in".

To design mobile pervasive applications one has to clearly outline the status of the different places and if they are suitable or not for a "second" layer of writing and reading. This has to do not only with comfort and access but also with the representations of the place: is it a suitable and legitimate place to write? It also means that the environment must be designed so as to produce signs and affordances that help understand that this is indeed the right place to read or write.

For the museums, what is at stake is obviously reconsidering the way they handle visitors. The problem is not to deprive the cultural institution of its legitimacy in terms of knowledge and heritage, but to pursue the reflection on how to impart this knowledge. Designing Plug: The Secrets of the Museum, we therefore discovered that not only should the environment be taken into consideration, but also the moves with their predefined meanings and their potential to include new significations as well as significant potential for trouble.

References

1. Björk, S.: Changing Urban Perspectives. In: Borries, F., Walz, S., Böttger, M. (eds.) Space Time Play. Birkhäuser, Basel (2007)
2. Cameron, D.: Un point de vue: le musée considéré comme système de communication et les implications de ce système dans les programmes éducatifs muséaux (1968). In: Desvallées, A. (ed.) Vagues: Une anthologie de la nouvelle muséologie, Éd. W/MNES, Mâcon/Savigny-le-Temple, vol. 1 (1992)
3. Davallon, J.: L'Exposition à l'œuvre: Stratégies de communication et médiation symbolique. L'Harmattan, Paris (1999)
4. De Bary, M.O., Tobelem, J.M.: Manuel de muséographie. Petit guide à l'usage des professionnels de musée, Séguier/Option culture, Biarritz (2000)
5. De Certeau, M.: L'invention du quotidien, Tome 1, Arts de Faire, Folio, Paris (1980)
6. Deguchi, A., Mizoguchi, H., Inagaki, S., Kusunoki, F.: A Next-Generation Audio-Guide System for Museums 'SoundSpot': An Experimental Study, LNCS. Springer, Heidelberg (2007)
7. Edson, G., Dean, D.: The Handbook for Museums. Routledge, Londres (1996)
8. Gentès, A., Jutant, C.: Pervasive gaming: Testing future context aware applications. Communications and strategies 73 (2009)
9. Gentès, A., Jutant, C., Guyot, A., Simatic, M.: RFID technology: fostering human interactions. IADIS Game and Entertainment Technologies (2009)
10. Gob, A., Drouguet, N.: La muséologie: histoire, développements, enjeux actuels, Armand Colin, Paris (2006)
11. Heumer, G., Gommlich, F., Jung, B., Müller, A.: Via Mineralia - A Pervasive Museum Exploration Game. In: Proceedings of Pergames 2007, Salzburg, AT (2009); Hsi, S.: The Electronic Guidebook: A Study of User Experiences using Mobile Web Content in a Museum. Institute of Electrical and Electronics Engineers. In: International Workshop on Wireless and Mobile technologies in Education. IEEE, Danvers (2002)
12. Hsi, S., Fait, H.: RFID: Tagging the World: Rfid Enhances Visitors' Museum Experience at the Exploratorium. Communications of the Association for Computing Machinery 48(9) (2005)
13. Le Marec, J.: Publics et musées: la confiance éprouvée. L'Harmattan, Paris (2007)
14. Liccope, C., Inada, Y.: Les usages émergents d'un jeu multijoueur sur terminaux mobiles géolocalisés ». Réseaux 5(133) (2005)
15. Montola, M., Stenros, J., Waern, A.: Pervasive Games. Theory and Design. Morgan Kaufmann, San Francisco (2009)
16. Sunier, S.: Le scénario d'une exposition. In: Publics et Musées, vol. 11-12, Actes Sud, Arles (1997)
17. Topalian, R.: Visite+: personnalisation de la visite et site mémoire des visites culturelles, Culture et Recherche, Paris, vol. 112 (2007)
18. Walther, B.K.: Atomic Actions – Molecular Experience: Theory of Pervasive Gaming. In: CIE. ACM, New York (2005)

Context-Aware Recommendations in Decentralized, Item-Based Collaborative Filtering on Mobile Devices

Wolfgang Woerndl[1], Henrik Muehe[1], Stefan Rothlehner[2], and Korbinian Moegele[1]

[1] Technische Universitaet Muenchen, 85748 Garching, Germany
{woerndl,muehe,moegele}@in.tum.de
[2] EXO Exhibition Overview, c/o Lehrstuhl fuer Baurealisierung und Bauinformatik, 80333 Muenchen, Germany
rothlehn@in.tum.de

Abstract. The goal of the work presented in this paper is to design a context-aware recommender system for mobile devices. The approach is based on decentralized, item-based collaborative filtering on Personal Digital Assistants (PDAs). The already implemented system exchanges rating vectors among PDAs, computes local matrices of item similarity and utilizes them to generate recommendations. We then explain how to contextualize this recommender system according to the current time and position of the user. The idea is to use a weighted combination of the collaborative filtering score with a context score function. We are currently working on applying this approach in real world scenarios.

Keywords: collaborative filtering, context, mobile guides, item-based collaborative filtering.

1 Introduction

Recommender systems are an established research area and have also been successfully introduced in commercial applications such as online shops. So far, most recommenders are centralized systems where all the information about users and items is stored on a server. Decentralized management and interpretation of data without a central server offers some advantages though. Decentralized recommender systems can be used without a permanent connection to a server and thus improve the portability of the system [7]. Recommendations can be made locally by considering other users in the current vicinity. In addition, decentralized recommender systems are regarded as a way of reducing privacy concerns and improving user control over their personal data such as ratings [6]. This is especially true in a mobile or ubiquitous environment. In addition, recommender systems used on mobile devices can be much more dependent on context, such as the current location of the user, than recommenders mainly used in desktop computer settings.

We have designed and implemented a system to recommend items like images on Personal Digital Assistants (PDAs) [9]. Our approach is an application of item-based collaborative filtering. The system exchanges rating vectors among PDAs, computes local matrices of item similarity and utilizes them to generate recommendations. Our innovation in comparison to existing systems includes improving the extensibility of the data model by introducing versioned rating vectors. The goal of the work presented in this paper is to extend this decentralized PDA recommender for context-aware recommendations in mobile scenarios. Context characterizes the situation of users [4]. In our case, context mostly means the physical location of a user when requesting a recommendation, or the current time. But the approach can be generalized to other context attributes such as temperature or the presence of other users in the current vicinity. The benefit of this model is that contextualized recommendations can be given in decentralized manner on a PDA.

The remainder of the paper is organized as follows. In Section 2 we provide some background on item-based collaborative filtering and an existing related approach for a decentralized recommender. Based on this existing approach, we explain our decentralized, item-based recommender system for PDAs in Section 3. Subsequently, we discuss the options for contextualization and then our applications scenarios. In Section 6 we briefly review related work and finally conclude with a short summary.

2 Background

2.1 Item-Based Collaborative Filtering

Recommender systems recommend items like products or services to an active user. Collaborative filtering (CF) utilizes the user-item matrix of ratings users have made to generate suggestions. The standard, user-based CF algorithm first determines a neighbourhood of similar users, i.e. users that have rated similar to the active user in the past. The algorithm then selects (new) items, which users in the neighborhood of the active user have preferred. The second alternative method of CF is item-based collaborative filtering.

Item-based CF does not consider the similarity of users, but of items [8]. Thus, the user-item matrix of ratings is not analyzed line by line, but column by column. The item similarities can be precomputed to build an item-item matrix. The item-item matrix is the model of the algorithm. Therefore, this type of recommendation algorithm is also called model-based collaborative filtering. One element of the item-item matrix expresses the similarity between two items, determined from the users' ratings. The rating vector of the active user is then used to recommend items that are similar to items that have been positively rated by the active user in the past.

Item-based CF has an advantage over user-based CF with regard to the complexity of the computation because the item-item matrix can be calculated as an intermediate result, independently from who the active user is. In addition, at the time of the recommendation, the algorithm also needs the rating vector of the active user only, not all ratings. This is promising for improving the privacy of personal user data. Therefore, item-based CF appears to be well suited for an implementation on PDAs.

2.2 PocketLens

In this subsection, we are explaining the PocketLens system because it serves as a foundation for our approach. The basic motivation behind the development of PocketLens is that characteristics of item-based recommenders are well suited for decentralized adoption [7]. Though some problems arise from utilizing the item-item matrix.

In particular, it is required to recompute the matrix when a new rating is inserted. The algorithm has to evaluate all ratings associated with the concerning items in this process, because only the computed similarity is stored in the matrix. This also raises privacy issues, because every peer in the system has to store the rating vectors of all users. In addition, the recalculation of the matrix every time a new rating is added is not very efficient and increases hardware demands. It is preferable to make use of the item similarity values calculated so far as intermediate results.

PocketLens modifies the basic item-based CF algorithm to compute the item-item matrix. Thereby, all the concerning elements of the matrix do not have to be completely recalculated. This is achieved by storing the intermediate results with every item similarity pair in the item-item matrix. PocketLens uses cosine similarity and stores the dot products and the lengths of the rating vectors for every item pair. Thus, it is possible to easily update the item-item matrix with new ratings: the new rating just has to be added to the dot products and the vector length. The whole data model does not have to be recomputed. Especially reapplying all rating vectors used so far is not necessary. Hence, the PocketLens algorithm allows to delete the rating vector of other users after integrating their ratings into the model.

3 A Decentralized Recommender System for PDAs

We have implemented and tested a decentralized, item-based CF approach for PDAs based on PocketLens [9]. The idea is that users rate items on mobile devices. Then rating vectors are exchanged with other peers and each PDA calculates the item-item matrix based on the received ratings. The system recommends items based on the model and the local user's ratings only. Our approach improves some problems of PocketLens that have become clear when applying the algorithm in our scenario.

While item-based CF does have advantages with regard to computational complexity to generate recommendation, storing the model – the item-item matrix – does demand more storage capacity in comparison to user-based CF.

Rating vector   Versioned rating vector

Item	Rating
A	4
C	1

Item	Versioned ratings
A	$(1,5) \leftarrow (4,4)$
B	$(2,4) \leftarrow (5,0)$
C	$(3,1)$

Fig. 1. Rating vector vs. versioned rating vector

Therefore, we have optimized the storage model for usage on PDAs. PocketLens does allow for the easy insertion of new users with their rating vectors into the model, but it does not account for updating the matrix with new or changed ratings from existing users. In a distributed scenario, though, users may frequently exchange ratings and add other users updated rating vectors to their model on the PDA.

To fulfill the second requirement of extensibility, we have extended the concept of a rating vector to reproduce older values. The rating vector consists of an item identifier and the rating value. Our new rating vector introduces a serial respective version number and the corresponding value for representing the rating of an item. We call this resulting data structure a versioned rating vector (Fig. 1). The version-rating tuples of an item are implemented as linked lists starting from the newest rating to allow efficient access.

A change of a rating results in a new entry in the values list of the corresponding item. The history of old values is preserved. When integrating a versioned rating vector into the model, the system saves the version of the rating vector. The version of the vector is defined as the highest version number of any rating's value. When updating the item-item matrix with a vector with new or additional ratings, the system can first undo the effects of the formerly used version of this vector and thus restore the old model. Then the new ratings can be integrated into the matrix just like a new rating vector [9]. The rating vectors of other users can be deleted after insertion into the model, it is still not necessary to store the rating vectors of each user on every PDA.

In the example in Figure 1, item A has received a new rating (4) with a higher version number and therefore an old rating (1) is overwritten but not lost. The rating for item B has been deleted, while the rating for item C has not been changed.

As far as the storage optimization is concerned, we have analyzed the properties of the algorithm and found many duplicate entries in the item-item matrix. That means the computed item similarity value was identical for many item pairs. The elements of the matrix in the PocketLens algorithm require at least 20 bytes of memory in a densely populated matrix. Replacing an element with a pointer (4 bytes) to a duplicate entry reduces the storage requirement for this matrix element by 16 bytes. We have evaluated this approach with the 'Movie-Lens' data set [5] containing 100000 ratings by 943 users for 1682 movies. We were able to reduce the storage requirement from about 45 megabytes to 32 megabytes.

Fig. 2. Rating and recommending images on a PDA

Besides the reduction by using pointers, we can additionally modify the definition of identical matrix elements. PocketLens uses float values for item similarity and vector size as matrix elements. If we assume that two elements are duplicates if they are very similar but not quite identical, we can increase the number of 'duplicates' found and thus further improve the storage efficiency. The resulting model quality suffers a little, but we have found that cutting the float value after some decimal points does indeed reduce the memory requirements while hurting recommendation quality only marginally if at all. It is possible to adapt this impreciseness of duplicate search dynamically to be able to handle much larger models than without this feature. If we compare the field elements of the item-item matrix with an inaccuracy of just 0.05, the data model can be further reduced to 20 megabytes. This is small enough to be completely loaded into the memory of our test PDAs.

The search for duplicates is done only when a memory threshold is exceeded. The insertion of a rating vector into the model took about 1.5 seconds on average on our PDAs. The search for duplicates needs the main share of this time, which can be done in the background. The algorithm needs only a fraction of the 1.5 seconds for actually inserting a rating vector into the item-item matrix.

The system has been implemented to recommend images as an application example on PDAs for the Windows Mobile operating system [9]. The system displays one image with the option to rate it on a scale from one to five stars (Fig. 2). The user can choose whether the system displays random or recommended images. Random images are used to let the user initially rate a couple of items. It is also possible to change the rating of an image, if it has been rated before. Afterwards, the user gets recommendations based on the algorithm explained above. The user can click on 'Weiter' (German for 'next') for the next — random or recommended — image.

The image recommender has been tested with 13 users and HTC P3600 PDA or similar mobile devices. The test also included a shared public display for group recommendations [9]. The item set consisted of 63 images. The users were told to initially rate about 15-20 randomly selected images on their PDAs. The corresponding rating vectors were subsequently exchanged, item-item matrices were calculated on the PDAs and users were able to generate recommendations on their devices. First of all, the test proved that our application ran smoothly and provided comprehensive recommendations in the explained scenario on PDAs. Both the computation of the item-item matrices and the generation of recommendations ran without performance problems. Users tested the generated models by judging the recommended images and gave positive feedback in a questionnaire [9].

4 Context-Aware Recommendations

In this section we will now explain the options we are currently working on to contextualize the existing PDA recommender that was explained in the last section. We are currently fine-tuning and implementing the approach in the application domains that follow in Section 5 of this paper.

4.1 Context Definition

First, it is important to define what context means in our scenario. We follow the context definition by Dey et al.: 'Context is any information that can be used to characterize the situation of entities (i.e. whether a person, place or subject) that are considered relevant to the interaction between a user and an application, including the user and the application themselves' [4]. That means, context is very dynamic and transient. By contrast, user profile information such as preferences or ratings are somewhat static and longerlasting. In a mobile application domain, a context model could include location, movement, lighting condition or current availability of network bandwidth of mobile devices, for example.

The assumption is that the generation of the model, i.e. the construction of the item-item matrix, is rather independent from context because the ratings express the preferences of the user. Nonetheless, to contextualize the construction of the item-item matrix, the idea is to compare the context attributes when a rating was made and use this information when calculating item similariry, i.e. item pairs that were rated similarly in a comparable context receive the highest values in the model.

In our mobile scenario, context is important when a recommendation is made, because a user wants a recommendation at a certain time in a certain location. To do so, a context snapshot is taken and used to recommend items that are best suited for the current context in combination with the model of item-based CF.

4.2 Context-Aware, Hybrid Item Recommendation

To generate recommendations, the algorithm calculates the predicted rating of candidate items based on the model and the rating vector of the active user. We will call this predicted rating the *cf-score*. The list of k items with the highest scores can be then shown to the user. For the image recommender (cf. Section 3 and Fig. 2), the system shows the image with then highest score of all non-rated images.

Our idea is to use the following linear combination of the *cf-score* with a 'context-score' (*ctx-score*) to contextualize the recommendations:

$$score = a * cf\text{-}score + b * ctx\text{-}score \tag{1}$$

a and *b* are the weights to balance the collaborative and context components. Different application scenarios may require different weights, examples are given in the next section (cf. Section 5). *cf-score* or *ctx-score* returns values in the range [-1 ... 1], so these functions have to be normalized according to these values. Items with negative scores will not be considered for recommendation. A *cf-score* or *ctx-score* of zero means that the item is either evaluated as moderate, or the algorithm has been unable to generate a score, for example when an item has not yet been rated. The linear combination of the scores was designed to allow for separate calculation of the two components and weigh them according to different requirements (see Section 5)).

In terms of hybrid recommender systems [3], this approach is a weighted hybrid recommender. Other options are possible, including a cascading hybrid recommender. In this case, an initial recommender would generate some intermediary results using only the available ratings. In a second step, the results are further filtered and ranked by considering the 3rd dimension of context. However this contextual pre- or postfiltering [1] can also be implemented with our weighted hybrid method.

4.3 Context Score Function

Context is modeled by a vector C whose elements represents the various context attributes $c_1, c_2, \ldots, c_3$. For example, (c_1, c_2) = (longitude, latitude) to model the geographical position using GPS coordinates or (c_3) = current time. Different attributes can model very different things. All items have associated context attributes, e.g. the location and opening times of a restaurant. That is similar to item attributes that characterize the item space in content- or case-based filtering, e.g. price, color, weight etc. of a product.

The goal of the context score function is to model the similarity of two context vectors, one representing the current context of the active user and the other one an item under consideration. What similarity constitutes depends on the considered attributes and the application domain. The context score function returns a normalized value with -1 representing the worst value and +1 the best value, i.e. the user context is identical to an item context, according to the model.

The context score function can also be used to rule out items that are not suitable at all in the current context. For example, if the user want a recommendation for a restaurant at a given time, restaurants that are closed should not be included in the list of recommended restaurants:

$$ctx - score = \begin{cases} +1, & \text{if restaurant is open} \\ -1, & \text{if restaurant is currently closed} \end{cases} \tag{2}$$

In this example, the context score function will return -1 and the overall score of the item will be negative if the restaurant is closed. The same can be done for location. Thereby some items will be reasonable close to the current user's context, so the context score function for two context vectors C1 and C2 will be a variant if the following in most cases:

$$ctx - score = \begin{cases} +1, & \text{C1 and C2 are (nearly) identical} \\ x, & x = \text{distance(C1, C2)} \\ -1, & \text{C1 and C2 are (too) different} \end{cases} \quad (3)$$

distance(C1, C2) is normalized to [-1 ... +1], with zero representing a mediocre value. A combined context score function to integrate different context attributes is also possible.

5 Application Scenarios

In this section, we explain two real world application scenarios we are currently working with start-up companies in two separate projects.

5.1 Mobile Tourist Guide

The first one is a mobile tourist guide developed by voxcity s.r.o. and jomedia s.r.o. The idea is to rent out a mobile device with GPS positioning capabilities to support tourists. The guide is currently available for the Czech city of Prague (see www.voxcity.de). The mobile application plays audio, video, pictures and (HTML) text of tourist attractions based on the current position, traveling direction and speed.

The plan is to extend this application with options to rate items (i.e. the multimedia files) and generate recommendations not only based on location but also the collaborative filtering model. Since the mobile devices are not permanently connected to a network in this scenario, our decentralized approach appears well suited. When a customer rents a device, it will have the model of item similarity according to past ratings on it. The tourists can use the guide and also rate items such as audio clips. These ratings can then be used to generate CF recommendations as described above. In this case, the context score function will have a higher weight, because we still want to recommend multimedia files that are related to attractions in the current vicinity. On the other hand, the CF recommendation process can point the user to items (in our case, tourist attractions) that are a little farther away but are recommended according to the CF model. When a user returns the mobile device, the model can be updated with new ratings on other devices and used for other users. One additional idea is to acquire implicit ratings according to observed behaviour when using the device, e.g. a user playing an entire audio clip instead of skipping over it would constitute a positive rating.

5.2 Mobile Exhibition Guide

The second application scenario is a navigation and information system that is being developed by EXO Exhibition Overview GbR. It allows the visitors of trade fairs, conferences and similar events to orient themselves at the venue. Visitors of exhibitions can search for products, exhibitors or places of interest and navigate there. Moreover, the visitor can use several additional functions on the device, such appointment schedule, virtual business cards or collecting electronic versions of available print media on the event. The indoor positioning is done by a bluetooth-based infrastructure, which is also used to transfer data between the mobile devices and a central data source. The system will be available at the Munich Trade Fair Center starting 2010.

In this scenario the context, e.g. the exact location of exhibition booths, is less important. Thus, the context score function will distinguish between exhibition halls and prefer items that are nearby when recommending, but not completely rule out any item. The collaborative part, i.e. the *cf-score* function, is more significant in this scenario. It allows the device recommend items might not have been aware of before, but which fit her previous rating profile.

On the other hand, for recommending exhibition events (e.g. talks), the temporal context is very important. This can be modeled in our approach using the context score function. In this case, the context score will return -1 for past events or expired information, and a high score for upcoming events.

6 Related Work

In addition to the already discussed PocketLens, there are a few related approaches with regard to decentralized CF. For example, Berkovsky et.al. propose a distributed recommender system that partitions the user-item matrix based on domain-specific item categories [11]. However, additional information about the application domain of the items is needed.

One fundamental solution to contexualize recommender systems is the multidimensional approach by Adomavicius et.al. [2]. This model enhances the user-item data model to a multidimensional setting where additional dimensions constitute different context attributes such as location and time. They propose a reduction-based approach with the goal to reduce the dimensions, ideally to n = 2. Then, traditional recommender techniques can be used to generate item lists. But to our knowledge, the approach has not been applied in a mobile scenario setting similar to ours.

Magitti is a recent system to recommend leisure activities on mobile devices [10]. Their approach is to combine and weigh very different recommenders such as CF, a distance model, content preference and others. However, they do not use item-based CF, which we believe is well suited for application on a mobile device.

7 Conclusion

The goal of this work is to design a context-aware recommender system for mobile devices in the explained application scenarios. The approach is based on decentralized, item-based collaborative filtering. Thereby, mobile devices exchange rating vectors of their respective users, calculate local matrices of item similarity and utilize them to generate recommendations. We are currently working on contextualizing the recommender by using a weighted combination of the collaborative filtering score with a context score function. It is important to note that in our solution, the context information about a user never leaves her personal mobile device, so the approach can provide context-aware recommendations without raising issues of location privacy, i.e. a server being able to track a user's movements.

References

1. Adomavicius, G., Tuzhilin, A.: Context-aware recommender systems. In: Tutorial at ACM Conference on Recommender Systems, RecSys 2008 (2008), http://ids.csom.umn.edu/faculty/gedas/talks/RecSys2008-tutorial.pdf
2. Adomavicius, G., Sankaranarayanan, R., Sen, S., Tuzhilin, A.: Incorporating contextual information in recommender systems using a multidimensional approach. ACM Transactions on Information Systems 23, 103–145 (2005)
3. Burke, R.: Hybrid web recommender systems. In: Brusilovsky, P., Kobsa, A., Nejdl, W. (eds.) Adaptive Web 2007. LNCS, vol. 4321, pp. 377–408. Springer, Heidelberg (2007)
4. Dey, K., Abowd, D., Salber, D.: A conceptual framework and a toolkit for supporting the rapid prototyping of context-aware applications. Human Computer Interaction 16, 97–166 (2001)
5. Herlocker, J.L.: An algorithmic framework for performing collaborative filtering. In: 22nd Annual international ACM SIGIR Conference on Research and Development in Information Retrieval, Berkeley, CA (1999)
6. Kobsa, A.: Privacy-enhanced personalization. Communications of the ACM 50(8), 24–33 (2007)
7. Miller, B.N., Konstan, J.A., Riedl, J.T.: PocketLens: Toward a personal recommender system. ACM Transactions on Information Systems 22(3), 437–476 (2004)
8. Sarwar, B., Karypis, G., Konstan, J.A., Riedl, J.T.: Item-based collaborative filtering recommendation algorithms. In: 10th International Conference on World Wide Web (WWW 10), Hong Kong, China (2001)
9. Woerndl, W., Muehe, H., Prinz, V.: Decentral item-based collaborative filtering for recommending images on mobile devices. In: Workshop on Mobile Media Retrieval (MMR 2009), MDM 2009 Conference, Taipeh, Taiwan (2009)
10. Ducheneaut, N., Partridge, K., Huang, Q., Price, B., Roberts, M., Chi, E.H., Bellotti, V., Begole, B.: Collaborative filtering is not enough? Experiments with a mixed-model recommender for leisure activities. In: Houben, G.-J., McCalla, G., Pianesi, F., Zancanaro, M. (eds.) UMAP 2009. LNCS, vol. 5535, pp. 295–306. Springer, Heidelberg (2009)
11. Berkovsky, S., Kuflik, T., Ricci, F.: Distributed collaborative filtering with domain specialization. In: ACM Conference on Recommender Systems (RecSys 2007), Minneapolis, MN (2007)

deSCribe: A Personalized Tour Guide and Navigational Assistant[⋆]

Dheeraj Kota[2], Neha Laumas[1], Urmila Shinde[1], Saurabh Sonalkar[1], Karthik Dantu[1], Sameera Poduri[1], and Gaurav S. Sukhatme[1,2]

[1] Computer Science Department
[2] Ming Hsieh Department of Electrical Engineering
University of Southern California, Los Angeles, CA 90089, USA
{dkota,laumas,ushinde,sonalkar,dantu,sameera,gaurav}@usc.edu

Abstract. Mobile phones have become ubiquitous in daily life. They are also becoming more powerful with more computation and a variety of sensors embedded in them. We have built *deSCribe*, a context-aware phone-based navigational aid. The application provides turn-by-turn directions from the user's current location to a requested destination. It provides information about the surroundings by processing the images taken by the camera on the phone. It enables a novel user interface with the ability of using the phone as a remote to point at buildings to get further information about a particular floor within the building. Lastly, the application uses user feedback to control how much information is presented to the user.

1 Introduction

Mobile phones are ubiquitous. Commodity mobile phones with powerful microprocessors (order of 500 MHz), high resolution cameras (3.2 megapixel camera), and a variety of embedded sensors (accelerometers, compass, GPS) are becoming the norm. Coupled with good network connectivity, they can be used as a great source of information and user context.

One can envision using such mobile phones as mobile tour guides in trade shows, university campuses, downtowns, national parks etc where the user is assisted with information based on her location and situational context to better assist her in the task she is doing. We have implemented *deSCribe*, a mobile tour guide application that encapsulates several features required of such an application. Currently, the application has been tailored to the University of Southern California (USC) but given the data for other locations, the application can easily be tailored for them. We will now describe the design of our application and its various components.

This paper is organized as follows. Sec. 2 motivates the need for such an application. Sec. 3 describes some related work. Sec. 4 elaborates on the design of our application. Sec. 5 shows some results. Sec. 6 lists some of the limitations of our application due to

---

[⋆] This work was supported in part by NSF grant CCR-0120778 (CENS: Center for Embedded Networked Sensing), and by a gift from the Okawa Foundation. It was initiated as a project for the graduate course CS 546: Intelligent Embedded Systems taught at USC in Spring 2009.

the tradeoffs we made and the contextual information at our disposal. Sec. 7 concludes our description followed by Sec. 8 which lists future directions.

2 Motivation

Every year University of Southern California gets hundreds of visitors who are not familiar with the campus and have trouble navigating and getting relevant information regarding the buildings on campus. The problem is further exacerbated by the fact that the campus spans a 226-acre area and has over a hundred buildings. This is not a problem unique to this or just other university campuses. We can imagine large downtowns, shopping complexes, fairs and exhibitions etc. posing a similar problem for a first time visitor. Such an application can be further extended using the user generated mapping in online sevices like wikimapia to further extract information about each building in any locality.

We hope to address this problem and have developed our application *deScribe* that would provide relevant information regarding the buildings in the vicinity of the users. The application also provides turn by turn directions to any destination on campus. Information of buildings is also provided in audio which makes it very user friendly, easy for hands free operation and provides better assistance for the visually impaired.

The input to our application was a detailed map of the USC campus at three resolutions and a database with building information and GPS coordinates. We use this to display the current location of the user on the USC campus and the point of the next hop in the route to the destination so that the user has an idea regarding her position on the map. Navigation is further aided by image processing algorithms to recognize side views of cars in the image which would help the visually handicapped to determine whether a car is going in front of them. This would aid in situations in case they need to decide whether they want to cross the road or not.

3 Related Work

There has been a lot of work in designing novel interfaces for mobile tourism and other interactive navigational services using a variety of embedded systems in the past few years. We cite a representative set here and draw from their experiences.

Schmidt-Belz et al. in [8] report their findings of a user validation study for the CRUMPET system which is a location aware mobile based tourism system focused on providing personalized tourism services. One of the findings of the study was that people need more textual/audio tour description as they find it difficult to interpret maps. Also, the study suggested that sights to visit are usually looked up both before traveling and also while on tour. The study also indicates that location based services would be very important for mobile based tourism applications. All these findings clearly indicate the need to provide users with relevant contextual information while on the go. Kray et al. in [5] review several navigation assisting services/devices available commercially or as research prototypes. The authors compare the various systems based on the services offered, positioning mechanism used, user interaction method, system architecture etc. Their observation is that more situational awareness is crucial for future mobile guides. The other key observation is that mobile guides need to adapt to real world situations like lack of network connectivity, visibility to GPS satellites etc. They also raise questions regarding the architecture to be used in the design of such guides and whether a client-server architecture is the most suitable. Gregory et al. in [4] talk about Cyberguide, a context aware tour guide. This is one of the first mobile applications to provide location aware services both indoor and outdoor. It also uses a past history of the user's locations to better assist the user.

[6] performed a multi-user mobile application game study. Their primary observation was that social interaction is a strong motive to participate in a multi-user service. They also concluded that it is equally influenced by three factors - context, communication and identification. The other novel outcome of their study was that user communication increased dramatically when the game was displayed on a large public display. This suggests that users are affected strongly by moving the viewership of their participation from their personal space to public space in some form.

[9] describes a location aware mobile tourist guide application where, it provides the users, in one of the mode they describe as a explorer mode, a way to provide details about buildings around the present location of the user. This data is provided on a map along with the current location of the user. We extend this to make the application more usable by utilizing the compass in the mobile phone to provide a mode intuitive way of providing information to the user, thereby giving the flexibility of using the mobile phone as a virtual tourist guide who explains more about the building when one points to a building an asks more details about it.

There are numerous studies on using the phones in novel ways. [7] develops an application that allows the use of the phone as a pen to jot down small pieces of information quickly by writing in air. The Zagat NRU [1] and acrossair NY subway [2] are examples of virtual reality applications that use the GPS and the compass sensor to provide a direction for restaurants closeby and the nearest subway stations respectively. Wikitude [3] is the closest to our work. It is an augemented reality application which uses HTC G-1 camera view to display annotated landscape, mountain names, landmark descriptions, and interesting stories.

4 Design

The *deSCribe* application has three key functions. These are distinguished with three tabs on the application as shown in Figs. 1, 2.

- **Contextual Information:** Allows a user to point to any building on the USC campus and hear the name and other relevant details regarding the building being spoken aloud. The user has the option of setting the granularity of the details that they would like to hear. On changing the angle of elevation of the phone the user hears the floor number being pointed to. On pointing the phone vertically towards the sky they hear the current weather information.
- **Turn-by-Turn Directions:** Allows a user to find the shortest walkable route to any building on campus from her current location. The user is told the distance and direction she needs to walk to reach the desired location. It also displays a map of the USC campus and shows the current location of the user (a red pin is displayed) and the next hop in the path (a blue pin is displayed) to the desired destination.
- **Phone as a remote:** The user can use the phone as a remote and point to buildings/surroundings of interest. She can then query for further information about the point of interest by the push of a button.

4.1 Platform

We used the Android Platform running on the HTC G1 phone for our development. The major factors that influenced the decision of choosing Android platform is the availabilty of various modes of localization available, with varying levels of power consumption and utilizing two different techniques which will come handy when the GPS direct signal is unavailable. Kray-C et al. in [5], provided a list of issues related to mobile application design after a survey of available tour guide applications. One of the criteria they mentioned is usage of application in a non networked mobile phone. We chose Android platform beacuse of the sqlite database they provide on the device itself. Considering the usability of the user, we provide the details of the building in text for which we required a text to speech library which is already available as an open source library on the android plaform. Since, the content regarding the buildings are saved on the local phone storage, it might become stale after some point, this problem can be solved once utilizing the already exisitng android market's infrastructures to provide applications and updates to the users. In addition to that, due to the open source licensing of android platform, more devices are likely to come, thereby increasing the target audience.

4.2 Turn by Turn Directions

As mentioned earlier, a database was provided with the GPS coordinates of all buildings on campus. The GPS sensor is switched on when the user requests route directions. This provides us updates regarding the user's current location after every two seconds. We also turn on the compass sensor on the phone to provide us with the user's orientation. We use the GPS to get her current location and then run the shortest path algorithm to find the appropriate route to the destination. On each location update from the GPS sensor, we determine the distance the user needs to walk by calculating her current distance to the GPS coordinate of the next hop in the route. A sample screenshot of the execution of the application is shown in Fig. 1.

Fig. 1. Turn-by-Turn Directions Initial Screen

We also determine the direction she should be walking in with the help of the onboard compass. We use this distance and direction to inform the user of where she should be heading in the form of a message displayed on the screen.

4.3 Contextual Information

We provide the user with contextual information by providing her the option to point to a particular building. The application determines the building and speaks aloud its name. The user can choose the granularity of details she hears about the building. The default option is *coarse* which only speaks aloud the name of the building. The user can switch to the *fine* option which would allow her to hear a detailed description of the purpose of the building along with any other interesting features such as names of cafes, restaurants etc., housed in the building. The text to speech library helps provide these details to the user in clear audio.

Fig. 2. Phone as a Remote Control Screen

As soon as the user clicks the option to determine the building ahead of her we query the GPS sensor on the phone to provide us with the user's current location and the compass to find his/her orientation. We have a database which has a GPS coordinate identifying each building on campus. It is stored in the android's sqllite database.

We use the database to filter out all the buildings within a specified range (currently the range consists of all buildings whose latitude and longitude is within 0.001 from our current latitude and longitude) from us. On obtaining this subset of buildings, we determine the distance from our current position to the GPS coordinates of each of these buildings. This distance is calculated in terms of Universal Transverse Mercator (UTM) coordinate system. The building description is obtained using the current location and the orientation of the compass. From the figure below consider the case where we are in between four buildings. The building to which the mobile device is currently pointing to is obtained by getting the angle the ends points of building makes to the current location of the device. We can conclude that we are pointing to that particular building if the compass is oriented within the range of this particular angle. For this purpose we need to assign angles dynamically from the current location to the buildings around. This would have been trivial in case we had the GPS co-ordinates of the four corners of the building. However, currently we have been able to obtain only one GPS co-ordinate of each building. Further, we think it would be the same case in many other scenarios, where we do not have complete information about the environment. We assume that the single GPS co-ordinate we have for each building, is at the center of the building. We now calculate the slope of the line joining the current location of the phone obtained via GPS to the building's latitude and longitude information available to us. Consider the building b1 in the figure which at a distance d1 which makes an angle of 135 degrees with the +x axis from the current view. Now we can tell for sure if the compass if oriented towards 135 degrees that we are pointing to b1. The building can be huge and we need identify the building even when the compass is pointing to any part of the building. In order to achieve this, we assign some range $(135 - dx, 135 + dx)$ to the building, which can be used as a measure to decide if we are pointing to that particular building. We obtain the value of dx based on the current distance $d1$ which is the distance between the current location of the phone to the GPS co-ordinate of the building. The value of dx varies with distance and the value with which it varies was obtained empirically by getting an estimate of about ten buildings at USC by calculating the angle made by each building from various distances. A larger value of dx is assigned to a smaller distance and vice a versa and we made a lookup table assigning the value of dx based on distances.

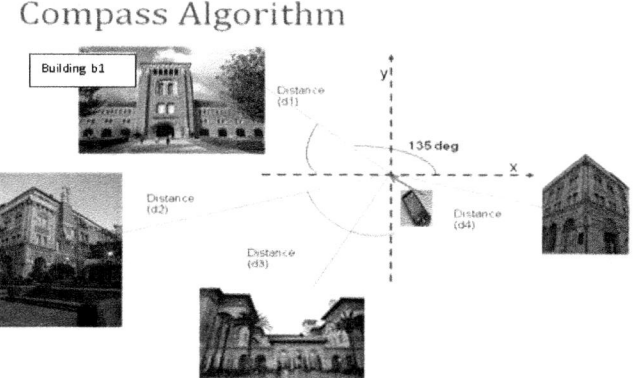

Fig. 3. Compass Algorithm

4.4 Phone as a Remote Control

Floor determination is done using the Y axis of the compass which indicates the pitch. When the phone is exactly horizontal, the Y axis indicates 0 degrees. When you start turning the phone upwards then angle becomes negative and is -90 degrees when completely vertical. When you take the phone downwards the angle increases from 0 to $+90$

degrees when the phone is completely pointing to the ground. To calculate which floor you are pointing to, one has to get the distance between the current location and the building. Then from the y axis angle and the distance you can get the height the phone is pointing to by the formula.

$$Height = \tan(angle) * Distance$$

The height of the floor is assumed constant at 3.7 meters. Based on this data we can calculate the floor number the phone is pointing to. One of the problems that we faced when calculating the height is that the GPS coordinates of the buildings we have is of the center of the building. Fig. 4 below illustrates the error introduced.

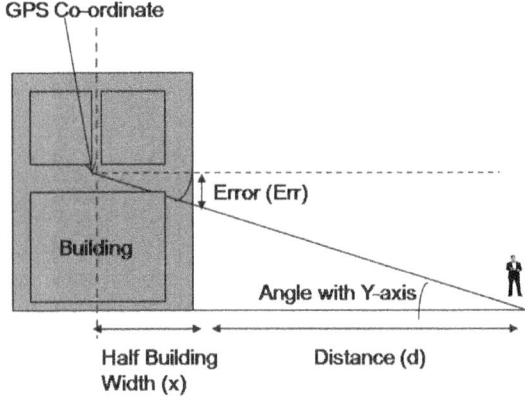

Fig. 4. Computing the Floor Number

As shown in the Fig. 4, the measured height is

$$H2 = \tan(angle) * (d + x)$$

when it should be $\tan(angle) * d$. So the error introduced is: $err = \tan(angle) * x$. Fig. 6 shows the plot of error versus angle in which the building width is assumed constant at 30 meters. Graph shows the increase in error as the angle increases. So the floor information will work best only for smaller angles which means buildings having less floors or lower floors of tall buildings. Also if we keep the angle constant, the error becomes proportional to the width of the building. So the floor works best only for the small buildings or we need to have the GPS coordinates of all the four corners of the building. Another factor that the floor determination depends on is the users view and the phone view. Sometimes the user thinks she is pointing to a particular floor based on her line-of-sight angle but the phone angle is slightly different. This also introduces significant error depending on the distance from the building. If the angle of pointing increases beyond 80° or beyond the number of floors in a building, then we get the weather information. We are currently using the Google weather API. The API returns an XML file which is parsed and the information such as current weather condition, the temperature, humidity and wind conditions are spoken out by the TTS engine.

4.5 Current Location Display

We determine our current location with the help of the GPS as shown in Fig. 5. As mentioned earlier, we have been provided with the USC map at three resolutions. We align the tiles such that it displays the user's current location. We also show the point on the map which in the next hop to the destination. This can provide the user an understanding of her relative position and a sense of orientation on the USC map. Determining the presence of a car We give the users an option to determine the presence of cars ahead of her. This work has been motivated to aid the visually handicapped. The user is required to take an image of the scene ahead of him. We then upload this image to the aludra server running in USC. We use the GentleBoost image recognition algorithm on the server to identify the number of cars detected in the image. We then send this information back to the phone and provide the user the number of cars that were detected in the image. This information is helpful since the user is aware that a car is going side ways in front of him, indicating that it is probably dangerous for him/her to cross the road.

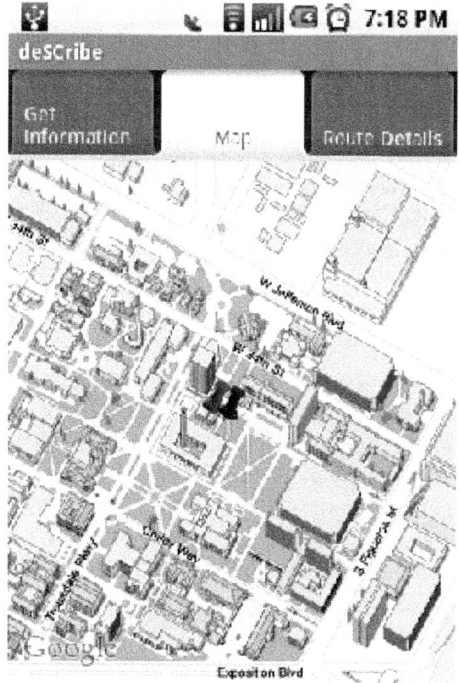

Fig. 5. Map Display with Current Location (Red) and Next Hop (Blue)

5 Results

We collected data from about 30 buildings on campus and our accuracy rate was 83.2%. We found that the buildings identified incorrectly were very big in size. We attribute our

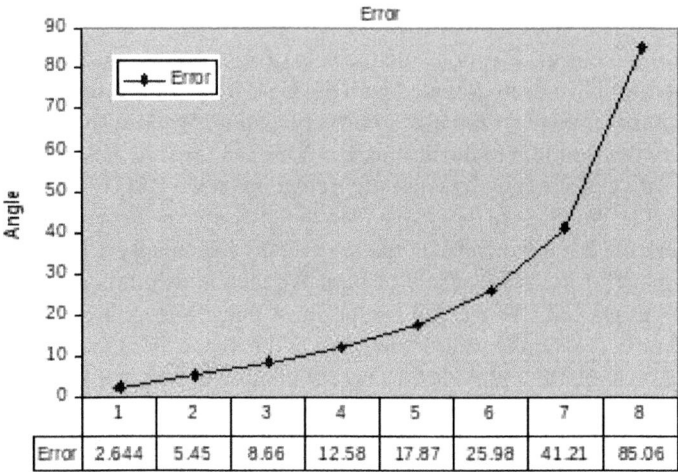

Fig. 6. Angular Error in Floor Computation

inaccuracy to the fact that our algorithm assigns angles based on the distance and for large buildings we would need to move very close to the building in order to get correct identification. In case we had been provided with GPS coordinates for all four corners of buildings we would not have to use an empirical angle based on distance and would have got more accurate readings.

6 Limitations

Our database currently identifies each building on campus by a single GPS coordinate. Also, these GPS coordinates are often not located in the center of buildings and building sizes are variable on campus. This sometimes leads to some inaccuracy in the estimation of the angle the phone is making with the buildings in the vicinity and leads to incorrect answers by the phone. The accuracy of our estimation of the building that is being pointed to by the phone would increase considerably in case we had GPS coordinates of all the four corners of each building since they would help us calculate the correct angle the phone is making with each building. There is an error introduced when estimating the height of a building since we do not have the GPS coordinate of corners of buildings. The GPS coordinate available being at the center of building introduces an error in the estimation of our distance from the building for the purpose of calculating the height of the building. We have assumed that the routing is point-to-point and that the roads are straight lines. This sometimes leads to somewhat ambiguous directions being given to the user. Both these limitations could be overcome by obtaining finer details in the map like the GPS coordinates of building corners and the size and coordinates of the roads.

Also, in our work we use GPS measurements both for finding turn by turn directions and finding the building that is being pointed to by the phone. GPS uses a lot of power and we find a significant reduction in the battery lifetime on using our application continuously. On an average the talk time on the phone is 5 hours and it reduces to nearly 3 hours.

7 Conclusion

We believe that this work is a pre cursor to develop such applications for schools, recreational parks etc. within which buildings and land marks are often not reachable by commercially available GPS devices like Garmin and Tom Tom. We believe that such an application is highly beneficial and in the long run will replace paper maps which can help find a route to a particular destination but do not provide any contextual information. Further, using the phone as a remote control is a novel user interface for such mobile devices and can see many other uses beyong mobile tourism.

8 Future Work

Currently the information about each and every building is saved on the mobile phones local storage. The information about a building can change. This new information has to be updated in the phones database. This has to be done with minimum amount of delay so that the users donot get incorrect information. The application can be extended to recognize more objects such as bicycles; pedestrians etc. that are present ahead of the user and integrate them to our application. This would make the application appealing for use for the visually impaired it would provide them information regarding what lies ahead of them.

References

1. http://www.zagat.com/Blog/Detail.aspx?SCID=42&BLGID=20939
2. http://www.acrossair.com/apps_newyorknearestsubway.htm
3. http://www.wikitude.org
4. Abowd, G.D., Atkeson, C.G., Hong, J., Long, S., Kooper, R., Pinkerton, M.: Cyberguide: A mobile context-aware tour guide. Baltzer/ACM Wireless Networks 3 (1997)
5. Baus, J., Kray, C., Cheverst, K.: A Survey of Map-based Mobile Guides. Springer, Heidelberg (2005)
6. Leikas, J., Stromberg, H., Ikonen, V., Suomela, R., Heinila, J.: Multi-user mobile applications and a public display: novel ways for social interaction. In: Fourth Annual IEEE International Conference on Pervasive Computing and Communications, PerCom 2006, pp. 5–70 (March 2006)
7. Sandip Agrawal, I.C., Gaonkar, S., Choudhury, R.R.: Phonepoint pen: Using mobile phones to write in air. In: ACM Workshop on Networking, Systems, Applications on Mobile Handhelds, Mobiheld 2009 (2009)
8. Schmidt-Belz, S.B., Schmidt-belz, B., Nick, A., Poslad, S., Zipf, A.: Personalized and location-based mobile tourism. In: Proceedings of Mobile-HCI (2002)
9. ten Hagen, K., Modsching, M., Kramer, R.: A location aware mobile tourist guide selecting and interpreting sights and services by context matching. In: 2nd Annual International Conference on Mobile and Ubiquitous Systems (MobiQuitous 2005), pp. 293–304. IEEE Computer Society, Los Alamitos (2005)

Author Index

Akselsen, Sigmund   142
Ankolekar, Anupriya   331
Annavaram, Murali   42

Baki, Omar Abdul   190
Bash, Cullen   160
Boda, Péter   107
Borning, Alan   92
Buthpitiya, Senaka   254

Chen, Han   72
Choudhury, Romit Roy   203
Clayton, Ben   307
Constandache, Ionut   203
Cox, Landon   203
Crepaldi, Riccardo   72

Dantu, Karthik   273, 343, 393
Das, Sajal K.   335
Dayal, Umeshwar   160
Dey, Anind K.   26
Dorman, Kyle   1
Duri, Sastry   72

Evjemo, Bente   142

Ferris, Brian   92

Ganihar, Sairabanu Z.   160
Gaonkar, Shravan   203
Gentès, Annie   339
Gentes, Annie   374
Goel, Diwakar   26
Gong, Michelle X.   59
Graham, Michelle   366
Gray, David   366
Griss, Martin   26, 190, 223, 254
Gupta, Manish   160
Guyot, Aude   374

Hao, Ming   160
Heu, Alfred   236
Hoelzl, Gerold   12
Huberman, Bernardo A.   331

Jayakumar, A.   160
Joag, Shriya   26
Joshi, Anand   273
Julien, Christine   335
Jutant, Camille   374

Kadmawala, Ritesh   273
Kaiser, William   1
Kher, Eisha   26
Kolli, Aalaya   343
Kommaraju, Sumalatha   254
Kota, Dheeraj   393
Krishnamachari, Bhaskar   42

Laumas, Neha   393
Le, Ken   343
Lee, Kyujoong   236
Lin, Tony   190
Liu, Zhigang   107
Luon, Yarun   331
Lyons, Kent   59

Madamanchi, Deepthi   254, 326
McCarthy, William   1
Melamed, Tom   307
Mikkonen, Tommi   123
Misra, Archan   335
Moegele, Korbinian   383
Mohan, Vani   160
Muehe, Henrik   383
Mujumdar, Veda   26
Munusamy, Ramesh   160

Nahapetian, Ani   1, 236
Naik, Deepa   160
Nyrhinen, Feetu   123

Pang, Hawk Yin   107
Parnandi, Avinash   343
Patel, Chandrakant   160
Pattan, Neha   223, 326
Pering, Trevor   59
Poduri, Sameera   273, 343, 393

Author Index

Policroniades, Calicrates   142
Pulli, Kari   293

Reason, Johnathan M.   72
Rivera, Alejandro   223
Rosario, Barbara   59
Rothlehner, Stefan   383
Roy, Nirmalya   335

Salminen, Arto   123
Sarrafzadeh, Majid   1, 236
Sayler, Matt   203
Sharma, Ratnesh   160
Shinde, Urmila   393
Simatic, Michel   172, 339, 374
Sonalkar, Saurabh   393
Sud, Shivani   59
Suh, Myung-kyung   1, 236
Sukhatme, Gaurav S.   273, 343, 393
Szabo, Gabor   331

Taivalsaari, Antero   123

Vaghela, Pradeep   343
Vennelakanti, Ravigopal   160

Wang, Xiaodong   355
Wang, Yi   42
Wang, Yun   355
Want, Roy   59
Watkins, Kari   92
Woerndl, Wolfgang   383

Xiong, Yingen   293

Yahyanejad, Marjan   1
Yang, Guang   107
Yang, Jun   107
Yang, Kathleen   223
Yin, Jun   355

Zhang, Joy   190
Zhang, Mi   273
Zhao, Qing   42

GPSR Compliance

The European Union's (EU) General Product Safety Regulation (GPSR) is a set of rules that requires consumer products to be safe and our obligations to ensure this.

If you have any concerns about our products, you can contact us on ProductSafety@springernature.com

In case Publisher is established outside the EU, the EU authorized representative is:

Springer Nature Customer Service Center GmbH
Europaplatz 3
69115 Heidelberg, Germany

Batch number: 09478804

Printed by Printforce, the Netherlands